T0136841

Studies in Computational Intelligence

Volume 827

Series Editor

Janusz Kacprzyk, Polish Academy of Sciences, Warsaw, Poland

The series "Studies in Computational Intelligence" (SCI) publishes new developments and advances in the various areas of computational intelligence—quickly and with a high quality. The intent is to cover the theory, applications, and design methods of computational intelligence, as embedded in the fields of engineering, computer science, physics and life sciences, as well as the methodologies behind them. The series contains monographs, lecture notes and edited volumes in computational intelligence spanning the areas of neural networks, connectionist systems, genetic algorithms, evolutionary computation, artificial intelligence, cellular automata, self-organizing systems, soft computing, fuzzy systems, and hybrid intelligent systems. Of particular value to both the contributors and the readership are the short publication timeframe and the world-wide distribution, which enable both wide and rapid dissemination of research output.

The books of this series are submitted to indexing to Web of Science, EI-Compendex, DBLP, SCOPUS, Google Scholar and Springerlink.

More information about this series at http://www.springer.com/series/7092

Oscar Castillo · Patricia Melin
Editors

Hybrid Intelligent Systems in Control, Pattern Recognition and Medicine

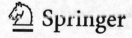
Springer

Editors
Oscar Castillo
Division of Graduate Studies and Research
Tijuana Institute of Technology
Tijuana, Baja California, Mexico

Patricia Melin
Division of Graduate Studies and Research
Tijuana Institute of Technology
Tijuana, Baja California, Mexico

ISSN 1860-949X   ISSN 1860-9503   (electronic)
Studies in Computational Intelligence
ISBN 978-3-030-34137-4   ISBN 978-3-030-34135-0   (eBook)
https://doi.org/10.1007/978-3-030-34135-0

This Springer imprint is published by the registered company Springer Nature Switzerland AG
The registered company address is: Gewerbestrasse 11, 6330 Cham, Switzerland

Preface

We describe in this book, recent advances on fuzzy logic, neural networks and optimization algorithms, as well as their hybrid combinations, and their application in areas, such as intelligent control and robotics, pattern recognition, medical diagnosis, time series prediction and optimization of complex problems. This book is organized into five main parts, which contain a group of papers around a similar subject. The first part consists of papers with the main theme of type-1 and type-2 fuzzy logic, which basically consists of papers that propose new concepts and algorithms based on type-1 and type-2 fuzzy logic and their applications. The second part contains papers with the main theme of pattern recognition and applications, which are basically papers dealing with new concepts and algorithms in neural networks and fuzzy logic applied in recognition. The third part contains papers that present theory and practice of meta-heuristics in different areas of application. The fourth part presents diverse applications of fuzzy logic, neural networks and hybrid intelligent systems in medical applications. Finally, the fifth part contains papers describing applications of fuzzy logic, neural networks and meta-heuristics in robotics problems.

In the first part of this book with four papers, we refer to theoretical aspects and applications of type-1 and type-2 fuzzy logic, which basically consists of papers that propose new concepts and algorithms based on type-1 and type-2 for different applications. The aim of using type-2 fuzzy logic is to provide better uncertainty management in problems of control, pattern recognition and prediction.

In the second part of pattern recognition theory and applications, there are three papers that describe different contributions that propose new models, concepts and algorithms for recognition applications. The aim of using neural networks and fuzzy logic is to provide learning and adaptive capabilities to intelligent pattern recognition systems.

In the third part of the theory and practice of meta-heuristics in different areas of application, there are six papers that describe different contributions that propose new models and concepts, which are also applied in diverse areas of application. The nature-inspired methods include variations of different methods as well as new nature-inspired paradigms.

The fourth part presents diverse applications of fuzzy logic, neural networks and hybrid intelligent systems in medical applications, and there are five papers that describe different contributions on the application of these kinds of systems to solve complex real-world medical problems.

In the fifth part of fuzzy logic, neural networks and meta-heuristic applications in robotics, there are six papers that describe different contributions on the application of these kinds of intelligent models to solve complex real-world robotic problems.

In conclusion, the edited book comprises papers on diverse aspects of fuzzy logic, neural networks and nature-inspired optimization meta-heuristics and their application in areas, such as intelligent control and robotics, pattern recognition, time series prediction and optimization of complex problems. There are theoretical aspects as well as application papers.

Tijuana, Mexico   Oscar Castillo
December 2018   Patricia Melin

Contents

Robotic Applications

Type-1 and Type-2 Fuzzy Logic

Intuitionistic and Neutrosophic Fuzzy Logic: Basic Concepts and Applications

Amita Jain and Basanti Pal Nandi

Abstract Fuzzy set proposed by Zadeh states that belongingness of an element in a set is a matter of degree unlike classical set where membership is a matter of affirmation or denial. Fuzzy set theory provides more natural representation for real world problems. Intuitionistic fuzzy set (IFS) is the generalization of fuzzy set, proposed by Atanassov, in 1986 (Fuzzy Sets Syst 20(1):87–96, 1986 [1]). It assigns two values called membership degree and a non-membership degree respectively. Later Florentin Smarandache introduced an additional parameter for neutrality which generalise Intuitionistic Fuzzy Set as Neutrosophic Fuzzy Set (NFS). The speciality lies in the 3D Neutrosophic space where each logical statement is evaluated with 3 components namely truth, falsity and indeterminacy. IFS and NFS revolve around these divisions of degree of belongingness to their component structure and so generate different variations. In this chapter we discuss the properties of these two variants of fuzzy set based on their different extension, propositional calculus, predicate calculus, degree of dependence of each component, geometric representation and various application areas of both the sets.

Keywords Intuitionistic fuzzy · Neutrosophic cube · Neutrosophic fuzzy · Predicate calculus · Propositional calculus · Research statistics

1 Introduction

Elements of classical fuzzy set has a membership value assigned to it. In Fuzzy set each element is mapped with some real value between 0 and 1. There are various ways to choose membership functions for classical fuzzy set like triangular function, trapezoid, L-function etc. Intuitionistic fuzzy set (IFS) [2, 3] is an extension to the

A. Jain (✉)
Ambedkar Institute of Advanced Communication Technologies and Research, Delhi, India
e-mail: amitajain@aiactr.ac.in

B. Pal Nandi
Guru Tegh Bahadur Institute of Technology, New Delhi, India
e-mail: basanti_pal@yahoo.com

© Springer Nature Switzerland AG 2020
O. Castillo and P. Melin (eds.), *Hybrid Intelligent Systems in Control,
Pattern Recognition and Medicine*, Studies in Computational Intelligence 827,
https://doi.org/10.1007/978-3-030-34135-0_1

classical fuzzy set where each element assigns a membership value and a nonmembership value contrast to the Zahed's fuzzy set [4, 5] where each element assigns only a membership value. Degree of Indeterminacy or degree of Hesitancy is the parameter which is calculated from the two varying parameter mentioned for membership and non-membership category. This concept was extended by Florentin Smarandache [1] which includes a three way decisions. This three way decision provides more natural representation for real world problems. It can be understood by considering some real world examples like for games: win/loss/draw, for voting: pro/cont/blank, for numbers: positive/negative/zero etc. In all these real world problems there are three decisions. To deal with this tri-component logic, Florentin proposed Neutrosophic logic [6–8] which means "The knowledge of neutral thought". The main difference between Intuitionistic fuzzy set (IFS) and Neutrosophic fuzzy set (NFS) is the middle/neutral/indeterminant component. In Neutrosophic Fuzzy set when the summation of truth, indeterminacy and falsity is ≤ 1 then the set is both IFS and NFS. On the other hand when the values of truth, falsity and indeterminacy overlap i.e., the summation of these three membership value is ≥ 1 then the set is NFS not IFS [9].

In this chapter a picture of comparison on the basis of variations of IFS and NFS is represented. Geometrical interpretations as well as the applications of these two logics in different fields are also discussed. IFS and its variants like 'Interval valued Intuitionistic fuzzy set', 'Intuitionistic L-fuzzy set', 'Temporal Intuitionistic fuzzy set' etc. have been applied on various fields like electoral system, medical pattern recognition, medical diagnosis, sociometry, petrochemical farm, pneumatic transportation process etc. [10–12]. On the other hand Neutrosophic Fuzzy logic has been applied in fields like physics, robotics, image segmentation, generating score of a neutral word etc. [13–16]. It gives a precise way to capture inconsistent or imprecise factors of a given problem. It continues to evolve new variations like 'Interval Neutrosophic set', 'Single valued Neutrosophic set', 'Refined Neutrosophic set'. The elements of Intuitionistic fuzzy logics give the relationship between the membership values with the non-membership value. The basic elements of Intuitionistic fuzzy which are covered in this chapter are 'Intuitionistic fuzzy propositional calculus', 'Intuitionistic fuzzy predicate logic'. Neutrosophic elements analogous to Intuitionistic fuzzy set express its propositional calculus and predicate calculus on an instance of neutrosophic theory called 'Interval valued Neutrosophic logic'. To show the differences in these two extensions of classical fuzzy system geometric interpretations give a visual idea which was shown by J. Dezert in 2002 [17] using a Neutrosophic cube. He has shown the difference between absolute and relative neutrosophic values and shown the regional partition for Intuitionistic and Neutrosophic ranges. As the summation of membership and non-membership values are different based on interdependency, in each case the degree of dependency has also been discussed by Florentin [18].

2 Intuitionist Fuzzy Set (IFS)

An Intuitionist Fuzzy set A in the domain E is defined according to the following form

$A = \{< x, \mu_a(x), \nu_a(x) > | x \in E\}$, where $(\mu_a(x))$ is the degree of membership and $(\nu_a(x))$ is the degree of non-membership which lies in the range [0, 1] and $0 \leq \mu_a(x) + \nu_a(x) \leq 1$.

This logic differs from classical fuzzy when the term indeterministic or hesitancy comes and defined in Intuitionistic fuzzy as $\pi_a = 1 - \mu_a(x) - \nu_a(x)$, It is degree of hesitancy of x to A. When the term π_a becomes 0 the set becomes classical fuzzy set [2].

3 Neutrosophic Fuzzy Set (NFS)

In Neutrosophic set three standard and non standard subset T (Truth), I (Indeterminacy), F (False) are defined for each element of the set where the ranges of these components lies in $]^-0\ 1^+[$. An element x in universe U within a set M has three components with (t, i, f) which interprets as the belongingness of x in M represents with t% truth, i% indeterminacy, and f% falsehood. The components vary from 0 to 1, even can be less than 0 or greater than 1. The components are not necessarily a number but can be a subset of type discrete or continuous set. The other categories can be open, closed, half open, half closed sets and it can also be union or intersection of previous sets [19].

4 Operations on Intuitionistic and Neutrosophic Fuzzy Set

4.1 Some Operations on Intuitionistic Fuzzy Set

A, B are two Intuitionistic Fuzzy Sets of the set E where $\mu(x)$ is the degree of membership and $v(x)$ is the degree of non-membership [2] then,

1. $A \subset B\ iff\ (\forall x \in E)(\mu_a(x) \leq \mu_b(x)\ and\ v_a(x) \geq v_b(x))$
2. $A = B\ iff\ (\forall x \in E)(\mu_a(x) = \mu_b(x)\ and\ v_a(x) = v_b(x))$
3. $A \wedge B = \{< x, \min(\mu_a(x), \mu_b(x)), \max(v_a(x), v_b(x)) > | x \in E\}$
4. $A \vee B = \{< x, \max(\mu_a(x), \mu_b(x)), \min(v_a(x), v_b(x)) > | x \in E\}$
5. $A + B = \{< x, (\mu_a(x) + \mu_b(x) - \mu_a(x).\mu_b(x)), v_a(x).v_b(x) > | x \in E\}$
6. $A . B = \{x, \mu_a(x) . \mu_b(x), (v_a(x) + v_b(x) - v_a(x) . v_b(x)) > | x \in E\}$

4.2   Some Operations on Neutrosophic Fuzzy Set

A,B are the sets over Universe U, where $x(T_1, I_1, F_1) \in A$ represents element x in set A having T_1 as neutrosophic membership, I_1 as neutrosophic indeterminacy and F_1 as neutrosophic non-membership values [9].

1. *If $x(T_1, I_1, F_1) \in A$ and $x(T_2, I_2, F_2) \in B$ then,*

$$A \cup B = x(T_1 \oplus T_2 \ominus T_1 \odot T_2, I_1 \oplus I_2 \ominus I_1 \odot I_2, F_1 \oplus F_2 \ominus F_1 \odot F_2)$$

2. *If $x(T_1, I_1, F_1) \in A$ and $x(T_2, I_2, F_2) \in B$ then,*

$$A \cap B = x(T_1 \odot T_2, I_1 \odot I_2, F_1 \odot F_2)$$

3. *If $x(T_1, I_1, F_1) \in A$ and $x(T_2, I_2, F_2) \in B$ then,*

$$A/B = x(T_1 \ominus T_1 \odot T_2, I_1 \ominus I_1 \odot I_2, F_1 \ominus F_1 \odot F_2)$$

4. *If $x(T_1, I_1, F_1) \in A$ and $y(T', I', F') \in B$ then,*

$$A \times B = \left(x(T_1, I_1, F_1), y(T', I', F') \right)$$

5   Variants of Intuitionistic and Neutrosophic Fuzzy Set

5.1   Extensions of Intuitionistic Fuzzy Set

There are few extensions of Intuitionistic fuzzy set which change the universe or extent of this fuzzy and sometimes derived from the classical fuzzy.

5.1.1   Interval Valued Intuitionistic Fuzzy Set

Interval valued Intuitionistic Fuzzy is a combination to both Intuitionistic fuzzy and Interval valued fuzzy set. It is defined on the basic set E where its membership function is M_a and non-membership function is N_a[20],

$M_a: E \to INT([0, 1])$ and $N_a: E \to INT([0, 1])$ and $INT([0, 1])$ is the set of all subsets of the unit interval.

When each of the intervals M_a and N_a contains exactly one element for each element then the set becomes ordinary Intuitionistic fuzzy set. On the other hand if $N_a = \phi$ for each element then the set becomes simple Interval valued fuzzy set.

5.1.2 Intuitionistic L-Fuzzy Set

Intuitionistic L-Fuzzy set is another variation which is derived from L-fuzzy set, where L may be a complete Lattice, complete chain or a complete ordered semi-ring [21]. It is an object in E defined as

$A^* = \{< x, \mu_a(x), \nu_a(x) > | x \epsilon E\}$, where, membership value $\mu_a(x) : E \to L$, Non-membership value $\nu_a(x) : E \to L$ where $x \epsilon E$ and for every $x \epsilon E$
$0 \leq \mu_a(x) \leq (\nu_a(x))'$, Where $': L \to L$ is an order reserving operation in $(L, \leq)$.

5.1.3 Temporal Intuitionistic Fuzzy Set

An instance of Temporal Intuitionistic Fuzzy Set A(T) is defined over non empty set E and T where elements of T is called 'Time-moment' [22].

$$A(T) = \{< x, \mu_a(x, t), \nu_a(x, t) > | (x, t) \epsilon E \times T\},$$

where:

(a) $A \subset E$ is a fixed set,
(b) $\mu_a(x, t) + \nu_a(x, t) \leq 1$ for every $(x, t) \epsilon ExT$
(c) $\mu_a(x, t)$ and $\nu_a(x, t)$ are the degree of membership and non-membership value of the element $x \epsilon E$ at the time $t \epsilon T$.

5.1.4 Intuitionistic Fuzzy Set of Second Type

Intuitionistic fuzzy set of second type [1] is another extension of IFS and is defined with varied degree of dependence of membership and non-membership values. Let A be the object of IFS of second type such that

$A = \{< x, \mu_a(x), \nu_a(x) > | x \in E\}$, in which functions $\mu_a(x) : E \to [0, 1]$ and $\nu_a(x) : E \to [0, 1]$ is in the relation then [1],
$0 \leq \mu_a(x)^2 + \nu_a(x)^2 \leq 1$ and $\pi_a(x) = \sqrt{(1 - \mu_a(x)^2 - \nu_a(x)^2)}$, here $\pi_a(x)$ is the degree of non-determinacy of the element $x \in E$.

5.2   Extensions of Neutrosophic Fuzzy Set

5.2.1   Interval Neutrosophic Set

Interval Neutrosophic set [23] A in the space X for element $x \in X$ is characterised by Truth-membership function which is denoted by T_a, Indeterminacy-membership function denoted by I_a, Falsity-membership function denoted by F_a. For each element $x \in X, T_a(x), I_a(x), F_a(x) \subseteq [0, 1]$.

$$\text{When } X \text{ is continuous} = \int_X < T(x), I(x), F(x) > /x, x \in X$$

$$\text{When } X \text{ is discrete, } A = \sum_{i=1}^{n} < T(x_i), I(x_i), F(x_i) > /x_i, \; x_i \in X$$

For each point x in X we can define T(x) = [inf T(x), sup T(x)], I(x) = [inf I(x), sup I(x)], F(x) = [inf F(x), sup F(x)] $\subseteq$ [0, 1] [24].

5.2.2   Single Valued Neutrosophic Set

Single Valued Neutrosophic set [25, 26] A in the space X for element $x \in X$ is characterised by Truth-membership function T_a, Indeterminacy-membership function I_a, Falsity-membership function F_a. For each element $x \in X, T_a(x), I_a(x), F_a(x) \in [0, 1]$. It is a generalization of classical set, fuzzy set, interval valued fuzzy set, intuitionistic fuzzy set and paraconsistent set.

A Single valued neutrosophic set A over a finite domain X is represented as

$$\text{When } X \text{ is continuous } A = \int_X < T(x), I(x), F(x) > /x, x \in X$$

$$\text{When } X \text{ is discrete } A = \sum_{i=1}^{n} < T(x_i), I(x_i), F(x_i) > /x_i, x_i \in X$$

Both of these variants of Neutrosophic set differs from the ordinary Neutrosophic set as the truth-membership value, Indeterminacy-membership value and the False-membership value for Interval Neutrosophic Set (INS) and Single Valued Neutrosophic Set (SVNS) lies in the range of [0, 1] but for Neutrosophic set it is non-standard value $]^-0, 1^+[$.

5.2.3 Refined Neutrosophic Set

n-valued refined neutrosophic set [16] introduced by Florentin [27] has a general definition where the Truth, Indeterminacy and Falsehood are subdivided into parts of the same components. An element x of A is comprised of $x\left(T_1, T_2, \ldots, T_p; I_1, I_2, \ldots, I_r; F_1, F_2, \ldots, F_s\right) \in A$

$$where\ p, r, s \geq 1\ and\ p + r + s \geq 3.$$

$T_1, T_2, \ldots, T_p; I_1, I_2, \ldots, I_r; F_1, F_2, \ldots, F_s$ are sub components of membership degrees, Indeterminacy degrees and non-membership degrees respectively.

6 Propositional and Predicate Calculus for Intuitionistic and Neutrosophic Fuzzy Set

6.1 Propositional and Predicate Calculus Defined Over Intuitionistic Fuzzy Logic

6.1.1 Propositional Calculus for Intuitionistic Fuzzy

For each proposition in Intuitionistic Fuzzy logic [28] a "truth-degree" and a "falsity-degree" is assigned to $\mu(p)$ and $v(p)$, here p is the proposition having a relation $\mu(p) + v(p) \leq 1$. Assignment of the proposition p to a function $V(p)$ can be defined as

$$V(p) = <\mu(p), v(p)>$$

The evaluation function over propositional logic is discussed below:

1. The evaluation of negation of proposition p i.e. $\sim p$ can be written as:

$$V(\sim p) = <v(p), \mu(p)>$$

2. The evaluation function for AND operator on proposition p and q

$$V(p \wedge q) = <\min(\mu(p), \mu(q)), \max(v(p), v(q))>$$

3. The evaluation function for OR operator on proposition p and q

$$V(p \vee q) = <\max(\mu(p), \mu(q)), \min(\nu(p), \nu(q))>$$

4. The evaluation of $p \supset q$ defined as

$$V(p \supset q) = <\max(\nu(p), \mu(q)), \min(\mu(p), \nu(q))>$$

6.1.2 Predicate Calculus for Intuitionistic Fuzzy

Predicate Calculus is the application of propositional operators over quantifiers (Existential and Universal). Predicate logic formulae are the applications of quantifiers using propositional operations "~"," $\wedge$ "," $\vee$ ","$\supset$". If A is a formula and x-A are variables then $\forall \times$ A and $\exists \times$ A are formulae. The function V defines on x ranges over E:

$$V(\forall xA) = < \min \mu(A(i(x) = a)), \max \nu(A(i(x) = a))> \; where \; a \in E$$
$$V(\exists xA) = < \max \mu(A(i(x) = a)), \min \nu(A(i(x) = a))> \; where \; a \in E$$

6.2 Propositional and Predicate Calculus Defined Over Neutrosophic Fuzzy Logic

In the case of Neutrosophic Fuzzy set propositional and predicate calculus [23] are defined over its variant Interval Neutrosophic set. We have already discussed the basics of Interval Neutrosophic set. In this section we will discuss only the Propositional Calculus and Predicate Calculus on Interval Neutrosophic Logic (INL). The proposition p in Interval Neutrosophic Fuzzy consists of $<t(p), i(p), f(p)>$ where $t(p), i(p), f(p) \in [0, 1]$ and it comprises of a syntax and a semantics to define the well-formed formulae.

6.2.1 Propositional Calculus for Neutrosophic Fuzzy

The set of formulae (well-formed formulae) on Interval Neutrosophic Propositional Calculus defined by its semantics are as follows in Table 1.

Table 1 Propositional calculus for Neutrosophic Fuzzy

Connectives	Semantics
$INL(\sim p)$	$<f(p), 1 - i(p), t(p)>$
$INL(p \wedge q)$	$< \min(t(p), t(q)), \max(i(p), i(q)), \max(f(p), f(q))>$
$INL(p \vee q)$	$< \max(t(p), t(q)), \min(i(p), i(q)), \min(f(p), f(q))>$
$INL(p \rightarrow q)$	$< \min(1, 1 - t(p) + t(q)), \max(0, i(q) - i(p)), \max(0, f(q) - f(p)) >$

Table 2 Predicate calculus for neutrosophic fuzzy

Connectives	Semantics
$INP(\forall x F)$	$< \min t(F(E(x))), \min i(F(E(x))), \max f(F(E(x))) >, E(x) \in D$
$INP(\exists x F)$	$< \max t(F(E(x))), \max i(F(E(x))), \min f(F(E(x))) >, E(x) \in D$

6.2.2 Predicate Calculus for Neutrosophic Fuzzy

The semantics in first order predicate logic on Interval Neutrosophic set gives the meaning of well formed formulae. The predicate calculus semantics are same as propositional calculus for the four connectives "~", "∧", "∨", "→". The semantics of qualifiers on the propositions of Interval Neutrosophic Predicate Logic (INP) are given below. Here the interpretation function (or interpretation) of a formula F in the first order interval neutrosophic predicate logic consists of a nonempty domain D (Table 2).

7 Degree of Dependence of Each Component of Neutrosophic and Intuitionistic Fuzzy

For Single valued neutrosophic set the sum of the components $(T + I + F)$ lies between 0 to 3 [18] when all three components are independent i.e, $0 \leq T + I + F \leq 3$.

When two components are dependent and the third component is independent with respect to the other two then the summation of the three lies between 0 to 2 i.e, $0 \leq T + I + F \leq 2$.

If each component of the three is dependent on each other, then the sum is between 0 to 1 i.e, $0 \leq T + I + F \leq 1$.

When the summation is greater than 1 i.e, when three or two components among T, I, F are independent then the information may be incomplete information ($sum < 1$) or para-consistent or contradictory information ($sum > 1$) or complete information ($sum = 1$).

If T, I, F are dependent then the information is either incomplete ($sum < 1$) or complete ($sum = 1$).

Three sources of T, I, F are independent if they do not influence each other and gives a max summation value 3. If they are fully dependent then it gives a max summation value 1.

In the case of Intuitionistic Fuzzy set the properties of dependence or independence for the two valued fuzzy set is applicable. In Intuitionistic fuzzy two components $\mu(x)$ and $\nu(x)$ that vary in the unit interval [0, 1] may be dependent or independent with a dependence degree $d°(\mu(x), \nu(x))$ with a relation: $0 \leq \mu(x) + \nu(x) \leq 2 - d°(\mu(x), \nu(x))$ where $d°(\mu(x), \nu(x))$ is called "degree of dependence". It is 0 when $\mu(x)$ and $\nu(x)$ are completely independent and 1 if both of them are completely dependent. The values of $d°(\mu(x), \nu(x))$ lie between [0, 1].

Therefore $2 - d°(\mu(x), \nu(x))$ is degree of independence between $\mu(x)$ and $\nu(x)$.

8 Geometrical Interpretation of Neutrosophic and Intuitionistic Fuzzy

A Neutrosophic cube introduced by J. Desert in 2002 [17] is a useful tool to visualize the concept of relative and absolute Neutrosophic set [29] and has the ability to differentiate between Intuitionistic and Neutrosophic ranges. It is drawn in the 3D Cartesian coordinate system where T is the truth axis value ranges from $]^-0\ 1^+[$, I is the indeterminacy axis value ranges from $]^-0\ 1^+[$ and F is the false axis value ranges from $]^-0\ 1^+[$.

In Fig. 1 we have taken a Neutrosophic cube with ranges [0, 1]. The cube can be extended in the more positive and more negative directions to get the range $]^-0\ 1^+[$.

The triangle ACH has the side of $\sqrt{2}$ units and has the summation of locus is 1 for any point situated on it. So, for any point p on or inside the triangle ACH gives $t_p + i_p + f_p = 1$ and it represents Atanassov-Intuitionistic fuzzy set.

Whereas points inside the pyramid ADCH including its side ADH,ADC and DCH but excluding the side ACH gives summation of locus value less than 1.

So for p in ADCH we have $t_p + i_p + f_p \leq 1$ and gives incomplete information.

The solid on the opposite side of pyramid ADCH with respect to the side ACH gives partially or fully independent information with $t_p + i_p + f_p \geq 1$.

9 Applications of Intuitionistic and Neutrosophic Fuzzy

IFS and NFS differ from classical fuzzy by incorporating undecidable and non-membership factors in the element of the set. The systems where these uncertain and non-membership elements lies underlying there the application of such fuzzy systems are successful. Liu and Wang [30] has applied IFS in Multi-criteria decision making problem. They have divided the uncertain portion of IFS into affirmative, dissent and abstention part in order to get decision by an evaluation function based on

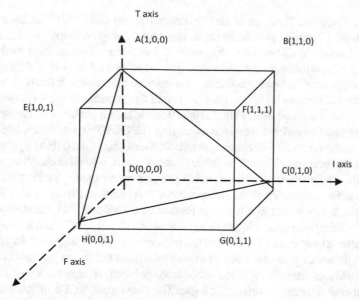

Fig. 1 Absolute Neutrosophic cube

Intuitionistic fuzzy logic. De et al. [10] determined the disease on the basis of symptoms by evaluation of max-min-max composition of IFS element with Intuitionistic Fuzzy relation. In this decision making problem non-membership function has more importance than membership function as there always exists a non-zero hesitant part. Fuzzy methods have been successfully applied for image segmentation. Huang et al. [31] applied Intuitionistic fuzzy to segment MRI images using C-means clustering techniques. Dengfeng and Chuntian [11] applied a new similarity measure between two Intuitionistic fuzzy set, as the previously proposed methods for similarity measure of a classical fuzzy is not applicable to IFS. They have used this measure for pattern recognition where maximum degree of similarity decides in which pattern class the sample belongs to. As introduction to the hesitant part gives better prediction level, the application of IFS in Financial interference is quite successful. This has been shown by Hajek and Olej [32], who have introduced a new de-fuzzification method (MOM) and applied in Intuitionistic neuro-fuzzy network trained with Particle swarm optimization for a financial inference system. They have shown that this method outperforms other Neuro-fuzzy inference techniques. In the network system also IFS has been applied by Dutta and Sait [12] for routing. The system is for both dynamic and static routing. The resource management is less costly in the system with overall gain of the performance of the network. Expert system is another application of Intuitionistic fuzzy. Degree of hesitancy can simulate better result for the decision of expert system as shown by Atanassov.

The neutrality component of Neutrosophic fuzzy logic gives information about the indeterminacy which can be captured into a mathematical formula to make certain decision. Deli et al. [16] makes feasible to use this uncertain information to get a medical diagnosis. This type of medical diagnosis was done on IFS earlier but here they have used Neutrosophic fuzzy logic. Hamming distance, Normalized hamming distance, Euclidean distance and Normalized Euclidean distance used on Neutrosophic set of symptoms to determine the type of disease. Liu et al. [30] suggested some new operator on Neutrosophic set and showed their result in multiple attribute group decision making problem. They proposed the generalized neutrosophic number Hamacher weighted averaging (GNNHWA) operator, generalized neutrosophic number Hamacher ordered weighted averaging (GNNHOWA) operator, and generalized neutrosophic number Hamacher hybrid averaging (GNNHHA) operator and studied their properties like union, intersection, t-norm, t-conorm etc. They showed these multi-attribute group decision making logic over air-quality ranking with more flexible results on single valued Neutrosophic information. Zhang et al. [24] used Interval Neutrosophic set which is an extension of SVNS and SNS, for multi-criteria decision making using some aggregation operator. NS has application to image processing also. Zhang et al. [13] decided the homogeneity of a image by using Neutrosophic logic which gives the value of degree of a pixel being a object pixel or edges or a background pixel. Kavitha et al. [33] have reduced the uncertainty of Intrusion detection system using Neutrosophic logic. They have used KDDcup'99 dataset and used Neutrosophic logic classifier to classify the dataset and used Genetic algorithm to make more precise Neutrosophic rules. Smarandache and Vladareanu [14] applied Neutrosophic logic to control a robot. A robot uses fusion of information from various sensors. The fuzziness or conflicting information optimized by Neutrosophic logic controls the kinematics of the robot. Ansari et al. [34] proposed a Neutrosophic classifier to classify Iris dataset. Besides the above mentioned application Neutrosophic logic has applied for detecting the neutrality of a word in a corpus also. Colhon et al. [15] determined the degree of neutrality of a neutral word. He classified the neutral word in three classes namely "pure neutral word", "half positive half negative word" and "Positive negative balanced word" using Neutrosophic logic.

10 Statistics of Research on Intuitionistic Fuzzy

As compared to Intuitionistic Fuzzy the effort of research has put into Neutrosophic Fuzzy are very less till today. So the picture we depict on the comparison of research publication are mainly based on Intuitionistic Fuzzy. We have tried to give a statistics of research work done in Intuitionistic Fuzzy on the basis of area of research, countries of the interest in Intuitionistic Fuzzy, fields where applied and the journals of publications.

Out of 1762 publications it shows that India is the second leading country after China to move ahead with the research in this topic. The following graph shows top 10 countries of the research publication in Intuitionistic Fuzzy (Fig. 2).

Statistics shows that Intuitionistic Fuzzy is the most explored area in the field of Computer Science. The other top four fields are Mathematics, Engineering, Operations research management science and Science. The application of Intuitionistic

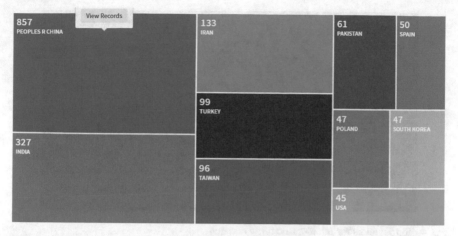

Fig. 2 Top 10 countries on the research interest in Intuitionistic fuzzy

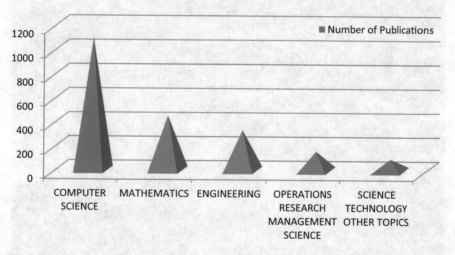

Fig. 3 Top 5 research areas for Intuitionistic fuzzy

Fuzzy in various Computer Science fields shows a new direction to handle uncertainty of data in corresponding domain (Figs. 3, 4 and 5; Table 3).

11 Conclusion

The differences of Neutrosophic and Intuitionistic fuzzy set gives the idea that both have their own importance in the field of mathematical logic, graphical, lattice, ring, field and other application areas. The implementation has changed diversely

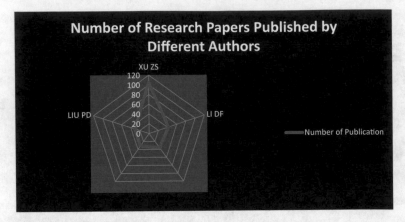

Fig. 4 Top 5 authors for Intuitionistic fuzzy research

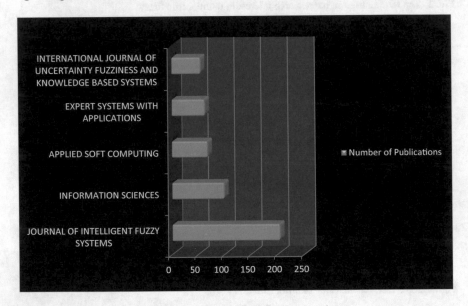

Fig. 5 Top 5 journals for publications of Intuitionistic fuzzy research

Table 3 Number of publications on Top 5 research areas

Research area	Number of publications
Computer Science	1096
Mathematics	454
Engineering	335
Operations Research Management Science	161
Science Technology and Other Topics	97

with the generalization or specialization of these two fuzzy sets. As the philosophy has changed so the application of these two types of fuzzy has also been changed. Neutrosophic fuzzy can be made fit to analyze fields where uncertainty plays a vital role. The undetermined factors give a mathematical notion to be included in the calculation. Specified degree of dependence and independence gives the idea of complete or incomplete or contradictory information. So in the age of information technology the data analysis can give new directions with the study of these fuzzy techniques over huge, uncertain and incomplete information.

References

1. F. Smarandache, Neutrosophic logic-generalization of the intuitionistic fuzzy logic. arXiv preprint math/0303009 (2003)
2. K.T. Atanassov, Intuitionistic fuzzy sets. Fuzzy Sets Syst. **20**(1), 87–96 (1986)
3. K.T. Atanassov, (2003, September). Intuitionistic fuzzy sets: past, present and future, in *EUSFLAT Conference*, pp. 12–19
4. L.A. Zadeh, Fuzzy sets. Inf. Control **8**, 3 (1965)
5. L.A. Zadeh, The concept of a linguistic variable and its application to approximate reasoning— I. Inf. Sci. **8**(3), 199–249 (1975)
6. C. Ashbacher, *Introduction to Neutrosophic Logic*. Infinite Study (2002)
7. F. Smarandache, *Classical Logic and Neutrosophic Logic*. Answers to K. Georgiev. Infinite Study (2016)
8. F. Smarandache, Neutrosophic logic—a generalization of the intuitionistic fuzzy logic. Multispace Multistructure Neutrosophic Transdisciplinarity (100 Collected Papers of Science) **4**, 396 (2010)
9. F. Smarandache, Neutrosophic set—a generalization of the intuitionistic fuzzy set. Int. J. Pure Appl. Math. **24**(3), 287 (2005)
10. S.K. De, R. Biswas, A.R. Roy, An application of intuitionistic fuzzy sets in medical diagnosis. Fuzzy Sets Syst. **117**(2), 209–213 (2001)
11. L. Dengfeng, C. Chuntian, New similarity measures of intuitionistic fuzzy sets and application to pattern recognitions. Pattern Recogn. Lett. **23**(1–3), 221–225 (2002)
12. A.K. Dutta, A.R.W. Sait, An application of intuitionistic fuzzy in routing networks. Editorial Preface **3**(6) (2012)
13. M. Zhang, L. Zhang, H.D. Cheng, A neutrosophic approach to image segmentation based on watershed method. Sig. Process. **90**(5), 1510–1517 (2010)
14. F. Smarandache, L. Vlădăreanu, Applications of neutrosophic logic to robotics: an introduction, in 2011 IEEE *International Conference on Granular Computing (GrC)* (IEEE, 2011), pp. 607–612
15. M. Colhon, Ş. Vlăduţescu, X. Negrea, How objective a neutral word is? A neutrosophic approach for the objectivity degrees of neutral words. Symmetry **9**(11), 280 (2017)
16. I. Deli, S. Broumi, F. Smarandache, On neutrosophic refined sets and their applications in medical diagnosis. J. New Theory **6**, 88–98 (2015)
17. J. Dezert, Open questions in neutrosophic inferences. Multiple Valued Logic Int. J. **8**(3), 439–472 (2002)
18. F. Smarandache, *Degree of Dependence and Independence of the (Sub) Components of Fuzzy Set and Neutrosophic Set*. Infinite Study (2016)
19. F. Smarandache, *Neutrosophic Theory and Its Applications. Collected Papers, I*. Neutrosophic Theory and Its Applications (2014), p. 10
20. K.T. Atanassov, Operators over interval valued intuitionistic fuzzy sets. Fuzzy Sets Syst. **64**(2), 159–174 (1994)

21. D. Çoker, Fuzzy rough sets are intuitionistic L-fuzzy sets. Fuzzy Sets Syst. **96**(3), 381–383 (1998)
22. K.T. Atanassov, Temporal intuitionistic fuzzy relations, in *Flexible Query Answering Systems* (Physica, Heidelberg, 2001), pp. 153–160
23. H. Wang, F. Smarandache, R. Sunderraman, Y.Q. Zhang, *Interval Neutrosophic Sets and Logic: Theory and Applications in Computing: Theory and Applications in Computing*, vol. 5. Infinite Study (2005)
24. H.Y. Zhang, J.Q. Wang, X.H. Chen, Interval neutrosophic sets and their application in multicriteria decision making problems. Sci. World J. (2014)
25. H. Wang, F. Smarandache, Y. Zhang, R. Sunderraman, Single valued neutrosophic sets, in *Proceeding of the 10th 476 International Conference on Fuzzy Theory and Technology* (2005)
26. P. Majumdar, S.K. Samanta, On similarity and entropy of neutrosophic sets. J. Intell. Fuzzy Syst. **26**(3), 1245–1252 (2014)
27. F. Smarandache, *n-Valued Refined Neutrosophic Logic and Its Applications to Physics*. Infinite Study (2013)
28. K.T. Atanassov, Elements of Intuitionistic Fuzzy Logics, in *Intuitionistic Fuzzy Sets* (Physica, Heidelberg, 1999), pp. 199–236
29. F. Smarandache, *A Geometric Interpretation of the Neutrosophic Set—A Generalization of the Intuitionistic Fuzzy Set*. arXiv preprint math/0404520 (2004)
30. H.W. Liu, G.J. Wang, Multi-criteria decision-making methods based on intuitionistic fuzzy sets. Eur. J. Oper. Res. **179**(1), 220–233 (2007)
31. C.W. Huang, K.P. Lin, M.C. Wu, K.C. Hung, G.S. Liu, C.H. Jen, Intuitionistic fuzzy c-means clustering algorithm with neighborhood attraction in segmenting medical image. Soft. Comput. **19**(2), 459–470 (2015)
32. P. Hájek, V. Olej, Intuitionistic neuro-fuzzy network with evolutionary adaptation. Evolving Syst. **8**(1), 35–47 (2017)
33. B. Kavitha, S. Karthikeyan, P.S. Maybell, An ensemble design of intrusion detection system for handling uncertainty using Neutrosophic Logic Classifier. Knowl. Based Syst. **28**, 88–96 (2012)
34. A.Q. Ansari, R. Biswas, S. Aggarwal, Neutrosophic classifier: an extension of fuzzy classifer. Appl. Soft Comput. **13**(1), 563–573 (2013)

Study of the Relevance of Polynomial Order in Takagi-Sugeno Fuzzy Inference Systems Applied in Diagnosis Problems

Emanuel Ontiveros-Robles, Patricia Melin and Oscar Castillo

Abstract Fuzzy Logic has been implemented successfully for different kind of problems. However, there is an opportunity for these methods to be improved in the realm of classifications problems. The present paper is focused in a specific application of classification problems, the diagnosis systems, this problem consists in training an intelligent system to learn the relationship between symptoms and diagnosis, this kind of problems are usually based in powerful non-linear methods for example Modular Neural-Networks or complex hybrids models, however, in this paper are applied the Type-1 Takagi Sugeno Fuzzy Systems (TSK) but analyzing the improvement of their performance by increasing the order of the Sugeno polynomial. The conventional Takagi-Sugeno Fuzzy Systems are based in the aggregation of first-order polynomial but it is interesting to observe the effect of increase the order of this polynomial, the TSK Fuzzy Diagnosis Systems are evaluated by their accuracy obtained in ten benchmark dataset of the UCI Dataset Repository, for different kind of diseases and different difficult levels.

Keywords Diagnosis systems · Sugeno fuzzy models · Neuro-fuzzy system

1 Introduction

Nowadays, the Fuzzy Logic representation is well accepted in different kind of applications, for example, control applications [1–12], image processing [13–16], industrial applications [17–19], and others, however, an interesting field of computational system is the diagnosis system, that is a classification problem, in the current paper it is evaluated the accuracy of Fuzzy Systems in diagnosis problems, for example the problems presented in [20–25], it is interesting to observe that the Fuzzy Systems are not commonly used for this kind of problems, usually, the diagnosis systems are solved helped with powerful algorithms or methods for example, Modular Neural

E. Ontiveros-Robles · P. Melin · O. Castillo (✉)
Tijuana Institute of Technology, Tijuana, BC, Mexico
e-mail: ocastillo@tectijuana.mx

© Springer Nature Switzerland AG 2020
O. Castillo and P. Melin (eds.), *Hybrid Intelligent Systems in Control,
Pattern Recognition and Medicine*, Studies in Computational Intelligence 827,
https://doi.org/10.1007/978-3-030-34135-0_2

19

Networks or hybrid methods, but, in this paper, we explore the performance of the Takagi-Sugeno Fuzzy Systems for this problems.

The Takagi-Sugeno Fuzzy Systems are frequently used for modelling non-lineal problems, and offer versatility to be combined with different kind of methods or systems. For example, they are the used in the Adaptive Neuro Fuzzy Inference Systems (ANFIS) that is a hybrid method of Fuzzy Systems with Artificial Neural Networks.

The organization of the present paper is explained as follows: Sect. 2 a brief introduction to Takagi-Sugeno Fuzzy Systems, that are the fundament of the paper, Sect. 3 talks about the Takagi-Sugeno Fuzzy Diagnosis Systems, that are the proposed approach to diagnosis based in fuzzy logic, Sect. 4 present the experimental results realized with ten datasets of UCI Dataset repository and finally Sect. 5 presents the conclusion of the work.

2 Takagi-Sugeno Fuzzy Systems

Fuzzy Logic is a logic model that expands the conventional logic [26], that handle two membership values, and introduces the concept of membership degree, this degree is within [0, 1]. This is expressed in Eq. (1).

$$A = \{(x, \mu_A(x))|x \in X\} \tag{1}$$

where $\mu_A(x):X \rightarrow [0, 1]$ is the membership degree of x to the set A.

This evolution of conventional logic allows to model the human knowledge in special systems called Fuzzy Inference Systems through rules that are usually called *knowledge base*, these rules are related with special logical operators called *T-Norm* and *S-Norm* and relates the membership degree associated with the input signals features (*big, small, close, far*) with output signals features (*go forward, go back*).

There is a different kind of Fuzzy Inference Systems, for example, Mamdani FIS, Tsukamoto FIS and Takagi-Sugeno FIS. The present paper is focused in the Takagi-Sugeno FISs, this kind of FIS had been implemented successfully in several problems because their output is polynomial.

The TSK Fuzzy rules are expressed as follows Eq. (2).

$$R^i : If\, x_1\, is\, A_1^i\, and\, x_2\, is\, A_2^i, \ldots, x_n\, is\, A_n^i$$

$$then\, f^i = f_i\left(x_1, x_2, \ldots, x_n; \vec{a}^i\right) = f_i(\vec{x}; \vec{a}^i) = a_0^i + a_1^i x_1 + \cdots + a_n^i x_n \tag{2}$$

The output of the fuzzy system is computed by the following Eq. (3)

$$\hat{y} = \frac{\sum_{i=1}^c R^i f^i}{\sum_{i=1}^c R^i} \tag{3}$$

where R^i is the firing force of the i_{th} rule.

Although the conventional form of TSK output denoted for (3) is a first-order polynomial, this approach can be expanded to a high-order polynomial output, that are proposed in [27] and their expression is described in (4).

$$f_i\left(\vec{x}; \vec{a}^i\right) = a_0^i + a_1^i x_1 + \cdots + a_n^i x_n + a_{n+1} x_1^2$$
$$+ \ldots a_{2n} x_n^2 + a_{(m-1)n+1} x_1^m + \cdots + a_{mn} x_n^m \tag{4}$$

That can be simplified as follows (5)

$$f_i\left(\vec{x}; \vec{a}^i\right) = a_0^i + \sum_{k=1}^m \sum_{s=1}^n a_{(k-1)n+s} x_1^k \tag{5}$$

3 Takagi-Sugeno Fuzzy Diagnosis System

For the present paper, it is proposed to use the architecture of the Neuro Fuzzy Inference System proposed for Jang in [28], the architecture consist on a neural network with five layers and it is illustrated in Fig. 1.

Layer 0: Inputs

$$X = \left[x_{p,1}, x_{p,2}\right] \tag{6}$$

Layer 1: Fuzzification

$$\mu_{F_i^l}\left(x_{p,i}\right) = e^{\frac{\left(x_{p,i} - m_{l,i}\right)^2}{\partial_{l,i}^2}} \tag{7}$$

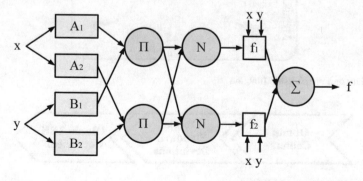

Fig. 1 ANFIS architecture

Layer 2: Compute firing strengths (T-Norm as Product)

Remembering the rules are defined by the number of input membership functions and are the combination of these.

$$\alpha_{p,R} = \mu_{F_1^{l(R)}}(x_{p,i}) * \mu_{F_2^{l(R)}}(x_{p,i}) * \mu_{F_3^{l(R)}}(x_{p,i}) \tag{8}$$

Layer 3: Normalization

$$\phi_i = \frac{\alpha_{p,R}}{\sum_{i=1}^{m} \alpha_{p,R}} \tag{9}$$

Layer 4: Sugeno Polynomial

$$f_i(\vec{x}; \vec{a}^i) = a_0^i + \sum_{k=1}^{m} \sum_{s=1}^{n} a_{(k-1)n+s} x_1^k \tag{10}$$

It is proposed to be used Gaussian membership functions in the inputs. This kind of membership function are widely used in classification approaches with fuzzy logic and the graphic representation can be observed in Fig. 2, the mathematical representation can be appreciated in Eq. (11),

$$\text{gaussmf}(x, \sigma, c) = e^{-\frac{1}{2}\left(\frac{x-c}{\sigma}\right)^2} \tag{11}$$

The process to generate the TSK FDS is divided in three steps and these are presented in Fig. 3.

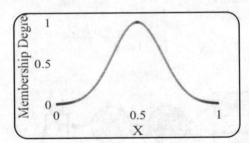

Fig. 2 Gaussian membership function

Fig. 3 Steps for generate TSK FDS

This process consists in obtaining the parameters of the proposed TSK FDS, the first step is obtaining the centers, this centers can be obtained by the implementation of clustering algorithms, for example, k-nearest neighbors, Fuzzy C-Means or Subtractive method, the centers of the Gaussian membership functions are the center of the clusters obtained in this step, in this case we proposed to use FCM.

The second step is obtaining the standard deviations of the Gaussian membership functions, these parameters can be obtained with statistical methods for example, in this case, we obtain the standard deviations helped in a non-lineal regression.

Finally, the third step is obtaining the TSK Coefficients, these values can be obtained with optimization methods or mathematical methods, in this case we proposed to use least-square error.

4 Experimental Results

This Section introduces the experiments realized in order to evaluate the impact of the Sugeno polynomial order in the accuracy of the fuzzy diagnosis system.

The experiments in evaluates the accuracy obtained for the FDS with different orders of Sugeno polynomial, and reporting the average of 30 experiments, on the other hand, there are also documented the performance in the training data in order to observe if there exist any relationship for training data and the accuracy for test data, it is important to note that the 70% of data are used for training and 30% is used for test the FDS.

The experiments are used as benchmark problems ten datasets that can be founded in the UCI Dataset Repository, these datasets correspond to different kind of diseases and are widely used for test different classification methods. The benchmark datasets that are proposed to be used are listed in Table 1.

Table 1 Benchmark datasets

Dataset	Attributes	Instances
Breast cancer Wisconsin (original) data set	10	699
Breast cancer Wisconsin (diagnostic) data set	32	569
Haberman's survival data	3	306
Mammographic mass data set	6	906
Pima Indians diabetes data set	9	768
Fertility data set	10	100
Immunotherapy data set	8	90
Cryotherapy data set	7	90
Breast cancer coimbra	10	116
Heart dataset	13	270

Figures 4, 5, 6, 7, 8, 9, 10, 11, 12 and 13 illustrate the obtained results, the results are measured in accuracy and the graphs represent the accuracy obtained for different order of Sugeno polynomial.

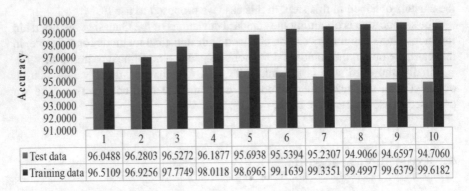

	1	2	3	4	5	6	7	8	9	10
■ Test data	96.0488	96.2803	96.5272	96.1877	95.6938	95.5394	95.2307	94.9066	94.6597	94.7060
■ Training data	96.5109	96.9256	97.7749	98.0118	98.6965	99.1639	99.3351	99.4997	99.6379	99.6182

Fig. 4 WBCD dataset—results

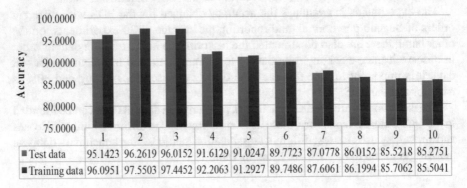

	1	2	3	4	5	6	7	8	9	10
■ Test data	95.1423	96.2619	96.0152	91.6129	91.0247	89.7723	87.0778	86.0152	85.5218	85.2751
■ Training data	96.0951	97.5503	97.4452	92.2063	91.2927	89.7486	87.6061	86.1994	85.7062	85.5041

Fig. 5 WDCB dataset—results

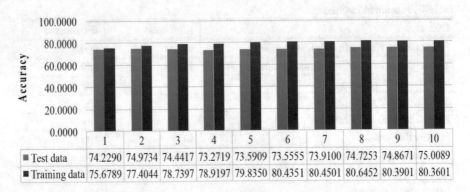

	1	2	3	4	5	6	7	8	9	10
■ Test data	74.2290	74.9734	74.4417	73.2719	73.5909	73.5555	73.9100	74.7253	74.8671	75.0089
■ Training data	75.6789	77.4044	78.7397	78.9197	79.8350	80.4351	80.4501	80.6452	80.3901	80.3601

Fig. 6 Haberman dataset—results

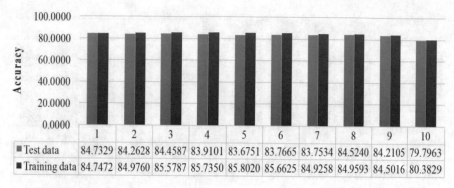

	1	2	3	4	5	6	7	8	9	10
■ Test data	84.7329	84.2628	84.4587	83.9101	83.6751	83.7665	83.7534	84.5240	84.2105	79.7963
■ Training data	84.7472	84.9760	85.5787	85.7350	85.8020	85.6625	84.9258	84.9593	84.5016	80.3829

Fig. 7 Mammograms dataset—results

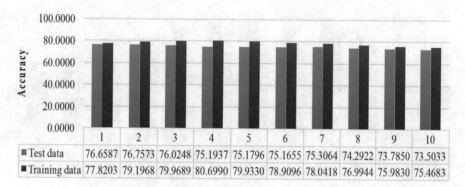

	1	2	3	4	5	6	7	8	9	10
■ Test data	76.6587	76.7573	76.0248	75.1937	75.1796	75.1655	75.3064	74.2922	73.7850	73.5033
■ Training data	77.8203	79.1968	79.9689	80.6990	79.9330	78.9096	78.0418	76.9944	75.9830	75.4683

Fig. 8 Diabetes dataset—results

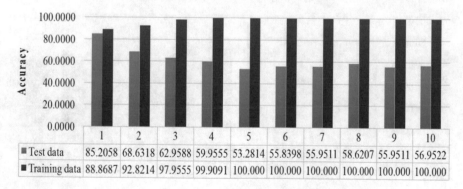

	1	2	3	4	5	6	7	8	9	10
■ Test data	85.2058	68.6318	62.9588	59.9555	53.2814	55.8398	55.9511	58.6207	55.9511	56.9522
■ Training data	88.8687	92.8214	97.9555	99.9091	100.000	100.000	100.000	100.000	100.000	100.000

Fig. 9 Fertility dataset—results

As you can observe, the performance does not change significantly for the different order of Sugeno polynomial, however, in several of the cases, the best performance is obtained for first-order polynomial, then, Fig. 14 illustrates the average of the evaluated datasets.

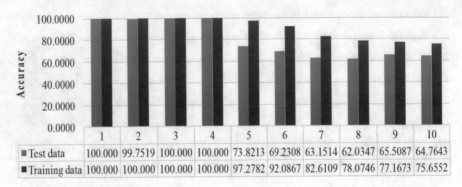

	1	2	3	4	5	6	7	8	9	10
Test data	100.000	99.7519	100.000	100.000	73.8213	69.2308	63.1514	62.0347	65.5087	64.7643
Training data	100.000	100.000	100.000	100.000	97.2782	92.0867	82.6109	78.0746	77.1673	75.6552

Fig. 10 Immunotherapy—results

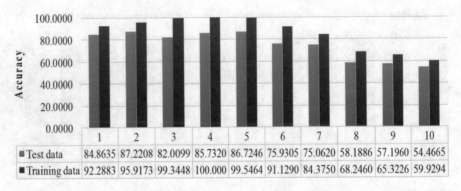

	1	2	3	4	5	6	7	8	9	10
Test data	84.8635	87.2208	82.0099	85.7320	86.7246	75.9305	75.0620	58.1886	57.1960	54.4665
Training data	92.2883	95.9173	99.3448	100.000	99.5464	91.1290	84.3750	68.2460	65.3226	59.9294

Fig. 11 Cryotherapy data set—results

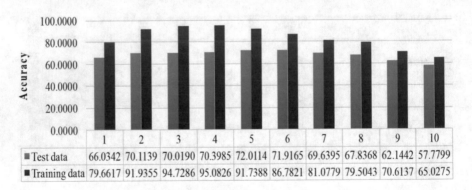

	1	2	3	4	5	6	7	8	9	10
Test data	66.0342	70.1139	70.0190	70.3985	72.0114	71.9165	69.6395	67.8368	62.1442	57.7799
Training data	79.6617	91.9355	94.7286	95.0826	91.7388	86.7821	81.0779	79.5043	70.6137	65.0275

Fig. 12 Breast cancer coimbra—results

It is interesting to observe that the better performance is obtained for first-order Sugeno polynomial and exist an over fitting effect, when the performance of the test data decrease, the performance of the training data grow up.

Study of the Relevance of Polynomial Order in Takagi-Sugeno ...

27

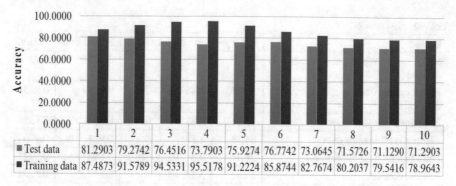

	1	2	3	4	5	6	7	8	9	10
■ Test data	81.2903	79.2742	76.4516	73.7903	75.9274	76.7742	73.0645	71.5726	71.1290	71.2903
■ Training data	87.4873	91.5789	94.5331	95.5178	91.2224	85.8744	82.7674	80.2037	79.5416	78.9643

Fig. 13 Heart dataset—results

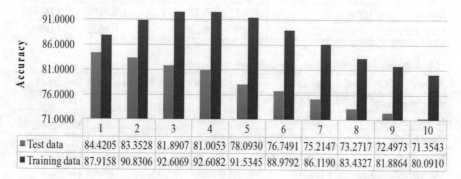

	1	2	3	4	5	6	7	8	9	10
■ Test data	84.4205	83.3528	81.8907	81.0053	78.0930	76.7491	75.2147	73.2717	72.4973	71.3543
■ Training data	87.9158	90.8306	92.6069	92.6082	91.5345	88.9792	86.1190	83.4327	81.8864	80.0910

Fig. 14 Results summary

Table 2 report the best performance obtained for the different datasets and document also the corresponding Sugeno polynomial order and two metrics related with the data, the fractal dimension and the data entropy, this metrics are documented in order to find a relation with them and the order of Sugeno polynomial.

Tables 3, 4, 5, 6, 7, 8, 9 and 10 documents some of the results reported in the literature, they are presented in order to compare the performance obtained with respect to other kinds of methods reported in the literature, for examples, decision trees, Support-Vector-Machine or Artificial Neural Networks.

5 Conclusions

Based in the realized experiments we can conclude that the use of low order of low-order polynomial in T1 TSK FDS is recommended. This can be explained because when the order of Sugeno polynomial is increased, the system is most complex and tends to *memorize* the diagnosis and it is not useful for another data.

Table 2 Best results

Dataset	Average	Std. dev.	Order	Fractal dim	Entropy
Breast cancer Wisconsin (original) data set	96.53	1.15	3	0.993	0.93
Breast cancer Wisconsin (diagnostic) data set	96.26	1.84	3	0.9628	0.9531
Haberman's survival data	75.01	1.82	10	0.9301	0.8326
Mammographic mass data set	84.73	1.82	1	0.917	0.9947
Pima Indians diabetes data set	76.76	2.5	2	0.8884	0.9311
Fertility data set	85.21	5.57	1	0.8924	0.5221
Immunotherapy data set	100	0	1	1	0.7186
Cryotherapy data set	87.22	7.39	2	1	0.9964
Breast cancer coimbra	72.01	5.72	5	0.9196	0.9916
Heart dataset	81.29	3.33	1	0.9522	0.9913

Table 3 Breast cancer Wisconsin (original) data set

Author	Method	Average	Std. dev.	Reference
Sang Won Yoon	SVM	97.1		[20]
L.B. Goncalves	Neuro-fuzzy	98.26		[36]
Eyad Elyan	Random forest	95.88		[37]
Eyad Elyan	Random forest	96.73		[23]
DiptenduSinha Roy	SMO	97.85		[24]
Sheng Weiguo	NN	97.71		[25]
Proposed approach		96.53	1.15	

Table 4 Breast cancer Wisconsin (diagnostic) data set

Author	Method	Average	Std. dev.	Reference
Sang Won Yoon	SVM	97.68		[20]
Diptendu Sinha Roy	SMO	98.77		[24]
E. Boros	LAD	96.9		[38]
Proposed approach		96.26	1.84	

Table 5 Haberman's survival data

Author	Method	Average	Std. dev.	Reference
Eyad Elyan	Random forests	72.82		[37]
Berk Ustun	SLIM	70.8		[39]
Morente-Molinera	Supervised classification	74.2		[40]
Hamdi Tolga Kahraman	ABC-based k-nn	87.28		[41]
William A. Young II	V-synth	79		[42]
Proposed approach			75.01	1.82

Table 6 Mammographic mass data set

Author	Method	Average	Std. dev.	Reference
Saritas I	ANN	85.5		[38]
Kuntoro Adi Nugroho	DT	83.3		[43]
Zadeh Shirazi A	SOM	94.5		[44]
Eyad Elyan	Random forests	82.19		[37]
Eyad Elyan	Random forests	83.83		[23]
R. Vidya Banua	PCA	82		[45]
Berk Ustun	SLIM	80.5		[39]
Proposed approach		84.73	1.82	

Table 7 Pima Indians diabetes data set

Author	Method	Average	Std. dev.	Reference
Sang Won Yoon	SVM	78.6		[36]
Pasi Luukka	ANFIS	75.29		[46]
Humar Kahramanli	FNN	84.2		[47]
Salih Güneşa	GDA–LS-SVM	82.05		[48]
Fatemeh Mansourypoor	RLEFRBS	84		[49]
Eyad Elyan	Random forests	78.20		[23]
R. Vidya Banua	PCA	78		[45]
Diptendu Sinha Roy	SMO	79.3		[24]
Sheng Weiguo	NN	79.6		[25]
E. Boros	LAD	71.9		[38]
Proposed approach		76.76	2.5	

Table 8 Immunotherapy data set

Author	Method	Average	Std. dev.	Reference
Sabita Khatri	DT	81		[46]
Sabita Khatri	DT	96.6		[46]
Selahaddin Batuhan Akben	DT	90		[50]
Khozeimeh F	FIS	83.33		[51]
Proposed approach		100	0	

As can be observed in the experiment, there not exist relationship between the accuracy obtained for training data and test data, so, optimize a FDS based in the training data produce an over fitting.

Another conclusion, based in the consulted literature, is that the fuzzy logic does not show the better results in comparison with respect another method designed for

Table 9 Cryotherapy data set

Author	Method	Average	Std. dev.	Reference
Sabita Khatri	DT	93.7		[46]
Sabita Khatri	DT	98.9		[46]
Selahaddin Batuhan Akben	DT	94.4		[50]
Khozeimeh F	FIS	80		[51]
Proposed approach		87.22	7.39	

Table 10 Heart dataset

Author	Method	Average	Std. dev.	Reference
Eyad Elyan	Random forests	83.96		[37]
Eyad Elyan	Random forests	82.64		[23]
Berk Ustun	SLIM	83.5		[39]
Sheng Weiguo	NN	83.6		[25]
Proposed approach		81.29	3.33	

classifications or most complex methods for example Artificial Neural Networks. However, the obtained results are similar than other complex methods and as future work it is possible to analyze the performance by using Type-2 Fuzzy Logic [29–32]. We also could consider other application areas like in [33–35].

References

1. C. Caraveo, F. Valdez, O. Castillo, Optimization of fuzzy controller design using a new bee colony algorithm with fuzzy dynamic parameter adaptation. Appl. Soft Comput. **43**, 131–142 (2016)
2. O. Castillo, L. Amador-Angulo, J.R. Castro, M. Garcia-Valdez, A comparative study of type-1 fuzzy logic systems, interval type-2 fuzzy logic systems and generalized type-2 fuzzy logic systems in control problems. Inf. Sci. **354**, 257–274 (2016)
3. O. Castillo, P. Melin, A. Alanis, O. Montiel, R. Sepulveda, Optimization of interval type-2 fuzzy logic controllers using evolutionary algorithms. Soft. Comput. **15**(6), 1145–1160 (2011)
4. N.R. Cazarez-Castro, L.T. Aguilar, O. Castillo, Designing type-1 and type-2 fuzzy logic controllers via fuzzy lyapunov synthesis for nonsmooth mechanical systems. Eng. Appl. Artif. Intell. **25**(5), 971–979 (2012)
5. L. Cervantes, O. Castillo, Type-2 fuzzy logic aggregation of multiple fuzzy controllers for airplane flight control. Inf. Sci. **324**, 247–256 (2015)
6. H. Chaoui, M. Khayamy, A.A. Aljarboua, Adaptive interval type-2 fuzzy logic control for PMSM drives with a modified reference frame. IEEE Trans. Ind. Electron. **64**(5), 3786–3797 (2017)
7. A.M. El-Nagar, M. El-Bardini, and N.M. EL-Rabaie, Intelligent control for nonlinear inverted pendulum based on interval type-2 fuzzy PD controller. Alex. Eng. J. **53**(1), 23–32 (Mar. 2014)

8. M.H. Khooban, T. Niknam, M. Sha-Sadeghi, Speed control of electrical vehicles: a time-varying proportional #x2013; integral controller-based type-2 fuzzy logic. IET Sci. Meas. Technol. **10**(3), 185–192 (2016)
9. E. Ontiveros-Robles, P. Melin, and O. Castillo, Comparative analysis of noise robustness of type 2 fuzzy logic controllers. Kybernetika 175–201 (Mar. 2018)
10. A.I. Roose, S. Yahya, H. Al-Rizzo, Fuzzy-logic control of an inverted pendulum on a cart. Comput. Electr. Eng. **61**, 31–47 (2017)
11. H. Zhou, H. Ying, J. Duan, Adaptive control using interval type-2 fuzzy logic for uncertain nonlinear systems. J. Cent. South Univ. Technol. **18**(3), 760 (2011)
12. P. Melin, E. Ontiveros-Robles, C.I. Gonzalez, J.R. Castro, and O. Castillo, An approach for parameterized shadowed type-2 fuzzy membership functions applied in control applications. Soft Comput. (Sep. 2018)
13. C.I. Gonzalez, J.R. Castro, O. Mendoza, A. Rodríguez-Díaz, P. Melin, and O. Castillo, Edge detection method based on interval type-2 fuzzy systems for color images, in *2015 Annual Conference of the North American Fuzzy Information Processing Society (NAFIPS) held jointly with 2015 5th World Conference on Soft Computing (WConSC)*, 2015, pp. 1–6
14. C.I. Gonzalez, P. Melin, J.R. Castro, O. Mendoza, Optimization of interval type-2 fuzzy systems for image edge detection. Appl. Soft Comput. **47**, 631–643 (2016)
15. P. Melin, C.I. Gonzalez, J.R. Castro, O. Mendoza, O. Castillo, Edge-detection method for image processing based on generalized type-2 fuzzy logic. IEEE Trans. Fuzzy Syst. **22**(6), 1515–1525 (2014)
16. O. Mendoza, P. Melin, and G. Licea, A new method for edge detection in image processing using interval type-2 fuzzy logic, in *2007 IEEE International Conference on Granular Computing (GRC 2007)*, (2007) pp. 151–151
17. M.H. Khooban, N. Vafamand, A. Liaghat, T. Dragicevic, An optimal general type-2 fuzzy controller for urban traffic network. ISA Trans. **66**, 335–343 (2017)
18. C.F. Juang, K.J. Juang, Circuit implementation of data-driven TSK-type interval type-2 neural fuzzy system with online parameter tuning ability. IEEE Trans. Ind. Electron. **64**(5), 4266–4275 (2017)
19. J. Debnath, D. Majumder, A. Biswas, Air quality assessment using weighted interval type-2 fuzzy inference system. Ecol. Inform. **46**, 133–146 (2018)
20. H. Wang, B. Zheng, S.W. Yoon, H.S. Ko, A support vector machine-based ensemble algorithm for breast cancer diagnosis. Eur. J. Oper. Res. **267**(2), 687–699 (2018)
21. E. Elyan, M.M. Gaber, A fine-grained random forests using class decomposition: an application to medical diagnosis. Neural Comput. Appl. **27**(8), 2279–2288 (2016)
22. N. MadhuSudana Rao, K. Kannan, X. Gao, D.S. Roy, Novel classifiers for intelligent disease diagnosis with multi-objective parameter evolution. Comput. Electr. Eng. **67**, 483–496 (Apr. 2018)
23. W. Sheng, P. Shan, S. Chen, Y. Liu, F.E. Alsaadi, A niching evolutionary algorithm with adaptive negative correlation learning for neural network ensemble. Neurocomputing **247**, 173–182 (2017)
24. I. Saritas, Prediction of breast cancer using artificial neural networks. J. Med. Syst. **36**(5), 2901–2907 (2012)
25. S. Khatri, D. Arora, A. Kumar, Enhancing decision tree classification accuracy through genetically programmed attributes for wart treatment method identification. Procedia Comput. Sci. **132**, 1685–1694 (2018)
26. L.A. Zadeh, Fuzzy sets. Inf. Control **8**(3), 338–353 (1965)
27. J.R. Castro, O. Castillo, M.A. Sanchez, O. Mendoza, A. Rodríguez-Diaz, P. Melin, Method for higher order polynomial Sugeno Fuzzy inference systems. Inf. Sci. **351**, 76–89 (2016)
28. J.-S.R. Jang, ANFIS: adaptive-network-based fuzzy inference system. IEEE Trans. Syst. Man Cybern. **23**(3), 665–685 (1993)

29. C. Leal Ramírez, O. Castillo, P. Melin, A. Rodríguez Díaz, Simulation of the bird age-structured population growth based on an interval type-2 fuzzy cellular structure. Inf. Sci. **181**(3), 519–535 (2011)
30. O. Castillo, P. Melin, Intelligent systems with interval type-2 fuzzy logic. Int. J. Innov. Comput. Inf. Control, **4**(4), 771–783 (2008)
31. G.M. Mendez, O. Castillo, Interval type-2 TSK fuzzy logic systems using hybrid learning algorithm. in *The 14th IEEE International Conference on, Fuzzy Systems, 2005. FUZZ'05.* (2005), pp. 230–235
32. E. Rubio, O. Castillo, F. Valdez, P. Melin, C.I. González, G. Martinez, An extension of the fuzzy possibilistic clustering algorithm using type-2 fuzzy logic techniques. Adv. Fuzzy Syst. **2017**, 7094046:1–7094046:23 (2017)
33. P. Melin, O. Castillo, Intelligent control of complex electrochemical systems with a neuro-fuzzy-genetic approach. IEEE Trans. Ind. Electr. **48**(5), 951–955
34. L. Aguilar, P. Melin, O. Castillo, Intelligent control of a stepping motor drive using a hybrid neuro-fuzzy ANFIS approach. Appl. Soft Comput. **3**(3), 209–219 (2003)
35. P. Melin, O. Castillo, Adaptive intelligent control of aircraft systems with a hybrid approach combining neural networks, fuzzy logic and fractal theory. Appl. Soft Comput. **3**(4), 353–362 (2003)
36. L.B. Goncalves, M.M.B.R. Vellasco, M.A.C. Pacheco, F.J. de Souza, Inverted hierarchical neuro-fuzzy BSP system: a novel neuro-fuzzy model for pattern classification and rule extraction in databases. IEEE Trans. Syst. Man Cybern. Part C Appl. Rev. **36**(2), 236–248 (Mar. 2006)
37. E. Elyan, M.M. Gaber, A genetic algorithm approach to optimising random forests applied to class engineered data. Inf. Sci. **384**, 220–234 (2017)
38. E. Boros, P.L. Hammer, T. Ibaraki, A. Kogan, E. Mayoraz, I. Muchnik, An implementation of logical analysis of data. IEEE Trans. Knowl. Data Eng. **12**(2), 292–306 (2000)
39. B. Ustun, C. Rudin, Supersparse linear integer models for optimized medical scoring systems. Mach. Learn. **102**(3), 349–391 (2016)
40. J.A. Morente-Molinera, J. Mezei, C. Carlsson, E. Herrera-Viedma, Improving supervised learning classification methods using multigranular linguistic modeling and fuzzy entropy. IEEE Trans. Fuzzy Syst. **25**(5), 1078–1089 (2017)
41. H.T. Kahraman, A novel and powerful hybrid classifier method: development and testing of heuristic k-nn algorithm with fuzzy distance metric. Data Knowl. Eng. **103**, 44–59 (2016)
42. W.A. Young, S.L. Nykl, G.R. Weckman, D.M. Chelberg, Using Voronoi diagrams to improve classification performances when modeling imbalanced datasets. Neural Comput. Appl. **26**(5), 1041–1054 (2015)
43. K.A. Nugroho, N.A. Setiawan, T.B. Adji, Cascade generalization for breast cancer detection, in *2013 International Conference on Information Technology and Electrical Engineering (ICITEE)*, (Yogyakarta, Indonesia, 2013), pp. 57–61
44. A. Zadeh Shirazi, S.J. Seyyed Mahdavi Chabok, Z. Mohammadi, A novel and reliable computational intelligence system for breast cancer detection. Med. Biol. Eng. Comput. **56**(5), 721–732 (May 2018)
45. R. Vidya Banu, N. Nagaveni, Evaluation of a perturbation-based technique for privacy preservation in a multi-party clustering scenario. Inf. Sci. **232**, 437–448 (May 2013)
46. P. Luukka, T. Leppälampi, Similarity classifier with generalized mean applied to medical data. Comput. Biol. Med. **36**(9), 1026–1040 (2006)
47. H. Kahramanli, N. Allahverdi, Design of a hybrid system for the diabetes and heart diseases. Expert Syst. Appl. **35**(1), 82–89 (2008)
48. K. Polat, S. Güneş, A. Arslan, A cascade learning system for classification of diabetes disease: generalized discriminant analysis and least square support vector machine. Expert Syst. Appl. **34**(1), 482–487 (2008)

49. F. Mansourypoor, S. Asadi, Development of a reinforcement learning-based evolutionary fuzzy rule-based system for diabetes diagnosis. Comput. Biol. Med. **91**, 337–352 (2017)
50. S.B. Akben, Predicting the success of wart treatment methods using decision tree based fuzzy informative images. Biocybern. Biomed. Eng. **38**(4), 819–827 (2018)
51. F. Khozeimeh, R. Alizadehsani, M. Roshanzamir, A. Khosravi, P. Layegh, S. Nahavandi, An expert system for selecting wart treatment method. Comput. Biol. Med. **81**, 167–175 (2017)

Adaptation of Parameters with Binary Cat Swarm Optimization Algorithm of Controller for a Mobile Autonomous Robot

Trinidad Castro Villa and Oscar Castillo

Abstract This paper describes a method to solve optimization problems based on the binary cat swarm optimization (BCSO) algorithm to optimize the trajectory of a mobile unicycle autonomous robot, with the aim of following an established trajectory with the lowest possible margin of error. Simulation results show that the proposed method achieves good results.

Keywords Binary cat optimization · Swarm · Cat swarm optimization · Soft computing · Fuzzy control · Mobile robot · Autonomous mobile robot · Generic algorithm

1 Introduction

In this paper, the application is shown, the parameters are configured, the functions of the controllers of an autonomous mobile robot are configured, to obtain a better trajectory of the robot, the results of the adjustment of the parameters and the results of an optimized diffuse in the points of Membership Functions, used to obtain a better trajectory with a minimum margin of error.

Chuan Chu observed the behavior of 32 types of wild cats proposing the idea of solving optimization problems. The result of his observation was: search and tracking modes. Sharafi et al. [1] took the bases of the investigation of the original algorithm for the binary algorithm of the pack of cats to improve the problems.

T. C. Villa (✉) · O. Castillo
Tijuana Institute of Technology, Tijuana, BC, Mexico
e-mail: trinidad.castro17@tectijuana.edu.mx

O. Castillo
e-mail: ocastillo@tectijuana.mx

© Springer Nature Switzerland AG 2020
O. Castillo and P. Melin (eds.), *Hybrid Intelligent Systems in Control,*
Pattern Recognition and Medicine, Studies in Computational Intelligence 827,
https://doi.org/10.1007/978-3-030-34135-0_3

1.1 Binary Cat Swarm Optimization Algorithm (BCSO)

The BCSO version of the algorithm has differences when compared to the original algorithm, and among the differences is that the vectors are composed of zeros and ones. BCSO is composed of two modes: search and tracking [1, 2].
The steps of the algorithm use the following:

Step 1: Create N cats in the process, the cats are randomly distributed in the m-dimensional solution space and randomly x_k, d, new values.
Step 2: Randomly initialize the speed of the cats V_x.
Step 3: According to MR, cats are randomly selected from the population and their labels are set to search and tracking mode.
Step 4: Evaluate the fitness of each cat.
Step 5: Re-select the number of cats and place them in tracking mode according to MR, then set the other cats in the search mode.
Step 6: Verify the termination condition, if satisfied, end the program, and repeat Step 3–Step 5 otherwise.

1.2 Fuzzy Logic Structure

The model is of a unicycle mobile robot mounted of the same axis formed by three wheels: two driving wheels and one free wheel.

The FBCSO is of Mamdani type, and as Fig. 1 shows us it has two inputs and two outputs. Where the inputs are the error in angular velocity (ew) and the error in linear velocity (ev). The outputs are torque one (T_1) and torque two (T_2) that need to have each wheel of the robot in this case the robot has two wheels with servomotors [3].

The IF-THEN rules are used in fuzzy inference systems to calculate the degree to which the input data relates to the condition of the fuzzy rule. For the fuzzy utilitarian controller, in the simulation, Table 1 shows that 9 rules were used, which are the following, where ev is the linear speed, ew angular velocity, T_1 and T_2 are the pairs of servomotors [4].

1.3 Mobile Autonomous Robot

The movement of the free wheel can be ignored according to Eq. (1) and the kinematic system is represented by Eq. (2) [5]:

$$M(q)\dot{v} + C(q, \dot{q})v + D_v = \tau + P(t) \tag{1}$$

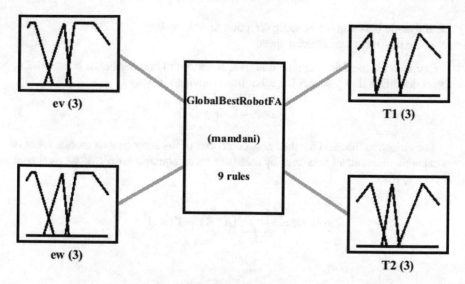

System GlobalBestRobotFA: 2 inputs, 2 outputs, 9 rules

Fig. 1 Type-1 FBCSO fuzzy controller for the simulation of the mobile autonomous robot

Table 1 Fuzzy rules

e_v/e_w then $T_1 T_2$	N	Z	P
N	N/N	N/Z	N/P
Z	Z/N	Z/Z	Z/P
P	P/N	P/Z	P/P

$q = (x, y, q)^t$ is the vector of the configuration coordinates.

$v = (v, w)^T$ is the vector of speeds.

$\tau = (T_1, T_2)$ is the vector of pairs applied to the wheels of the robot where T_1 and T_2 indicate the pairs of the right and left wheels.

$P \in R^2$ is the perturbation vector uniformly delimited.

$M(q) \in R^{2 \times 2}$ is the defined positive inertia matrix.

$C(q, \dot{q})\vartheta$ is the vector of centripetal and Coriolis forces.

$D \in R^{2 \times 2}$ is a diagonal positive defined damping matrix.

Equation (2): represents the evaluation of the direction that the mobile autonomous robot will take.

$$\dot{q} = \begin{vmatrix} \cos \theta & 0 \\ \sin \theta & 0 \\ 0 & 1 \end{vmatrix} \begin{vmatrix} v \\ w \end{vmatrix} \tag{2}$$

(x, y) is the position in the reference box X–Y (world).

θ is the angle between the heading direction and the x-axis.
v y w are the linear and angular speeds.

Equation (3) modifies the direction (non-holonomic) in the robot, it is an anti-slip wheel condition that prevents the robot from moving sideways.

$$\dot{y}\cos\theta - \dot{x}\sin\theta = 0 \tag{3}$$

The objective function for the fuzzy controller of the autonomous mobile robot is to calculate the control measures by means of mean squared error (MSE), as shown in Eq. (4) [6].

$$MSE = \frac{1}{N}\sum_{K=1}^{N}[X(K) - Y(K)]^2 \tag{4}$$

2 Proposed Method

To optimize the parameters of the membership functions in the fuzzy controller consists obtaining the smallest possible error with respect to a given reference, where the ideal minimum value is zero. In order to achieve this goal, the cat swarm algorithm was used to move dynamically the parameters of the membership functions of the fuzzy controller until the optimal value is reached. To optimize the parameters of the controller membership functions of a robot of different type, the robot plant was used [4].

In the Autonomous unicycle mobile robot, the problem is to control the tracking and stability of the desired path, and to solve problem we propose to use the cat swarm algorithm in the optimization of the trajectory of a unicycle mobile robot. In Fig. 2, the cat swarm algorithm is responsible for constructing the best Type 1 FBCSO, once selected and the best fuzzy system employ to the mobile robot model in order to follow an established path with the margin of minimal error.

3 Simulation Results

Table 3 shows the comparison of the optimization results of the BCSO algorithm with 30 experiments; using the values by the BCSO algorithm Table 2, against 30 of the same optimization problem with the metaheuristic called plant pollination algorithm (FPA) [4], and with an FLC of Type 1, the comparison between both methods, as it is a minimization problem, the closest to zero is the BCSO. The minimum MSE of the metaheuristics obtained in this work is 7.20E-05 and the highest value obtained in the 30 experiments is 1.20E-3. Table 3 shows that the results of the thirty experiments of

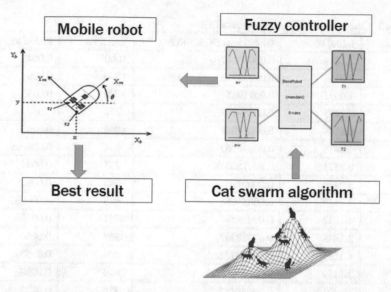

Fig. 2 Proposed optimization for the reference of the trajectory

Table 2 Parameters values by algorithm BCSO-ROBOT

Parameters	Value
No. experiments	30
Dimensions	10
SMP (memory set search)	3
SRD (searching the range of the selected dimension)	0.70
CMC (count of dimensions to change)	0.30
MR (mixture ratio)	0.33
Speed	10
Inertia weight	0.729
c1	2
r1	[0, 1]
tb	10
bitt	30
nPop	40
MaxIt	2000

the BCSO are the same in several experiments. In the case of the FPA the best MSE error is 8.99E-05 and the worst MSE error was 9.10E-03. The BCSO algorithm has the lowest error in all the experiments of this comparison and an experiment which is 0.000072, is show in Fig. 3.

Table 3 Comparison BCSO-Robot versus FPA

Exp.	Initial Pop.	Global best BCSO-Robot	Iterations	FPA-Robot
1	2.1848	0.00020247	1000	0.0046
2	2.0456	0.00038713	1000	0.009
3	4.0931	0.0011985	1000	0.0035
4	0.7295	0.00059102	976	0.006
5	6.8149	0.00051948	1960	0.0004
6	4.8476	0.00026267	651	0.000569
7	0.1573	0.0007503	737	0.0044
8	5.0469	0.000072	843	0.0062
9	2.7159	0.00054085	829	0.000089862
10	4.1832	0.0008838	257	0.0029
11	2.1848	0.00020247	359	0.0064
12	4.1153	0.00060722	294	0.0053
13	2.7159	0.00054085	829	0.0039
14	19.4919	0.00032387	336	0.0072
15	2.1848	0.00020247	359	0.000541
16	1.179	0.00028139	698	0.0073
17	2.0456	0.00038713	714	0.0065
18	2.7159	0.00054085	829	0.0065
19	2.1848	0.00020247	359	0.0091
20	5.8802	0.00029075	387	0.0082
21	1.3009	0.00051448	408	0.0085
22	5.5996	0.00040238	213	0.0010404
24	2.1951	0.000080012	989	0.0066
23	2.1848	0.00020247	359	0.0041
25	2.7159	0.00054085	829	0.0085
26	2.7159	0.00054085	829	0.002
27	2.1848	0.00020247	359	0.0032
28	2.1848	0.00020247	359	0.0054
29	2.7159	0.00054085	829	0.0058
30	2.7159	0.00054085	829	0.0014
	Best	0.000072		0.000089862
	Worst	0.0011985		0.0091

In the following fuzzy systems Global Best Robot FA belongs to experiment 8, the parameters to be considered in the fuzzy controller consists of two input variables and two output variables Fig. 4 shows a description of the fuzzy system:

Input variables: (ev) (ew). Each entry and exit has three membership functions, where they are triangular-type membership functions with trapezoidal to the sides.

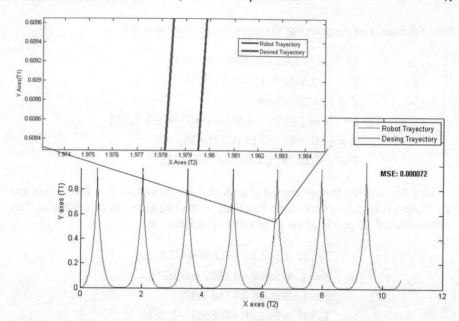

Fig. 3 Experiment 8 ROBOT-BCSO 7.2000E-05

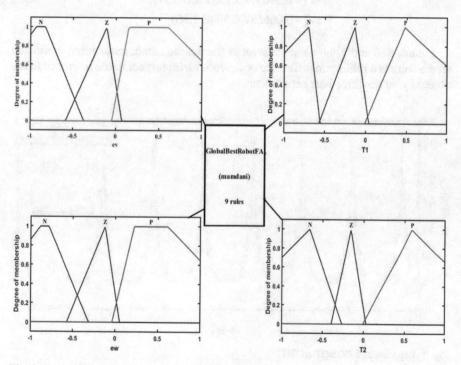

Fig. 4 Best experiment 8 fuzzy system 7.2000E-05

The distribution of the points of the membership functions are:

$$ev: N = [-1.72\ -0.9298\ -0.8242],$$
$$Z = [-0.5832\ -0.1047\ 0.07136],$$
$$P = [0\ 1250\ 2500].$$
$$ew : N = [-1.72\ -0.8818\ -0.7838\ -0.3096],$$
$$Z = [-0.5686\ -0.1182\ 0.05425],$$
$$P = [-0.03397\ 0.2205\ 0.6152\ 1.72].$$

Variable outputs: Torque applied to each of the servomotors (T_1: Torque one and T_2: Torque two). The membership functions in the two exits are all triangular. The distribution of the points of the membership functions are:

$$T_1 : N = [-1.80\ -0.6964\ -0.5519],$$
$$Z = [-0.5686\ -0.1322\ 0.04023],$$
$$P = [0.009766\ 0.4746\ 1.80].$$
$$T_2 : N = [-1.80\ -0.6567\ -0.26],$$
$$Z = [-0.3903\ -0.1537\ 0.03254],$$
$$P = [-0.006687\ 0.5946\ 1.80]$$

Figures 5, 6 and 7 show the behavior of the mobile autonomous robot optimized where there is a minimum difference of the robot with respect to the reference to the trajectory, of the three best experiments.

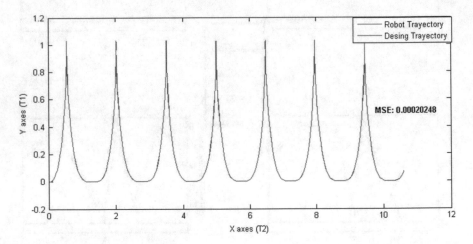

Fig. 5 Experiment 1 ROBOT-BCSO

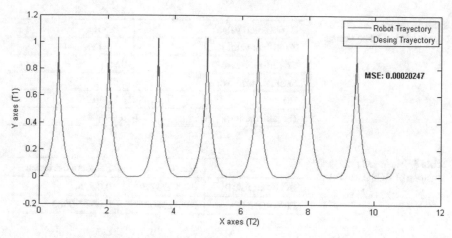

Fig. 6 Experiment 2 ROBOT-BCSO

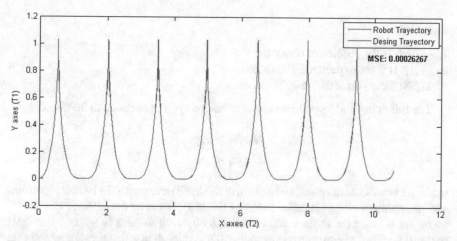

Fig. 7 Experiment 6 ROBOT-BCSO

4 Hypothesis Test BCSO Versus FPA

A statistical test was carried out to verify that the mean of the MSE errors of the population obtained with the FLC optimized with BCSO (μ_1) is lower than that obtained with the optimization results with FPA of investigation [7] (μ_2), 30 were performed experiments in each investigation. Table 4 shows the observed value Z which is obtained from Eq. 5. Table 5 shows the means and standard deviations of each of the optimization methods obtained from Table 2. The parametric test Z for two samples, which is used with the following Formula [8]:

Table 4 Parameters of the hypothesis test BCSO versus FPA

Z (observed value)	−8.661
Z (critical value)	−1.672
Confidence level	0.95
Significance level	0.05
H0	$\mu1 \geq \mu2$
Ha (affirmation)	$\mu1 < \mu2$

Table 5 Mean and standard deviation of BCSO and FPA

	Mean	Standard deviation
BCSO-ROBOT	0.000425179	0.000341
FPA-ROBOT	0.00483803	0.002779863

$$Z = \frac{(\bar{x}_1 - \bar{x}_2) - (\mu_1 - \mu_2)}{\sigma_{\bar{x}_1 - \bar{x}_2}} \tag{5}$$

where:

$\bar{x}_1 - \bar{x}_2$: It is the difference observed.
$\mu_1 - \mu_2$: It is the expected difference.
$\sigma_{\bar{x}_1 - \bar{x}_2}$: Standard error of the differences.

The following null hypothesis and alternative hypotheses need to be tested.

$$H_0: \mu_1 \geq \mu_2$$
$$H_a: \mu_1 < \mu_2$$

where μ_1 represents the population mean of the MSE errors with the BCSO algorithm and μ_2 represents the population mean of the MSE errors of the FPA.

Figure 8 based on Table 3 shows that the observed value z is −8.661 and falls within the critical region with a value of −1.672, which means that the null hypothesis is rejected, and this means that. There is enough information to support the alternative

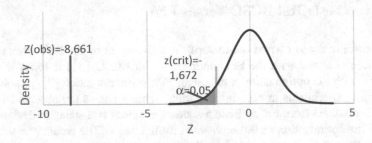

Fig. 8 Probability distribution chart for BCSO versus FPA

hypothesis that says the mean MSE error using the BCSO algorithm is less than the mean MSE error using FPA algorithms with a confidence level of 95%.

5 Conclusions

In this paper we present the BCSO algorithm with is new in the implementation of fuzzy parameter settings and fuzzy logic, that is, in order for the fuzzy system to dynamically move the parameters of this algorithm, and with this we obtained a faster convergence and better results than with the original method and others already exist in the literature. The fuzzy systems have two inputs and outputs of the robot for optimization with BCSO which served as a basis for testing the hypothesis against the FPA metaheuristics for the same optimization problem of which the hypothesis test performed in this research was sufficient to support with a level of confidence of 95%. As future work, will need to perform more experiments with variations in values of some parameters and we could use type-2 fuzzy logic as in [9–14]. In addition, we can deal with other types of applications, like in [15–21].

Acknowledgements The authors would like to express thank to the Consejo Nacional de Ciencia y Tecnología and Tecnológico Nacional de México/Tijuana Institute of Technology for the facilities and resources granted for the development of this research.

References

1. Y. Sharafi, M.A. Khanesar, M. Teshnehlab, Discrete binary cat swarm optimization algorithm, in *2013 3rd IEEE International Conference on Computer, Control and Communication, IC4 2013*, IEEE, (Sept. 2013)
2. S.K. Saha, S.P. Ghoshal, R. Kar, D. Mandal, Cat swarm optimization algorithm for optimal linear phase FIR filter design. ISA Trans. **52**(6), 781–794, Elsevier, (Aug. 2013)
3. J. Perez, P. Melin, O. Castillo, F. Valdez, C. Gonzalez, G. Martinez, J. Kacprzyk, M. Reformet, W. Melek, Trajectory optimization for an autonomous mobile robot using the bat algorithm, in *Fuzzy Logic in Intelligent Systems Design, Theory and Applications*, vol. 648 (Springer, Berlin, 2018), pp. 232–241
4. L. Astudillo, P. Melin, O. Castillo, Chemical optimization algorithm for fuzzy controller design, in *Briefs in Applied Sciences and Technology* (Springer, Berlin, 2014)
5. C. Caraveo, F. Valdez, O. Castillo, A new meta-heuristics of optimization with dynamic adaptation of parameters using type-2 fuzzy logic for trajectory control of a mobile robot. Algorithms **10**(3), 1–16 (2017)
6. C. Peraza, F. Valdez, P. Melin, Optimization of intelligent controllers using a Type-1 and interval Type-2 fuzzy harmony search algorithm. Algorithms MDPI **10**(3), 1–17, (July 2017)
7. O.R. Carvajal, O. Castillo, J. Soria, Optimization of membership function parameters for fuzzy controllers of an autonomous mobile robot using the flower pollination algorithm. J. Autom. Mob. Robot Intellegence Syst. **12**(1), 44–49 (2018)
8. M.G. Kendall, Elementary statistics. **155** (1945)

9. C. Leal Ramírez, O. Castillo, P. Melin, A. Rodríguez Díaz, Simulation of the bird age-structured population growth based on an interval type-2 fuzzy cellular structure. Inf. Sci. **181**(3), 519–535 (2011)
10. N.R. Cázarez-Castro, L.T. Aguilar, O. Castillo, Designing type-1 and type-2 fuzzy logic controllers via fuzzy lyapunov synthesis for nonsmooth mechanical systems. Eng. Appl. AI **25**(5), 971–979 (2012)
11. O. Castillo, P. Melin, Intelligent systems with interval type-2 fuzzy logic. Int. J. Innov. Comput. Inf. Control **4**(4), 771–783 (2008)
12. G.M. Mendez, O. Castillo, Interval type-2 TSK fuzzy logic systems using hybrid learning algorithm, Fuzzy Systems, 2005, in *The 14th IEEE International Conference on FUZZ'05*, 230–235
13. Claudia I. González, P. Melin, J.R. Castro, O. Mendoza, Castillo: an improved sobel edge detection method based on generalized type-2 fuzzy logic. Soft. Comput. **20**(2), 773–784 (2016)
14. C.I. González, P. Melin, J.R. Castro, O. Castillo, O. Mendoza, Optimization of interval type-2 fuzzy systems for image edge detection. Appl. Soft Comput. **47**, 631–643 (2016)
15. P. Melin, O. Castillo, Intelligent control of complex electrochemical systems with a neuro-fuzzy-genetic approach. IEEE Trans. Ind. Electr. **48**(5), 951–955
16. E. Rubio, O. Castillo, F. Valdez, P. Melin, C.I. González, G. Martinez, An Extension of the Fuzzy Possibilistic Clustering Algorithm Using Type-2 Fuzzy Logic Techniques. Adv. Fuzzy Syst. **2017**, 7094046:1–7094046:23 (2017)
17. P. Melin, A. Mancilla, M. Lopez, O. Mendoza, A hybrid modular neural network architecture with fuzzy Sugeno integration for time series forecasting. Appl. Soft Comput. **7**(4), 1217–1226 (2007)
18. P. Melin, O. Castillo, *Modelling, Simulation and Control of Non-Linear Dynamical Systems: An Intelligent Approach Using Soft Computing and Fractal Theory* (CRC Press, Boca Raton, 2001)
19. P. Melin, D. Sánchez, O. Castillo, Genetic optimization of modular neural networks with fuzzy response integration for human recognition. Inf. Sci. **197**, 1–19 (2012)
20. P. Melin, D. Sánchez, Multi-objective optimization for modular granular neural networks applied to pattern recognition. Inf. Sci. **460–461**, 594–610 (2018)
21. D. Sánchez, P. Melin, O. Castillo, Optimization of modular granular neural networks using a firefly algorithm for human recognition. Eng. Appl. of AI **64**, 172–186 (2017)

Comparison of Fuzzy Controller Optimization with Dynamic Parameter Adjustment Based on of Type-1 and Type-2 Fuzzy Logic

Marylu L. Lagunes, Oscar Castillo, Fevrier Valdez and Jose Soria

Abstract This paper presents the comparison of fuzzy controller optimization results using dynamic parameter adjustment Type-1 (T1) and Interval Type-2 (T2) fuzzy logic to the Firefly Algorithm (FA). The FA is used for optimizations parameters of the membership functions in the fuzzy controllers. The dynamic adjustment is applied to the randomness parameter of the search space, which represents the exploration of the method, avoiding stagnation or premature convergence. The FA generates the values that the parameters of the membership functions take for optimization use in the fuzzy systems for control. The control plants have one or more input variables that are processed and result in one or more output variables, it would be very difficult to model the human reasoning in equations to achieve a machine acquires the knowledge acquired by humans. For that reason the fuzzy logic that generates that insertity is used as if it were human reasoning.

Keywords Type-1 fuzzy logic · Interval type-2 fuzzy logic · FA · Dynamic adjustment

1 Introduction

The main contribution focuses on the performance of the FA [1, 2] to generate an efficient parameter data vector for the minimization of error in optimization problems, using T1 and T2 fuzzy logic for the dynamic adjustment of the alpha parameter (exploration). The generated values, optimize the parameters of the membership functions of fuzzy controllers. To perform the dynamic adjustment a simple fuzzy model of one input representing the iterations and one output that results in the values of the randomness parameter was developed. The parameters of the optimized membership functions are values of fuzzy controllers, used in plants to perform processes in general industrial, such as controlling temperatures, water flow, etc. In order to transmit human reasoning in this type of processes, fuzzy logic is used as

M. L. Lagunes · O. Castillo (✉) · F. Valdez · J. Soria
Tijuana Institute of Technology, Tijuana, BC, Mexico
e-mail: ocastillo@tectijuana.mx

© Springer Nature Switzerland AG 2020
O. Castillo and P. Melin (eds.), *Hybrid Intelligent Systems in Control,*
Pattern Recognition and Medicine, Studies in Computational Intelligence 827,
https://doi.org/10.1007/978-3-030-34135-0_4

a source of uncertainty for making decisions and evaluating the input and output variables of the plant. There are related works on optimization using metaheuristics such as [3–9]. This paper is structured as follows: Sect. 2 describes T1 and T2 fuzzy logic, and presents the Firefly Algorithm (FA), Sect. 3 describes the fuzzy dynamic adjustment model, Sect. 4 shows the fuzzy controllers to be optimized, Sect. 5 details the experimentation, Sect. 6 shows the results obtained from the fuzzy controller comparison and finally, Sect. 7 describes the conclusions.

2 Fuzzy Logic

Fuzzy logic Zadeh in [10, 11] was established with the principle of incompatibility between precision and complexity, when the analysis and the way of solving problems by humans, could not be interpreted easily, so that a machine would do the same. As some processes are not compatible with the complexity of human reasoning.

The T2 fuzzy logic is represented by an uncertainty trace (FOU) that can be visualized as the interval between these two T1 membership functions. This type fuzzy logic was described by Zadeh [12] and later on its use and popularity with Mendel and Liang [13, 14]. It is used to model uncertainty and imprecision [15, 16].

2.1 Firefly Algorithm

The FA has basic 3 rules [17, 18].

a. Fireflies are of the same genus, so any firefly could be attracted to any other firefly.
b. The attractiveness depends on the brightness and it is minimized depending on the distance between the fireflies, the firefly that shines the least will move towards the one that shines the most, if there is not one that shines more, they will move at random.
c. The glow of a firefly is given by its function of adequacy.

3 Model of Fuzzy Dynamic Parameter Adjustment

Dynamic parameter adjustment is applied to the FA to help improve the performance in the search space, with a good exploration the method has a good chance of finding the optimal objective function. For this reason a fuzzy system with a range [0, 1] was designed, with iteration as input, and alpha as output, with low, medium and high as linguistic variables respectively, in each of the triangular membership functions. Figure 1 shows the input and Fig. 2 the output of the T1 fuzzy system [18, 19].

Fig. 1 Input 1, T1 system
fuzzy logic

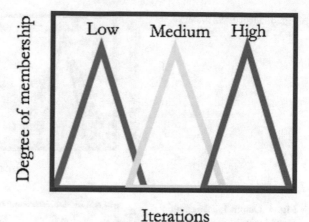

Fig. 2 Output 1, T1 system
fuzzy logic

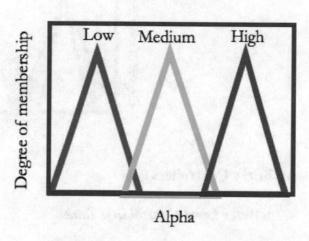

To compare the behavior of the method, we designed the same model with T2 fuzzy logic and observe which contributes more, in improving the behavior of firefly algorithm in the optimization. Figures 3 and 4 show the T2 fuzzy system.

The equation for the iteration variable is:

$$Iteration = \frac{Current\ Iteration}{Maximum\ of\ Iterations} \qquad (1)$$

Fig. 3 Input 1, T2 system
fuzzy logic

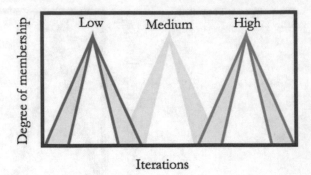

Fig. 4 Output 1, T2 system
fuzzy logic

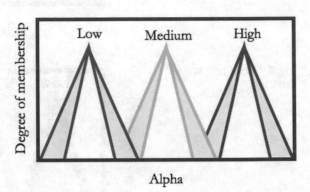

4 Fuzzy Controllers

4.1 Water Level Control in a Tank

With this model we control a water tank to be more specific the water level. Input
one is the water level error and the water level change rate as input two, and each of
these inputs uses 3 membership functions. It has an output that represents the speed
with which the control valve closes or opens, and uses 5 membership functions [20].
In Table 1 the rules of the fuzzy inference system are observed (Figs. 5 and 6).

Table 1 Water level control in the tank rules		
	1	If level is okay, then valve is noChange (1)
	2	If level is low, then valve is openFast (1)
	3	If level is high, then valve is closeFast (1)
	4	If level is okay, and rate is positive, then valve is closeSlow (1)
	5	If level is okay, and rate is negative, then valve is openSlow (1)

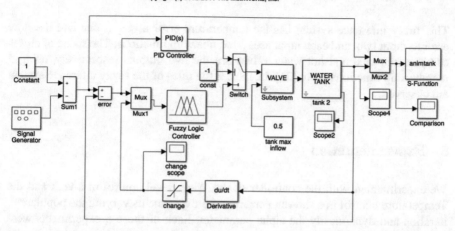

Fig. 5 Water level control in the tank plant

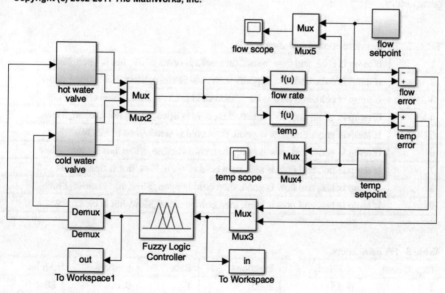

Fig. 6 Temperature control in a shower plant

4.2   Temperature Control in a Shower

This fuzzy inference system has the temperature error as input one and the flow error as input two, and each input uses 3 membership functions. The speed of closes or opens the cold and hot water valves are the two outputs respectively, using 5 membership functions each, [21]. In Table 2 the rules of the fuzzy inference system are observed.

5   Experimentation

We experimented with the controller of the Water level control in a tank and the Temperature control in a shower, performing 5 experiments varying the population, iteration and dynamically the alpha parameter. Each of these 5 experiments were executed 30 times, with the same parameters, to make a comparison of the behavior and results of the optimization using T1 and T2. As can be noticed in Table 3, the parameters used in the FA for the controller with T1 and T2 fuzzy logic are the same parameters.

Table 2 Temperature control in a shower rules

1	If temp is cold, and flow is soft, then cold is open_Slow, hot is open_Fast
2	If temp is cold, and flow is good, then cold is close_Slow, hot is open_Slow
3	If temp is cold, and flow is hard, then cold is close_Fast, hot is close_Slow
4	If temp is good, and flow is soft, then cold is open_Slow, hot is open_Slow
5	If temp is good, and flow is good, then cold is steady, hot is steady
6	If temp is good, and flow is hard, then cold is close_Slow, hot is close_Slow
7	If temp is hot, and flow is soft, then cold is open_Fast, hot is open_Slow
8	If temp is hot, and flow is good, then cold is open_Slow, hot is close_Slow
9	If temp is hot, and flow is hard, then cold is close_Slow, hot is close_Fast

Table 3 FA parameters

Experiment	Firefly	Iterations	Beta	Gamma	Alpha
1	15	2000	1	0.1	D
2	20	1500	1	0.1	D
3	25	1200	1	0.1	D
4	30	750	1	0.1	D
5	35	600	1	0.1	D

6　Simulations Results

The results obtained in the optimization of the parameters of the membership functions, the fuzzy controllers of the Water level control in a tank and Temperature control in a shower, using dynamic adjustment of the alpha parameter with T1 fuzzy logic, doing 5 experiments of 30 executions each, are present in Table 4.

The experimentation was done with T1 fuzzy logic, but to observe its behavior and comparison was also done with T2 fuzzy logic, for which the same parameters were used as with T1. Table 5 shows the results obtained with T2.

6.1　Statistical Test

With the results obtained, the statistical test was performed, for each of the optimized controllers, to compare the values that resulted from using T1 and T2 fuzzy logic for the dynamic adjustment of the alpha parameter in the FA method, as shows in Tables 6 and 7.

The equation for Z test is:

$$Z = \frac{(\bar{x}_1 - \bar{x}_2) - (\mu_1 - \mu_2)}{\sigma_{\bar{x}_1 - \bar{x}_2}} \tag{2}$$

Table 4 Type-1 results

Experiment	Water level control in a tank		Temperature control in a shower	
	T1 MSE	Average	T1 MSE	Average
1	0.0885	0.7178	0.0671	0.5463
2	0.0664	0.6413	0.0299	0.9274
3	0.0785	0.8104	0.0139	0.5085
4	0.0664	0.5758	0.0739	0.6930
5	0.0364	0.9274	0.0139	0.5085

Table 5 Type-2 results

Experiment	Water level control in a tank		Temperature control in a shower	
	T2 MSE	Average	T2 MSE	Average
1	0.0012	0.1921	0.0041	0.1292
2	0.0049	0.1410	0.0039	0.0353
3	0.0283	0.1638	0.0278	0.1960
4	0.0133	0.1592	0.0014	0.0634
5	0.0033	0.1399	0.0010	0.1907

Table 6 Parameters statistical test water level control in a tank

Parameter	Value
Level of significance	95%
H_a	$\mu_1 < \mu_2$
H_0	$\mu_1 \geq \mu_2$
Difference	-0.436
z (observed value)	-5.929
z (critical value)	-1.645
p-value (one-tailed)	<0.0001
Alpha	0.05

Table 7 Statistical test results obtained

Dynamic adjustment	Number of samples	Mean	Deviation standard	Alpha
T1	30	0.576	0.363	0.05
T2	30	0.140	0.175	0.05

Tables 8 and 9 show the parameters and results obtained in the statistical test with temperature control in a shower.

Table 8 Statistical test temperature control in a shower control in a tank

Parameter	Value
Level of significance	95%
H_a	$\mu_1 < \mu_2$
H_0	$\mu_1 \geq \mu_2$
Difference	-0.473
z (observed value)	-7.356
z (critical value)	-1.645
p-value (one-tailed)	<0.0001
Alpha	0.05

Table 9 Statistical test results obtained

Dynamic adjustment	Number of samples	Mean	Deviation standard	Alpha
T1	30	0.509	0.350	0.05
T2	30	0.035	0.040	0.05

7 Conclusions

In this paper we proposed fuzzy dynamic adjustment of the alpha parameter for, developing a fuzzy controller of one input and one output, with the objective of optimizing the fuzzy controller membership functions, these are optimized by a data vector generated and tested by the firefly algorithm.

As it was observed previously, the results in the comparison of type-1 and type-2 fuzzy logic are very similar, therefore it is intended to add noise to the plants to observe their behavior in the optimization. However, the results were favorable with the error that was generated before being optimized.

As future work, will need to perform more experiments with variations in values of some parameters and we could use general type-2 fuzzy logic [22–25] or use other applications as in [26–30].

References

1. X.S. Yang, X. He, Firefly algorithm: recent advances and applications. Int. J. Swarm Intell. **1**(1), 36 (2013)
2. X.-S. Yang, *Nature-Inspired Metaheuristic Algorithms*. (Luniver Press, 2010)
3. L. Amador-Angulo, O. Castillo, Comparative analysis of designing differents types of membership functions using bee colony optimization in the stabilization of fuzzy controllers, in *Nature Inspired Design of Hybrid Intelligent Systems*, vol. 667 (Springer, Berlin, 2017), pp. 551–571
4. M.L. Lagunes, O. Castillo, J. Soria, Methodology for the optimization of a fuzzy controller using a bio-inspired algorithm. Fuzzy Log. Intell. Syst. Des. **648**, 131–137 (2017). Springer
5. E. Bernal, O. Castillo, J. Soria, Imperialist competitive algorithm with dynamic parameter adaptation applied to the optimization of mathematical functions. Nat. Inspired Des. Hybrid Intell. Syst. **667**, 329–341 (2017). Springer
6. L. Rodríguez, O. Castillo, J. Soria, A study of parameters of the grey wolf optimizer algorithm for dynamic adaptation with fuzzy logic. Nat. Inspired Des. Hybrid Intell. Syst. **667**, 371–390 (2017). Springer
7. C. Peraza, F. Valdez, O. Castillo, Improved method based on type-2 fuzzy logic for the adaptive harmony search algorithm. Fuzzy Log. Augment. Neural Optim. Algorithms **749**, 29–37 (2018). Springer
8. M.L. Lagunes, O. Castillo, F. Valdez, J. Soria, P. Melin, Parameter optimization for membership functions of type-2 fuzzy controllers for autonomous mobile robots using the firefly algorithm. Fuzzy Inf. Process. **831**, 569–579 (2018). NAFIPS
9. M.L. Lagunes, O. Castillo, J. Soria, Optimization of membership function parameters for fuzzy controllers of an autonomous mobile robot using the firefly algorithm. Fuzzy Log. Augment. Neural Optim. Algorithms **749**, 199–206 (2018). Springer
10. L.A. Zadeh, Fuzzy sets. Inf. Control **8**(3), 338–353 (1965)
11. L.A. Zadeh, Fuzzy logic, *Computer (Long. Beach. Calif)*, vol. 21, no. 4, pp. 83–93, (Apr. 1988)
12. L.A. Zadeh, The concept of a linguistic variable and its application to approximate reasoning-III. Inf. Sci. (Ny) **9**(1), 43–80 (1975)
13. N.N. Karnik, J.M. Mendel, Q. Liang, Type-2 fuzzy logic systems. IEEE Trans. Fuzzy Syst. **7**(6), 643–658 (1999)
14. Q. Liang, J.M. Mendel, Interval type 2 fuzzy logic systems: theory and design. IEEE Trans. Fuzzy Syst. **8**(5), 535–550 (2000)

15. O. Castillo, P. Melin, J. Kacprzyk, W. Pedrycz, Type-2 fuzzy logic: theory and applications, in *2007 IEEE International Conference on Granular Computing (GRC 2007)*, (2007), pp. 145–145

16. J. Pérez, F. Valdez, O. Castillo, Modification of the bat algorithm using type-2 fuzzy logic for dynamical parameter adaptation. Nat. Inspired Des. Hybrid Intell. Syst. **667**, 343–355 (2017). Springer

17. C. Soto, F. Valdez, O. Castillo, A review of dynamic parameter adaptation methods for the firefly algorithm. Nat. Inspired Des. Hybrid Intell. Syst. **667**, 285–295 (2017). Springer

18. C. Solano-Aragón, O. Castillo, Optimization of benchmark mathematical functions using the firefly algorithm. Recent. Adv. Hybrid Approaches Des. Intell. Syst. **547**, 177–189 (2014). Springer

19. P. Ochoa, O. Castillo, J. Soria, Fuzzy differential evolution method with dynamic parameter adaptation using type-2 fuzzy logic, in *Intelligent Systems, 8th International Conference on, IEEE*, (2016), pp. 113–118

20. Water Level Control in a Tank—MATLAB & Simulink Example—MathWorks America Latina. [Online]. Available: https://la.mathworks.com/help/fuzzy/examples/water-level-control-in-a-tank.html. Accessed 04 Jul 2018

21. Temperature Control in a Shower—MATLAB & Simulink—MathWorks America Latina. [Online]. Available: https://la.mathworks.com/help/fuzzy/temperature-control-in-a-shower.html. Accessed 04 Jul 2018

22. P. Melin, C.I. González, J.R. Castro, O. Mendoza, O. Castillo, Edge-detection method for image processing based on generalized type-2 fuzzy logic. IEEE Trans. Fuzzy Syst. **22**(6), 1515–1525 (2014)

23. C.I. González, P. Melin, J.R. Castro, O. Castillo, O. Mendoza, Optimization of interval type-2 fuzzy systems for image edge detection. Appl. Soft Comput. **47**, 631–643 (2016)

24. C.I. González, P. Melin, J.R. Castro, O. Mendoza, O. Castillo, An improved sobel edge detection method based on generalized type-2 fuzzy logic. Soft. Comput. **20**(2), 773–784 (2016)

25. E. Ontiveros, P. Melin, O. Castillo, High order α-planes integration: a new approach to computational cost reduction of general Type-2 fuzzy systems. Eng. Appl. AI **74**, 186–197 (2018)

26. C. Leal Ramírez, O. Castillo, P. Melin, A. Rodríguez Díaz, Simulation of the bird age-structured population growth based on an interval type-2 fuzzy cellular structure. Inf. Sci. **181**(3), 519–535 (2011)

27. N.R. Cázarez-Castro, L.T. Aguilar, O. Castillo, Designing type-1 and type-2 fuzzy logic controllers via fuzzy lyapunov synthesis for nonsmooth mechanical systems. Eng. Appl. of AI **25**(5), 971–979 (2012)

28. O. Castillo, P. Melin, Intelligent systems with interval type-2 fuzzy logic. Int. J. Innov. Comput. Inf. Control **4**(4), 771–783 (2008)

29. G.M. Mendez, O. Castillo, Interval type-2 TSK fuzzy logic systems using hybrid learning algorithm, Fuzzy Systems, 2005, in *The 14th IEEE International Conference on FUZZ'05*, 230–235

30. P. Melin, O. Castillo, Intelligent control of complex electrochemical systems with a neuro-fuzzy-genetic approach. IEEE Trans. Ind. Electr. **48**(5), 951–955

Pattern Recognition

Particle Swarm Algorithm for the Optimization of Modular Neural Networks in Pattern Recognition

Beatriz Gonzalez, Patricia Melin and Fevrier Valdez

Abstract In this paper a Particle Swarm Algorithm (PSO) is applied for the optimization of modular neural networks. This method is used for optimizing modular neural network in medical image recognition (Echocardiograms). As echocardiograms, these images help to diagnose heart diseases by the specialists and so can reduce the number of deaths by disease. Simulation results show that the scaled conjugate gradient (trainscg method) offers much higher performance with an average 81.4% recognition rate and with the gradient descent with adaptive learning (traingda method) an average of 76.3% recognition rate.

Keywords Particle Swarm Algorithm · PSO · Medical images · Recognition · Echocardiogram · Disease · Modular neural networks · Optimizing

1 Introduction

One of the areas of greatest recent interest is medical image recognition [1–5] with images such as medical images [6], mammography [7, 8], X-ray [9, 10], biopsies [11, 12] and echocardiography [13, 14]. It is an important field of artificial intelligence, which has been applied in general to face recognition [15, 16], diagnosis diseases, Dental diagnosis [17], optimization of medical images [18] and among others.

Some works related with medical images are the optimization of modular neural networks [19, 20], images segmentation [21, 22], neural network [23], etc.

In these works related to optimization of modular networks in imaging recognition methods we can find: firefly algorithm [3, 24] inspired by the flashing behavior of fireflies, fuzzy logic in the gravitational search algorithm [19, 20] which is based on

B. Gonzalez · P. Melin (✉) · F. Valdez
Tijuana Institute of Technology, Tijuana, BC, Mexico
e-mail: pmelin@tectijuana.mx

B. Gonzalez
e-mail: betygm8@hotmail.com

F. Valdez
e-mail: fevrier@tectijuana.mx

© Springer Nature Switzerland AG 2020
O. Castillo and P. Melin (eds.), *Hybrid Intelligent Systems in Control,*
Pattern Recognition and Medicine, Studies in Computational Intelligence 827,
https://doi.org/10.1007/978-3-030-34135-0_5

the gravity rules, genetic algorithms [25] are inspired from Darwinian evolutionary theory, the grey wolf algorithm [26], the Adaptive Harmony Search Algorithm [27, 28], Galactic Swarm Optimization [29], etc.

In this case, we use Particle Swarm algorithm [30], which is a method that simulates the behavior of flock of bird for optimizing the modular neural networks.

This paper is organized as follows: in Sect. 2 the particle swarm optimization algorithm is described, in Sect. 3 the medical images are presented, in Sect. 4 the proposed method is explained, in Sect. 5 the results are presented, and in Sect. 6 the conclusion are presented.

2 Particle Swarm Algorithm

This section describes the Particle Swarm algorithm (PSO) that was developed by Eberhart R., and Kennedy J. in 1995 [30].

The PSO is a metaheuristic algorithm based on a population of particles. In this method, the main idea comes from the social behavior of flocks of birds.

Each particle moves through the d-dimensional problem space to search for new solutions. Each particle has a position and a velocity represented by the vectors $xi = (xi_1, xi_2, ..., xi_D)$ and $Vi = (vi_1, vi_2, ..., vi_D)$ for the i-th particle. At each iteration, particles are compared with each other. Each particle records its best position and is called the global best, and is represented as $G = (G_1, G_2, ..., G_D)$. The velocity of each particle is given by Eq. (1).

$$v_i(t+1) = wv_i(t) + c_1 r_1 \left(pbest_i - x_i(t) \right) + c_2 r_2 (gbest - x_i(t)) \qquad (1)$$

where

$w =$ is the inertia weight,
$v_i(t) =$ is the velocity of particle i,
$x_i(t) =$ is the position of particle i,
c_1 and $c_2 =$ the cognitive and social components,
r_1 and r_2 are random values between [0, 1].

Equation (2) shows the position which is determined by the sum of the previous position and the new velocity

$$x_{ij}(t+1) = x_{ij}(t) + v_{ij}(t+1) \qquad (2)$$

2.1 Pseudocode for PSO

The steps of the PSO are presented in the pseudo code below:

Algorithm Particle Swarm Optimization

Initialize the size of the particle swarm n
Initialize the positions and the velocities for
all the particles randomly
While end criterion false **do**

$$t = t + 1$$

Compute fitness value of each particle

$$x = \arg min_{t-1}^{n}(f(x(t-1)), f(x_1(t)), f(x_2(t)) \dots f(x_t(t)), \dots f(x_n(t)));$$

For i = 1 to n

$$x_t(t) = argmin_{t-1}^{n}(x_t(t-1)), f(x_t(t))$$

textbfFor j = 1 to Dimension
Update the j-th dimensión value of x_t and v_t

$$v_i(t+1) = wv_i(t) + c_1r_1(pbest_i - x_i(t)) + c_2r_2(gbest - x_i(t))$$

$$x_i(t+1) = x_i(t) + v_i(t+1)$$

$$v_{ij} = sign(v_{ij})\min(|v_{ij}|, v_{max})$$

End For

  End For

End while

2.2 Flowchart of the PSO

The flowchart of PSO is presented in Fig. 1.

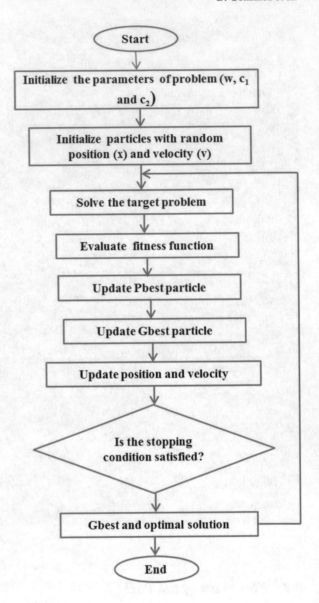

Fig. 1 Flowchart of Particle Swarm Optimization

3   Medical Images

Medical images can help diagnose diseases, and the types of medical images that we can have are magnetic resonance, ultrasound, mammography, echocardiography, etc.

Echocardiography is a diagnostic technique that is considered to be non-invasive and sure, which uses ultrasound for the study of the anatomy and functioning of the

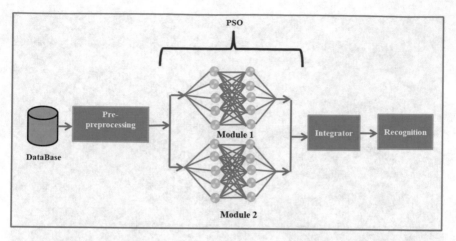

Fig. 2 Proposed method

heart. Research in echocardiogram imaging is very important because it can help diagnose various diseases.

As echocardiograms, these images help to diagnose heart diseases by the specialists and so reduce the number of deaths by disease.

4 Architecture of the Proposed Method

In Fig. 2 we can note that the objective is to find the optimal MNN architecture for finding the number layers and nodes with the PSO.

4.1 Database

In Fig. 3 we show the database include 90 patient echocardiograms, where there are ten images per patient, and with a total of 900 images in JPG format.

In Fig. 4 we show the step for images Preprocessing.

Table 1 shows the parameters used in method, which is the Particle Swarm Algorithm (PSO).

5 Results

In Table 2 we show results of PSO with the scaled conjugate gradient (trainscg method).

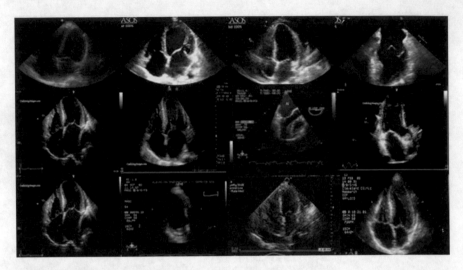

Fig. 3 Example of database images

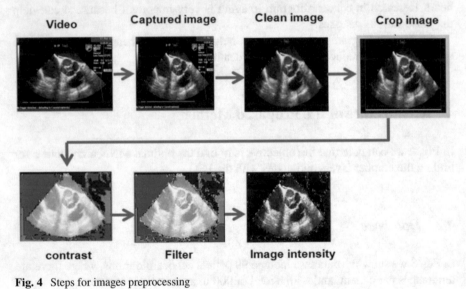

Fig. 4 Steps for images preprocessing

Table 1 Parameters used

Parameters	Value
Particles	120
Iterations	20
C_1	1
C_2	4-c1

Table 2 Results with the trainscg method

Training	Neurons	Epoch	Error	% Rec.
1	52, 100, 60	500	0.0001	100
2	77, 58, 92	500	0.0001	75
3	80, 96	500	0.0001	75
4	77, 59	500	0.0001	83.3
5	100, 67	500	0.0001	72.2
6	80, 75	500	0.0001	75
7	77, 67	500	0.0001	75
8	43, 50	500	0.0001	81
9	68, 60, 56	500	0.0001	75
10	63, 80, 100	500	0.0001	100
11	51, 58, 63	500	0.0001	75
12	42, 99	500	0.0001	83.3
13	39, 59, 73	500	0.0001	100
14	80, 96	500	0.0001	75
15	92, 69	500	0.0001	75
16	77, 108	500	0.0001	83.3
17	77, 63, 74	500	0.0001	72.2
18	84, 64, 96	500	0.0001	100
19	91, 50, 80	500	0.0001	75
20	97, 68, 60	500	0.0001	81
21	100, 57	500	0.0001	72.2
22	80, 96	500	0.0001	75
23	77, 57	500	0.0001	75
24	82, 78, 67	500	0.0001	100
25	77, 50, 90	500	0.0001	81
26	68, 48, 60	500	0.0001	72.2
27	107, 144	500	0.0001	83.3
28	123, 68	500	0.0001	81
29	77, 63, 74	500	0.0001	72.2
30	52, 91, 85	500	0.0001	100

In Table 3 we show results of PSO with the gradient descent with adaptive learning (traingda method).

Table 3 Results with the traingda method

Training	Neurons	Epoch	Error	% Rec.
1	63, 80	500	0.0001	75
2	94, 48, 69	500	0.0001	72.2
3	96, 79	500	0.0001	72.2
4	69, 82, 86	500	0.0001	81
5	100, 100, 85	500	0.0001	81
6	92, 49, 94	500	0.0001	75
7	69, 89	500	0.0001	75
8	100, 89	500	0.0001	72.2
9	89, 89, 73	500	0.0001	75
10	80, 80	500	0.0001	75
11	84, 77	500	0.0001	75
12	76, 100, 87	500	0.0001	75
13	89, 62, 66	500	0.0001	75
14	95, 95	500	0.0001	81
15	78, 89, 73	500	0.0001	75
16	93, 89	500	0.0001	81
17	89, 89, 73	500	0.0001	75
18	87, 93	500	0.0001	83
19	91, 50, 80	500	0.0001	75
20	76, 100, 87	500	0.0001	83
21	150, 33	500	0.0001	72.2
22	51, 80, 92	500	0.0001	72.2
23	90, 67, 81	500	0.0001	75
24	89, 92, 86	500	0.0001	81
25	80, 78, 87	500	0.0001	75
26	100, 82	500	0.0001	72.2
27	57, 49, 89	500	0.0001	75
28	79, 90	500	0.0001	81
29	80, 90	500	0.0001	75
30	78, 59, 73	500	0.0001	81

6 Conclusions

In this paper, we proposed Particle Swarm Algorithm for the Optimization of Modular Neural Networks in Echocardiogram Recognition. The proposed method was compared with the scaled conjugate gradient (*trainscg* method) and the gradient descent with adaptive learning (*traingda* method). The results indicated that the *trainscg* method offers much higher performance with an average of 81.4% with respect to the *traingda* method with an average of 76.3%. As future work we can apply our approach in other areas, like in the works of [31–33]. In addition we could consider using type-2 fuzzy logic like in [34–37].

Acknowledgements We like to express our gratitude to CONACYT, Tijuana Institute of Technology for the resources granted for the development of this research.

References

1. J. Zhang, K. Shao, X. Luo, Small sample image recognition using improved Convolutional Neural Network. J. Vis. Commun. Image Represent. **55**, 640–647 (2018)
2. A. Jalalvand, K. Demuynck, W. Neve, J. Martens, On the application of reservoir computing networks for noisy image recognition. Neurocomputing **277**, 237–248 (2018)
3. D. Sánchez, P. Melin, O. Castillo, Optimization of modular granular neural networks using a firefly algorithm for human recognition. Eng. Appl. Artif. Intell. **64**, 172–186 (2017)
4. P. Melin, D. Sánchez, Multi-objective optimization for modular granular neural networks applied to pattern recognition. Inf. Sci. **460–461**, 594–610 (2018)
5. F. Gaxiola, P. Melin, F. Valdez, J.R. Castro, Person recognition with modular deep neural network using the iris biometric measure, in *Fuzzy Logic Augmentation of Neural and Optimization Algorithms* (2018), pp. 69–80
6. M. Manchanda, R. Sharma, An improved multimodal medical image fusion algorithm based on fuzzy transform. J. Vis. Commun. Image Represent. **51**, 76–94 (2018)
7. H. Amer, F. Schmitzberger, B. Ingold-Heppner, J. Kussmaul, M. Tohamy, H. Tantawy, B. Hamm, M. Makowski, E. Fallenberg, Digital breast tomosynthesis versus full-field digital mammography—which modality provides more accurate prediction of margin status in specimen radiography? Eur. J. Radiol. **93**, 258–264 (2017)
8. K. Hu, W. Yang, X. Gao, Microcalcification diagnosis in digital mammography using extreme learning machine based on hidden Markov tree model of dual-tree complex wavelet transform. Expert Syst. Appl. **86**, 135–144 (2017)
9. K. Ishihara, T. Ogawa, M. Haseyama, Helicobacter Pylori infection detection from gastric X-ray images based on feature fusion and decision fusion. Comput. Biol. Med. **84**, 69–78 (2017)
10. C. Spampinato, S. Palazzo, D. Giordano, M. Aldinucci, R. Leonardi, Deep learning for automated skeletal bone age assessment in X-ray images. Med. Image Anal. **36**, 41–51 (2017)
11. D. D'Amario, A.M. Leone, M.L. Narducci, C. Smaldone, D. Lecis, F. Inzani, M. Luciani, A. Siracusano, F. La Neve, M. Manchi, G. Pelargonio, F. Perna, P. Bruno, M. Massetti, D. Pitocco, D. Cappetta, G. Esposito, K. Urbanek, A. Angelis, F. Rossi, R. Piacentini, Human cardiac progenitor cells with regenerative potential can be isolated and characterized from 3D-electro-anatomic guided endomyocardial biopsies. Int. J. Cardiol. **241**, 330–343 (2017)
12. J.L. McAfee, Ch. Warren, R. Prayson, Ultrastructural examination of skin biopsies may assist in diagnosing mitochondrial cytopathy when muscle biopsies yield negative results. Ann. Diagn. Pathol. **29**, 41–45 (2017)
13. V.A. Gupta, N.C. Nanda, V.L. Sorrell, Role of echocardiography in the diagnostic assessment and etiology of heart failure in older adults: opacify, quantify, and rectify. Heart Fail. Clin. **13**(3), 445–466 (2017)
14. X. Gao, W. Li, M. Loomes, L. Wang, A fused deep learning architecture for viewpoint classification of echocardiography. Inf. Fusion **36**, 103–113 (2017)
15. S. Wang, P. Phillips, Z. Dong, Y. Zhang, Intelligent facial emotion recognition based on stationary wavelet entropy and Jaya algorithm. Neurocomputing **272**, 668–676 (2018)

16. W. Song, Y. Lei, S. Chen, Z. Pan, Q. Wang, Multiple facial image features-based recognition for the automatic diagnosis of turner syndrome. Comput. Ind. **100**, 85–95 (2018)
17. L. Hoang, T. Manh, H. Fujita, N. Dey, D. Chu, Dental diagnosis from X-Ray images: an expert system based on fuzzy computing. Biomed. Signal Process. Control **39**, 4–73 (2018)
18. N. Jiang, Y. Zhuang, D. Chiu, Multiple transmission optimization of medical images in recourse-constraint mobile telemedicine systems. Comput. Methods Programs Biomed. **145**, 103–113 (2017)
19. B. González, F. Valdez, P. Melin, G. Prado-Arechiga, Fuzzy logic in the gravitational search algorithm enhanced using fuzzy logic with dynamic alpha parameter value adaptation for the optimization of modular neural networks in echocardiogram recognition. Appl. Soft Comput. **37**, 245–254 (2015)
20. B. González, F. Valdez, P. Melin, G. Prado-Arechiga, Fuzzy logic in the gravitational search algorithm for the optimization of modular neural networks in pattern recognition. Expert Syst. Appl. **42**(14), 5839–5847 (2015)
21. S. Abdel-Khaled, A.B. Ishak, A. Osama, A.-S.F. Omer, Obada: A two-dimensional image segmentation method based on genetic algorithm and entropy. Neurocomput. Opt. **131**, 414–422 (2017)
22. P. Ghosh, P. Mitchell, J. Tanyi, A. Hung, Incorporating priors for medical image segmentation using a genetic algorithm. Neurocomputing **195**, 181–194 (2016)
23. A. Qayyum, S.M. Anwar, M. Awais, M. Majid, Medical image retrieval using deep convolutional neural network. Neurocomputing **266**, 8–20 (2017)
24. M. Lagunes, O. Castillo, F. Valdez, J. Soria, P. Melin, Parameter optimization for membership functions of Type-2 fuzzy controllers for autonomous mobile robots using the firefly algorithm, in *NAFIPS* (2018), pp. 569–579
25. D. Sánchez, P. Melin, O. Castillo, Optimization of modular granular neural networks using a hierarchical genetic algorithm based on the database complexity applied to human recognition. Inf. Sci. **309**, 73–101 (2015)
26. L. Rodríguez, O. Castillo, J. Soria, P. Melin, J. Soto, A fuzzy hierarchical operator in the grey wolf optimizer algorithm. Appl. Soft Comput. **57**, 315–328 (2017)
27. C. Peraza, F. Valdez, O. Castillo, Improved method based on Type-2 fuzzy logic for the adaptive harmony search algorithm in *Fuzzy Logic Augmentation of Neural and Optimization Algorithms* (2018), pp. 29–37
28. C. Peraza, F. Valdez, P. Melin, Optimization of intelligent controllers using a Type-1 and Interval Type-2 fuzzy harmony search algorithm. Algorithms **10**(3), 82 (2017)
29. E. Bernal, O. Castillo, J. Soria, F. Valdez, Galactic swarm optimization with adaptation of parameters using fuzzy logic for the optimization of mathematical functions, in *Fuzzy Logic Augmentation of Neural and Optimization Algorithms* (2018), pp. 131–140
30. R. Eberhart, J. Kennedy, A new optimizer using particle swarm theory, in *Proceedings of the Sixth International Symposium on Micro Machine and Human Science*, vol. 1 (IEEE, New York, 1995), pp. 39–43
31. Patricia Melin, Alejandra Mancilla, Miguel Lopez, Olivia Mendoza, A hybrid modular neural network architecture with fuzzy Sugeno integration for time series forecasting. Appl. Soft Comput. **7**(4), 1217–1226 (2007)
32. P. Melin, O. Castillo, *Modelling, Simulation and Control of Non-linear Dynamical Systems: An Intelligent Approach Using Soft Computing and Fractal Theory* (CRC Press, Boca Raton, 2001)
33. P. Melin, D. Sánchez, O. Castillo, Genetic optimization of modular neural networks with fuzzy response integration for human recognition. Inf. Sci. **197**, 1–19 (2012)
34. C. Leal Ramírez, O. Castillo, P. Melin, A. Rodríguez Díaz, Simulation of the bird age-structured population growth based on an interval type-2 fuzzy cellular structure. Inf. Sci. **181**(3), 519–535 (2011)
35. N.R. Cázarez-Castro, L.T. Aguilar, O. Castillo, Designing Type-1 and Type-2 fuzzy logic controllers via fuzzy lyapunov synthesis for nonsmooth mechanical systems. Eng. Appl. of AI **25**(5), 971–979 (2012)

36. O. Castillo, P. Melin, Intelligent systems with interval type-2 fuzzy logic. Int. J. Innov. Comput. Inf. Control **4**(4), 771–783 (2008)
37. G.M. Mendez, O. Castillo, Interval type-2 TSK fuzzy logic systems using hybrid learning algorithm, Fuzzy Systems, 2005, in *The 14th IEEE International Conference on FUZZ'05*, 230–235

Optimal Recognition Model Based on Convolutional Neural Networks and Fuzzy Gravitational Search Algorithm Method

Yutzil Poma, Patricia Melin, Claudia I. González and Gabriela E. Martinez

Abstract In this paper we propose the optimization of a convolutional neural network (CNN) using the Fuzzy Gravitational Search Algorithm method (FGSA). The FGSA is inspired in extension of the Gravitational Search Algorithm (GSA) using fuzzy logic and this method is used to obtain the number of images per block that will enter in the training phase. The optimized CNN is applied for pattern recognition using the 10 handwritten numbers of the MINIST database. The model of the CNN model presented in this paper can be applied for any recognition or image classification application. In addition, the recognition rate achieved with the CNN optimized by the FGSA was compared against the results obtained with the non-optimized CNN.

Keywords Convolutional neural network · FGSA · Neural network · Classification · Optimization

1 Introduction

The convolutional neural network (CNN) help to identify and classify images, which are adapted to process data in multidimensional arrays. One of the main advantages when using these neural networks is reducing the number of connections and the number of parameters to train in comparison with the fully connected neural network. The first time that the convolutional neuronal network was used was for the

Y. Poma · P. Melin (✉) · C. I. González · G. E. Martinez
Tijuana Institute of Technology, Tijuana, Mexico
e-mail: pmelin@tectijuana.mx

Y. Poma
e-mail: yutpoma@hotmail.com

C. I. González
e-mail: cgonzalez@tectijuana.mx

G. E. Martinez
e-mail: gmartinez@tectijuana.mx

© Springer Nature Switzerland AG 2020
O. Castillo and P. Melin (eds.), *Hybrid Intelligent Systems in Control,*
Pattern Recognition and Medicine, Studies in Computational Intelligence 827,
https://doi.org/10.1007/978-3-030-34135-0_6

71

recognition of handwritten digits using a neural network with back-propagation [1]. In the last years fully connected neural networks have been used in various applications such as the optimization of the modular neural network (MNN) using a particle swarm with a fuzzy dynamic parameter [2], also using MNN for pattern recognition using the paradigm Ant Colony for network optimization [3], or like when using traditional Neural Networks (NN) for face recognition with fuzzy edge detector [4], also the integration with MNN based in Choquet integral with type-1 and type-2 applied to face recognitions [5], other application is provide the hypertension risk diagnosis of a person designed a hybrid model using modular neural networks and fuzzy logic [6], or the hybridation with methods like Particle Swarm Optimization (PSO) and Genetic Algorithms (GAs) using Fuzzy Logic to integrate the results applied in a NN optimized [7], others works are like using the genetic optimization of MNN with fuzzy response integration [8], or the modular neural network optimization based on a Multi-objective Hi-erarchical Genetic Algorithm [9], the optimization of modular granular neural network (MNN) using the firefly algorithm [10], the edge detection method like is based on Sobel technique and generalized tipe-2 fuzzy logic system [11], and the optimization of interval type-2 fuzzy systems [12]. Recently CNNs have been used, like, in the reading of system checks where character recognizers are used, combined with global training techniques [13]; they have also been used in the automatic detection and blurring of plates and faces in order to protect privacy in Google Street View [14]. There are some experimental applications in which these networks have been used as the detection of obstacles at a great distance, using a deep hierarchical network trained to extract significant characteristics of an image, where the classifier can predict the transfer capacity in real time, detecting roads and obstacles between 100 or to more and 5 m and it is adaptive [15]. The main contribution of the CNN is the extraction of characteristics of the images, trying to improve the results in pattern recognition. It has been decided to use the fuzzy gravitational search algorithm FGSA method [16] which is a variation of the Gravitational Search Algorithm (GSA) [17], unlike its predecessor, it obtains the *Alpha* variable through a fuzzy system which tends to increase or decrease, in comparison with other methods in which GSA is used where the *Alpha* is a static value [18–21].

The main contribution in this paper is the optimization of a convolutional neural network, together with the FGSA method, which obtains the number of images per block (Bsize) for the training phase in the CNN.

The paper is structured as follows: Sect. 2 shows the background where the basic foundations of convolutional neural networks are known. In Sect. 3, the proposed method to optimize CNN is shown. Section 4 shows the results achieved when the Bsize value of the CNN is changing in a manual way and when this parameter is obtained by using the FGSA method in the convolutional neural network. Finally, in Sect. 5 we can find the conclusions reached in the experimentation of the case study.

2   Literature Review

This section presents the basic concepts necessary to understand the proposed method.

2.1   Convolutional Neural Networks

CNN or ConvNet is a neural network that performs the extraction of characteristics from the input data. It is constituted by different kinds of layers, which each of them obtain important features. At the end it classifies the characteristics of the image, resulting in the corresponding recognition [22]. The CNNS have gone through a phase of evolution where some publications have established more efficient ways to train these networks using GPUs [23, 24].

Within the convolutional neural network we can define the number of times it will be trained (Batch), as well as the number of images per block that will enter the training process (Bsize).

Figure 1 shows the classical architecture of a CNN, which consists in three main layers: convolution, pooling and classification.

2.1.1   Convolution Layer

This layer generates new images called "Map of features", this accentuates the unique characteristics of the input data. This layer contains filters (kernels) that convert the images into new images, these are called "Convolution Filters", and these consist of two-dimensional arrays of between 5 * 5, and in recent applications have been used up to 1 * 1.

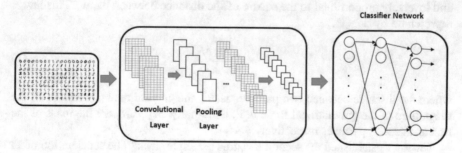

Fig. 1 Basic architecture of CNN

2.1.2 Non-linearity Layer

Several activation functions are applied to this layer. Which are more used non-linearity, in multilayer networks. The most commonly used activation functions are *Tanh, Sigmoide as* and rectified linear units (ReLU). Compared to other functions ReLU is preferable because networks train faster several times [25].

2.1.3 Pooling Layer

Also called "grouping" layer, it is responsible for reducing the size of the image and combines the nearby pixels in a certain area, taking small blocks of the convolution layer and sub-samples to obtain an output, or a single representative value [26, 27], which consists of a set of pixels, of which the average or maximum is calculated [28] as is in the corresponding case may be.

2.1.4 Classifier Layer

After the convolution and reduction layers, a completely connected layer is used, in which each pixel is a separate neuron as a multilayer perceptron. This layer has as many neurons as the number of classes to predict, in this layer the neural network recognizes or classifies the images that will obtain as output [29, 30, 31].

2.2 GSA

It is inspired on the law of gravity proposed by Newton in 1685, "The gravitational force between two particles is directly proportional to the product of their masses and inversely proportional to the square of the distance between them." This law is represented by Eq. (1):

$$F = G\frac{M_1 M_2}{R^2} \tag{1}$$

where the distance between two particles is R^2, the gravitational constant is G, the magnitude of the gravitational force is F, and finally, M_1 y M_2 are the mass of the first and second particles respectively.

Newton's established the second law, this principle says: "The acceleration of an object, is directly proportional to the net force acting on it, and inversely proportional to its mass" [32]. It is represented with Eq. (2):

$$a = \frac{F}{M} \tag{2}$$

In where the M is the mass of the object, F is the magnitude of gravitational force and the magnitude of acceleration is a.

2.3 FGSA

It is an agent-based method, this method has been used in various applications [33, 34].

In this method agents are objects, which are determined for their masses. All objects are attracted to each other, by this force of gravity, in turn, causes a global movement of all objects and maintains direct communication with the masses.

As the *Alpha* value is changing, different gravitation and acceleration can be obtained for each agent, which improves its performance.

The *Alpha* parameter was optimized by means of a fuzzy system, where the membership functions are triangular, and their the ranges were decided in order to give a wider value to look for the α [5].

It was decided to use the linguistic values: Not Large, Medium and so High with triangular membership functions, which are the following:

Not Large : [−50 0 50], Medium[0 50 100], So High [50 100 150]

The fuzzy system that carries out the modification of the *Alpha* variable has 3 rules which are:

1. *If the repetition is Not Large then α is Small*
2. *If the repetition is Medium and then α is medium*
3. *If the repetition is So High then α is tall.*

In Fig. 2 the flow chart of FGSA method is illustrated [5].

3 Proposed Method

The proposed model begins with the input of images of handwritten numbers from the MINIST database (Mixed National Institute of Standards and Technology) [5]. In Fig. 3 we can find the proposed method where CNN and FGSA work together to have a higher recognition percentage.

These are the steps:

1. The FGSA obtains the Bsize parameter.
2. The FGSA passes the Bsize parameter to the CNN.
3. The database is entered into in the CNN.
4. The CNN get the Bsize.
5. The Bsize partition the images number of blocks for training.

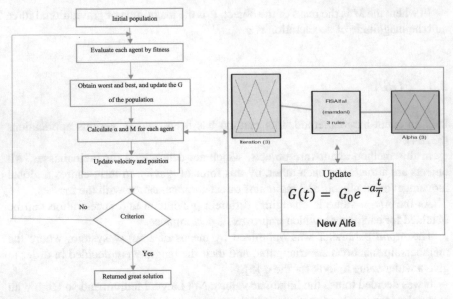

Fig. 2 Flow chart of FGSA [16]

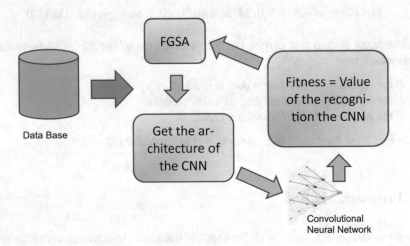

Fig. 3 Proposed method where the CNN is optimized with FGSA

6. The CNN training the data images.
7. The CNNobtains the value of the recognition.

The images will be recognized and classified by the CNN where going through the various layers of convolution, pooling and ending with the classification layer, in turn, the FGSA method will be responsible for optimizing the Bsize value of the neural network. This value is responsible for partitioning the block of images that will enter the Batch, which repeats the training process of the various layers of the network.

4 Results and Discussion

As a case study, the MNIST database was used, which contains 70,000 images, from which a sample of 10,000 images was taken, using 8000 for training and 2000 for testing. In Fig. 4 some examples of the numbers used in the database are illustrated, and each image is in black and white with a size of 28 * 28 pixels.

Experiments were performed by manually modifying the Bsize value. This variable separates the block of images that will enter the Batch (number of times the CNN training will take place), where it applies the phases of convolution, ReLU, Pooling, ReLU and classification.

Subsequently, the different images of numbers written by hand are identified and classified, obtaining 10 resulting outputs (0, 1, 2, 3, 4, 5, 6, 7, 8, 9). In Table 1, the results obtained when performing the variation of the Bsize value manually from 10 to 10 until 100, are shown until reaching the best value of the percentage of recognition of the images.

The best solution of this manual experiment is when the Bsize value is 10 achieving a recognition value of 97.65.

Subsequently, the Bsize value of the convolutional neural network was optimized with the FGSA method (where better results were obtained). The fuzzy system allows the *Alpha* value to decrease and increase, in this case we can control the exploitation and exploration the agent, when *Alpha* parameter decreases, the agents explore the search space and by increasing the *Alpha* we get a better exploitation of the area found. Table 2 shows the results obtained where the CNN is optimized with the FGSA method using the fuzzy system when the *Alpha* increases or decreases.

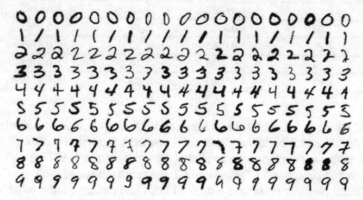

Fig 4 Images from the MNIST database

Table 1 Bsize modified manually without using the FGSA in the convolutional neuronal network

Bsize/Block	Recognition rate
10	0.9765
20	0.972
30	0.9715
40	0.965
50	0.962
60	0.9615
70	0.955
80	0.943
90	0.944
100	0.9465

Table 2 Bsize of the CNN optimized with the FGSA method

Experiment	FGSA—decrement *Alpha*		FGSA—increment *Alpha*	
	Number blocks	Recognition rate	Number blocks	Recognition Rate
1	29	0.9725	33	0.9675
2	10	0.9765	10	0.9765
3	10	0.9765	10	0.9765
4	10	0.9765	10	0.9765
5	10	0.9765	10	0.9765
6	10	0.9765	10	0.9765
7	10	0.9765	10	0.9765
8	10	0.9765	10	0.9765
9	10	0.9765	10	0.9765
10	10	0.9765	10	0.9765
11	10	0.9765	10	0.9765
12	10	0.9765	10	0.9765
13	10	0.9765	10	0.9765
14	10	0.9765	10	0.9765
15	10	0.9765	10	0.9765

Based on the results obtained where the FGSA method is used optimizing the Bsize value of the CNN, it was determined that the highest recognition percentage was obtained when the Bsize has a value of 10 and its fitness is 0.9765.

We can be appreciate, in Tables 1 and 2, the results obtained with the Convolutional Neural Network optimized with the FGSA method are similar to when manually calculated. In both cases 0.9765 of recognition rate was obtained with the best configuration using 10 Blocks (variable Bsize), but the optimized CNN get the optimal value in an "automatic" way, for the two cases where the *Alpha* increases and where the *Alpha* decreases.

5   Conclusions

We concluded that the best rate recognition was obtained when the CNN is optimized, the results show that optimizing the variable (Bsize), which is responsible for partitioning the images into blocks that later enter the CNN training, obtains a higher rate recognition.

As future work it is expected that by optimizing other parameters, such as the number of filters and the number of nodes in hidden layer of CNN, a higher percentage of recognition of the images will be obtained. In addition, we can perform more experiments with variations in values of some parameters and we could use type-2 fuzzy logic as in [35–38, 43]. In addition, we can deal with other types of applications, like in [39–42].

Acknowledgements We thank our sponsor CONACYT and the **Tijuana Institute of Technology** for the financial support provided with the scholarship number 816488.

References

1. L.D. Le Cun Jackel, B. Boser, J.S. Denker, D. Henderson, R.E. Howard, W. Hubbard, B. Le Cun, J. Denker, D. Henderson, Handwritten digit recognition with a back-propagation network. Adv. Neural Inf. Process. Syst., pp. 396–404 (1990)
2. D. Sánchez, P. Melin, O. Castillo, Fuzzy adaptation for particle swarm optimization for modular neural networks applied to iris recognition. NAFIPS **648**, 104–114 (2017)
3. F. Valdez, O. Castillo, P. Melin, Ant colony optimization for the design of Modular Neural Networks in pattern recognition. Proc. Int. Jt. Conf. Neural Netw. pp. 163–168 (2016 Oct)
4. C.I. Gonzalez, J.R. Castro, O. Mendoza, P. Melin, General Type-2 fuzzy edge detector applied on face recognition system using neural networks, in *2016 IEEE International Conference on Fuzzy Systems (FUZZ-IEEE)*, pp. 2325–2330 (2016)
5. G.E. Martínez, P. Melin, O.D. Mendoza, O. Castillo, Face recognition with a sobel edge detector and the Choquet integral as integration method in a modular neural networks, in *Design of Intelligent Systems Based on Fuzzy Logic, Neural Networks and Nature-Inspired Optimization*, ed. by P. Melin, O. Castillo, J. Kacprzyk (Springer International Publishing, Cham, 2015), pp. 59–70
6. P. Melin, I. Miramontes, G. Prado-Arechiga, A hybrid model based on modular neural networks and fuzzy systems for classification of blood pressure and hypertension risk diagnosis. Expert Syst. Appl. **107**, 146–164 (2018)
7. F. Valdez, P. Melin, O. Castillo, Modular Neural Networks architecture optimization with a new nature inspired method using a fuzzy combination of Particle Swarm Optimization and Genetic Algorithms Inf. Sci. (Ny) **270**, 143–153 (2014)
8. P. Melin, D. Sánchez, O. Castillo, Genetic optimization of modular neural networks with fuzzy response integration for human recognition. Inf. Sci. (Ny) **197**, 1–19 (2012)

9. P. Melin, D. Sánchez, Multi-objective optimization for modular granular neural networks applied to pattern recognition. Inf. Sci. (Ny) **460–461**, 594–610 (2018)
10. D. Sánchez, P. Melin, O. Castillo, Engineering Applications of Artificial Intelligence Optimization of modular granular neural networks using a firefly algorithm for human recognition. Eng. Appl. Artif. Intell. **64**(Oct 2016), 172–186 (2017)
11. C.I. Gonzalez, P. Melin, J.R. Castro, O. Mendoza, O. Castillo, An improved sobel edge detection method based on generalized type-2 fuzzy logic. Soft Comput. J. **20**(2), 773–784 (2016)
12. C.I. Gonzalez, J.R. Castro, O. Mendoza, P. Melin, O. Castillo, Optimization by cuckoo search of interval type-2 fuzzy logic systems for edge detection. Stud. Fuzziness Soft Comput. **342**, 141–154 (2016)
13. Y. LeCun, L. Bottou, Y. Bengio, P. Haffner, Gradient-based learning applied to document recognition. Proc. IEEE **86**(11), 2278–2323 (1998)
14. A. Frome, G. Cheung, A. Abdulkader, M. Zennaro, B. Wu, A. Bissacco, H. Adam, H. Neven, L. Vincent, Large-scale privacy protection in google street view. Evaluation, pp. 2–9 (2009)
15. R. Hadsell, P. Sermanet, J. Ben, A. Erkan, M. Scoffier, K. Kavukcuoglu, U. Muller, Y. LeCun, Learning long-range vision for autonomous off-road driving. J. F. Robot. **26**(2), 120–144 (2009)
16. A. Sombra, F. Valdez, P. Melin, O. Castillo, A new gravitational search algorithm using fuzzy logic to parameter adaptation, in *2013 IEEE Congress on Evolutionary Computation no. 3*, pp. 1068–1074 (2013)
17. E. Rashedi, H. Nezamabadi-pour, S. Saryazdi, GSA: a gravitational search algorithm. Inf. Sci. (Ny) **179**(13), 2232–2248 (2009)
18. O.P. Verma, R. Sharma, Newtonian gravitational edge detection using gravitational search algorithm. Int. Conf. Commun. Syst. Netw. Technol., pp. 184–188 (2012)
19. A. Hatamlou, S. Abdullah, Z. Othman, Gravitational search algorithm with heuristic search for clustering problems. Conf. Data Min. Optim., pp. 190–193, (2011 June)
20. S. Mirjalili, S.Z. Mohd Hashim, H. Moradian Sardroudi, Training feedforward neural networks using hybrid particle swarm optimization and gravitational search algorithm. Appl. Math. Comput. **218**(22), 11125–11137 (2012)
21. S. Mirjalili, S.Z.M. Hashim, A new hybrid PSOGSA algorithm for function optimization, in *Proceedings of ICCIA 2010–2010 International Conference on Computer and Information Application*, no. 1, pp. 374–377, 2010
22. Y. LeCun, Y. Bengio, Convolution networks for images, speech, and time-series. Igarss **2014**(1), 1–5 (1998)
23. Y. Bengio, P. Lamblin, Greedy layer-wise training of deep networks. Adv. Neural Inf. Process. Syst. (1), 153–160 (2007)
24. K. Chellapilla, S. Puri, P. Simard, High performance convolutional neural networks for document processing. Int. Work. Front. Handwrit. Recognit. (2006)
25. V. Nair, G.E. Hinton, Rectified linear units improve restricted Boltzmann machines, in *Proceedings of the 27th International Conference on Machine Learning*, no. 3, pp. 807–814 (2010)
26. M. Ranzato, F.J. Huang, Y.-L. Boureau, Y. LeCun, Unsupervised learning of invariant feature hierarchies applications to object recognition, in IEEE Conference on Computer Vision and Pattern Recognition, pp. 1–8 (2007)
27. J. Yang, K. Yu, Y. Gong, T.H. Beckman, Linear spatial pyramid matching using sparse coding for image classification, in *IEEE Computer Society Conference* on Computer Vision and *Pattern Recognition*, pp. 1794–1801 (2009)
28. T. Wang, D.J. Wu, A. Coates, A.Y. Ng, End-to-end text recognition with convolutional neural networks, in *ICPR, International Conference on Pattern Recognition*, pp. 3304–3308 (2012)
29. P. Kim, *MATLAB Deep Learning* (2017)
30. R. Venkatesan, B. Li, *Convolutional Neural Networks in Visual Computing: A Concise Guide* (CRC Press, Boca Raton, 2017)
31. L. Lu, Y. Zheng, G. Carneiro, L. Yang, *Deep Learning and Convolutional Neural Networks for Medical Image Computing* (2017)
32. J. Walker, R. Resnick, D. Halliday, *Fundamentals of Physics* (Wiley, New York, 2008)

33. B. González, F. Valdez, P. Melin, G. Prado-Arechiga, Fuzzy logic in the gravitational search algorithm for the optimization of modular neural networks in pattern recognition. Expert Syst. Appl. **42**(14), 5839–5847 (2015)
34. B. González, F. Valdez, P. Melin, G. Prado-Arechiga, Fuzzy logic in the gravitational search algorithm enhanced using fuzzy logic with dynamic alpha parameter value adaptation for the optimization of modular neural networks in echocardiogram recognition. Appl. Soft Comput. J. **37**, 245–254 (2015)
35. C. Leal Ramírez, O. Castillo, P. Melin, A. Rodríguez Díaz, Simulation of the bird age-structured population growth based on an interval type-2 fuzzy cellular structure. Inf. Sci. **181**(3), 519–535 (2011)
36. N.R. Cázarez-Castro, L.T. Aguilar, O. Castillo, Designing Type-1 and Type-2 Fuzzy logic controllers via Fuzzy Lyapunov synthesis for nonsmooth mechanical systems. Eng. Appl. AI **25**(5), 971–979 (2012)
37. O. Castillo P. Melin, Intelligent systems with interval type-2 fuzzy logic. Int. J. Innov. Comput. Inf. Control **4**(4), 771–783 (2008)
38. G.M. Mendez, O. Castillo, Interval type-2 TSK fuzzy logic systems using hybrid learning algorithm, in *The 14th IEEE International Conference on Fuzzy Systems, 2005. FUZZ'05*, pp. 230–235
39. P. Melin, O. Castillo, Intelligent control of complex electrochemical systems with a neuro-fuzzy-genetic approach. IEEE Trans. Ind. Electron. **48**(5), 951–955
40. P. Melin, A. Mancilla, M. Lopez, O. Mendoza, A hybrid modular neural network architecture with fuzzy Sugeno integration for time series forecasting. Appl. Soft Comput. **7**(4), 1217–1226 (2007)
41. P. Melin, O. Castillo, Modelling, Simulation and Control of Non-Linear Dynamical Systems: an Intelligent Approach Using Soft Computing and Fractal Theory (CRC Press, Boca Raton, 2001)
42. P. Melin, *G Prado-Arechiga: New Hybrid Intelligent Systems for Diagnosis and Risk Evaluation of Arterial Hypertension* (Springer, Switzerland, 2018)
43. P. Melin, C.I. González, J.R. Castro, O. Mendoza, O. Castillo, Edge-detection method for image processing based on generalized Type-2 fuzzy logic. IEEE Trans. Fuzzy Syst. **22**(6), 1515–1525 (2014)

Optimal Number of Clusters Finding Using the Fireworks Algorithm

Juan Barraza, Fevrier Valdez, Patricia Melin and Claudia González

Abstract The main goal of this paper is to find the optimal number of clusters for data set ungrouped using an optimization algorithm; in this case, we are using Fireworks Algorithm (FWA). The optimal number of clusters will be finding based on centroids and to determine the maximum K centroids of the interval where to search the Fireworks Algorithm, we are introducing two statistics rules: Sturges Law and the square root of N, also we are introducing two metrics to evaluate the clusters, which are Intra-clusters and Inter-clusters.

Keywords Fireworks · Algorithm · Clustering · Optimization

1 Introduction

In computational science the more important functions to imitate for machines are the learning and solve problems, for consequent, there are methodologies as neural networks [1], fuzzy logic [2] or algorithms based on nature, swarm intelligence [3], or the physical [4], which are used on the artificial intelligence to imitate the functions of humans mentioned above [5, 6].

At the end of the 80s and the beginning of the 90s the concept of swarm intelligence was created, which was introduced by Gerardo Beni and Jing Wang; nowadays, it is a concept mentioned a lot in the artificial intelligence area, and, swarm intelligence

J. Barraza · F. Valdez · P. Melin (✉) · C. González
Tijuana Institute of Technology, Tijuana, Mexico
e-mail: pmelin@tectijuana.mx

J. Barraza
e-mail: Jbarrazagerardo@gmail.com

F. Valdez
e-mail: fevrier@tectijuana.mx

C. González
e-mail: cgonzalez@tectijuana.mx

© Springer Nature Switzerland AG 2020
O. Castillo and P. Melin (eds.), *Hybrid Intelligent Systems in Control,*
Pattern Recognition and Medicine, Studies in Computational Intelligence 827,
https://doi.org/10.1007/978-3-030-34135-0_7

is considered as a branch of artificial intelligence. Swarm intelligence studies the collective behavior of decentralized, self-organized natural or artificial systems [7–9].

On the other hand, in the clustering algorithms [10–12], we need to know the optimal number of clusters or centroids for a certain data set, with the finally that the algorithm will have a performance satisfactory, on the contrary, if we does not know the optimal number of clusters or centroids, the probability of the clustering algorithm fails with the clusters or locations of centroids is higher and is a disadvantage of the clustering algorithm that we are going to attack [13, 14].

This paper is organized as follows: Sect. 2 shows the Conventional Fireworks algorithm (FWA). Our proposed method is in Sect. 3, i.e. we present C-FWA, in Sect. 4 describes the rules and metrics to evaluate the performance of the C-FWA; Sect. 5 shows the experiments and results, and Sect. 6 the conclusions.

2 Fireworks Algorithm (FWA)

The conventional Fireworks Algorithm (FWA) is a swarm intelligence algorithm as mentioned above, and it is composed for 4 general steps: initialization of locations, calculation of the number of sparks, calculation of the explosion amplitude for each firework and selection of the best location [15, 16].

In this Section, the main equations of the algorithm are presented:

Number of Sparks
The number of sparks is calculated with the following two Equations (Eqs. 1 and 2).

$$Minimize\ f(x_i) \in R,\ x_i\_min \le x_i \le x_i\_max \tag{1}$$

where $x_{i_{min}} \le x_i \le x_{i_{max}}$ represents the bounds of the search space

$$S_i = m. \frac{y_{max} - f(x_i) + \epsilon}{\sum_{i=1}^{n}(y_{max)} - f(x_i)) + \epsilon} \tag{2}$$

where m is a constant parameter, y_{max} is the worst value of the objective function and ϵ is a smallest number in the computer.

Explosion Amplitude
The explosion amplitude for each firework is calculated as follows:

$$A_i = \hat{A}. \frac{f(x_i) - y_{min} + \epsilon}{\sum_{i=1}^{n}(f(x_i) - y_{min}) + \epsilon} \tag{3}$$

where $\hat{A}$ is a constant parameter that controls the maximum amplitude of each firework, $y_{min} = \min(f(x_i))(i = 1, 2, 3, \ldots, n)$, indicate the minimum value (best) of the objective function among n fireworks and ϵ indicates the smallest constant in the computer, and it is utilized with the goal that an error of division by zero does not occur.

Selection of Locations

The equations that are used in this algorithm to select the best current location are the following:

$$R(x_i) = \sum_{j \in K} d(x_i, x_j) = \sum_{j \in K} \|x_i - x_j\| \tag{4}$$

where K is the set of all current locations from both fireworks. Then the probability for selection of a location at x_i is defined as:

$$p(x_i) = \frac{R(x_i)}{\sum_{j \in K} R(x_j)} \tag{5}$$

This is explained in more detail in [17–19].

3 Proposed Method (C-FWA)

In this Section, the modification of the Fireworks Algorithm (FWA) to find the optimal number of clusters is presented.

We implemented the Algorithm (FWA) to automatically find the optimal number of clusters, and we decided called as Centroid Fireworks Algorithm and we denoted as C-FWA.

The reason why we adapted FWA to find the optimal number clusters is because in almost every clustering algorithm, the number of clusters is initialized in a manually way and is maintaining constant while the algorithm is executed. To avoid the inconvenience to initialized the arbitrary way every execution of the algorithm, we modified the FWA with the main goal to find the optimal number of cluster as we mentioned above, and is explained in detail in the operation with the sequence of the following steps: 1. Select n initial locations, 2. set off n fireworks at n locations, 3. Obtain the locations of sparks, 4. Evaluate the quality of locations and 5. Obtain the new locations. These steps are presented in Fig. 1.

To select n initial locations, the first step is generating the initial swarm for the algorithm (C-FWA) to start working; the initial swarm is generated with the following equation:

$$Swarm_{ij} = LB_j + (UB_j - LB_j) * r_{ij}, \quad i = 1, 2, 3, \ldots n \tag{6}$$

where LB and UB are the lower and upper bounds for individual i in the dimension j, the dimensions of individuals is depending on the features of the data set given and r is a random value between 0 and 1.

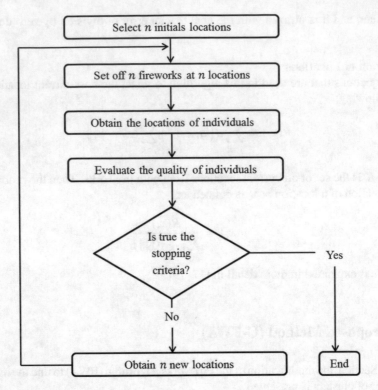

Fig. 1 Flow chart of C-FWA

Each individual, in this algorithm, is a spark or a firework, both are represented in a vector divided in two parts, in Fig. 2 we illustrate an example.

Where K is an integer number that is representing the number of centroids that one possible solution has and the rest of the vector $(d_1, d_2, d_3, \ldots, d_{nd} * K)$ are real numbers that are representing the features of data set given and nd are the features per each data.

To avoid problems with mathematics calculus we need to convert the second part of vector, i.e., the real numbers; these are shown in Fig. 3 represented of general way.

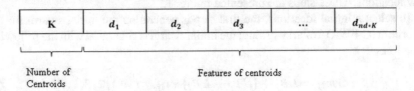

Fig. 2 Representation of the individual in C-FWA

d_1	d_2	d_3	d_4	...	d_{nd*K}
$d_{1,1}$	$d_{1,2}$	$d_{1,3}$	$d_{1,4}$	...	$d_{1,nd}$
$d_{2,1}$	$d_{2,2}$	$d_{2,3}$	$d_{2,4}$	...	$d_{2,nd}$
$d_{3,1}$	$d_{3,2}$	$d_{3,3}$	$d_{3,4}$	...	$d_{3,nd}$
$d_{4,1}$	$d_{4,2}$	$d_{4,3}$	$d_{4,4}$	...	$d_{4,nd}$
$d_{K,1}$	$d_{K,2}$	$d_{K,3}$	$d_{K,4}$	...	$d_{K,nd}$

Fig. 3 General representation of vector solution converts to matrix solution in C-FWA

We decided introducing two statistic rules with the goal of creating bounds for the number of clusters, i.e., the rules will allow to obtain a K_{max} to help the performance of the FWAC. Both rules are explained in the following Section.

4 Rules and Metrics

Rules

To approximate the optimal number of classes depending the numbers of data (N) are used. One of these rules is called the Sturges law [20], and it is shown in Eq. (7):

$$K = 1 + 3.322 \log N \tag{7}$$

Other law to find the optimal number of clusters is the square root of N, which is indicated in the following Equation.

$$K = \sqrt{N} \tag{8}$$

where K is the number of classes, but to our problem is the number that representing the maximum number of centroids (K_{max}) and N is the number of data in both Eqs. (7 and 8) [20, 21].

Metrics

There are two ways to evaluate the clusters: intra-cluster and inter-cluster, Intra-cluster is the value obtained, by adding the distances of all points with respect to their centroid, this metric is represented for the following Equation:

$$Intra = \sum_{i=1}^{n} dist(c_i, C) \tag{9}$$

where c_i are the data that belong to centroid C [22].

Inter-cluster is the value obtained, adding up the distances among the centroids. In Eq. (10), we show a general form to calculate the inter-cluster distance:

$$Inter = \sum_{\substack{i, j = 1, \\ i \neq j}}^{k} dist(C_i, C_j) \tag{10}$$

where i and j are the number of the centroids, C_i and C_j are different centroids and k is the maximum number of centroids [23].

5 Experiments and Results

In this section we present the results obtained by the experiments realized with Centroid Fireworks Algorithm (C-FWA). We perform 31 independents runs using the following parameters in the algorithm:

- 5 Fireworks.
- Amplitude Coefficient = 40.
- Amplitude Sparks = 50.
- Dimensions = features of data set.
- 15,000 function evaluations.

We tested with three datasets the performance of the C-FWA, the first is the Iris data set, with 3 K-optimal (centroids). Wine data set is the second with 3 K-optimal (centroids) and the third data set is the WBDC with 2 K-optimal (centroids).

The features, number of data and K-Optimal for each data set is illustrates in Table 1.

Tables 2 and 3 shows the results obtained of average of the 31 independent runs with 15,000 function evaluations using intra-cluster for minimum distance with square root of N and Sturges rule, respectively.

Table 1 Data sets

Data set	Number of data	Features	K-optimal
Iris	150	4	3
Wine	178	13	3
WBDC	569	30	2

Table 2 Results of C-FWA for minimum distance using intra-cluster with square root of N

Data set	K optimal	Mean	Standard deviation
Iris	3	5.13	2.13
Wine	3	8.94	3.23
WBDC	2	11.1	6.98

Table 3 Results of C-FWA for minimum distance using intra-cluster with Sturges rule

Data set	K optimal	Mean	Standard deviation
Iris	3	3.45	1.59
Wine	3	5.61	2.16
WBDC	2	6.16	2.41

For the results of Tables 2 and 3, when we are using intra-cluster as metric evaluation, the best combination is with Sturges, although, the C-FWA does not have good performance with the others combinations, while intra-cluster is used.

The results obtained using intra-cluster with maximum distances for C-FWA is presented in Tables 4 and 5. In Table 4 is shown with the rule square root of N and in Table 5 is shown with Sturges rule, respectively.

The combinations used in the Tables 4 and 5 was not present the results satisfactory.

In Tables 6 and 7, we show the results of the C-FWA using inter-cluster with minimum distance, in Table 6 with the rule Square root of N and with Sturges rule is presented in Table 7.

Table 4 Results of C-FWA for maximum distance using intra-cluster with square root of N

Data set	K optimal	Mean	Standard deviation
Iris	3	7	2.66
Wine	3	8.71	2.80
WBDC	2	14.94	6.21

Table 5 Results of C-FWA for maximum distance using intra-cluster with Sturges rule

Data set	K optimal	Mean	Standard deviation
Iris	3	5.06	1.97
Wine	3	5.13	1.63
WBDC	2	6.68	2.45

Table 6 Results of C-FWA for minimum distance using inter-cluster with square root of N

Data set	K optimal	Mean	Standard deviation
Iris	3	4.03	2.01
Wine	3	4.03	2.03
WBDC	2	6.77	5.44

Table 7 Results of C-FWA for minimum distance using inter-cluster with Sturges rule

Data set	K optimal	Mean	Standard deviation
Iris	3	2.81	0.95
Wine	3	3.39	1.69
WBDC	2	4.45	2.51

The results of the C-FWA obtained using Inter-cluster but with maximum distance are shown in Tables 7 and 8, in the same way in the Table 7 are with the rule Square root of N and in the Table 8 are with Sturges rule (Table 9).

In Table 10 we made a comparison of the means among Intra-cluster versus Inter-cluster and the rules: square root of N versus Sturges using the minimum distance in the metric evaluation.

We also made a comparison of the means among Intra-cluster versus Inter-cluster and the rules: square root of N versus Sturges but using the maximum distance in the metric evaluation.

Table 8 Results of C-FWA for maximum distance using inter-cluster with square root of N

Data set	K optimal	Mean	Standard deviation
Iris	3	4.10	1.68
Wine	3	4.35	2.18
WBDC	2	9.61	5.69

Table 9 Results of C-FWA for maximum distance using inter-cluster with Sturges rule

Data set	K optimal	Mean	Standard deviation
Iris	3	3.10	1.22
Wine	3	3.48	1.36
WBDC	2	3.87	1.59

Table 10 Comparison of means with minimum distance in C-FWA

Data	Optimal	Metric	FWAC	
Set	K	Evaluation	$\sqrt{n}$	S
WBDC	2	Intra	11.10	6.16
		Inter	6.77	**4.45**
IRIS	3	Intra	5.13	**3.45**
		Inter	4.03	**2.81**
WINE	3	Intra	8.94	5.61
		Inter	4.03	**3.39**

Bold indicates Best results

Table 11 Comparison of means with maximum distance in C-FWA

Data	Optimal	Metric	FWAC	
Set	K	Evaluation	$\sqrt{n}$	S
WBDC	2	Intra	14.94	6.68
		Inter	9.61	**3.87**
IRIS	3	Intra	7	5.06
		Inter	4.1	**3.10**
WINE	3	Intra	8.71	5.13
		Inter	4.35	**3.48**

Bold indicates Best results

The reason why we are using minimum and maximum distance is because we are trying to find the K-optimal number of clusters based on centroid and we are not to finding the locations of the centroids (Table 11).

The numbers in bold shows the better result for the *K*-Optimal clusters with its respective metric evaluation and rules, the results are the average obtained of the 31 independent runs for Centroid Fireworks Algorithm (C-FWA).

6 Conclusions

A way to conclude this work, we can say that used the Fireworks Algorith for finding the optimal number of clusters to three different datasets.

As we mentioned before, the main objective of this work is to find the optimal number cluster and not the locations of the clusters, for these reason, we used Inter-cluster that calculate the distance among all points of centroids and Intra-cluster that calculate the distances of all points with respect to their centroid. The decision to use Inter and Intra-cluster is because we believe that the cluster validation is a multi-objective problem [24], i.e., we should to calculate the separation and compact of the cluster.

Based on the results, the performance of C-FWA was not good i.e., the optimizations of the number of clusters with FWAC are very variable when the square root of N is used, although, in some cases the optimizations is satisfactory, for example, when the evaluations of FWAC are with Inter-cluster and Sturges rule, we achieve an approximation for the Iris and Wine data sets with minimum and maximum distance.

To future work could be test the adjust of parameters into the Centroid Fireworks Algorithm (C-FWA) using Type-1 Fuzzy Logic and Interval Type-2 Fuzzy Logic [1, 2, 13–15, 23, 24], the parameters could be controlling the amplitude and spark coefficient [25–28] exploration and explotation of the algorithm [29, 30], and so, the C-FWA has more precision. In addition, other areas of application could be considered [16, 19, 20, 22]

References

1. G.M. Mendez, O. Castillo, Interval type-2 TSK fuzzy logic systems using hybrid learning algorithm, in *The 14th IEEE International Conference on Fuzzy Systems, 2005. FUZZ'05*, pp. 230–235
2. C. Leal Ramírez, O. Castillo, P. Melin, A. Rodríguez Díaz, Simulation of the bird age-structured population growth based on an interval type-2 fuzzy cellular structure. Inf. Sci. **181**(3), 519–535 (2011)
3. Y. Tan, Y. Zhu, *Fireworks Algorithm for Optimization* (Springer, Berlin Heidelberg, 2010), pp. 355–364
4. J. Soler, F. Tencé, L. Gaubert, C. Buche, Data clustering and similarity, in *Proceedings of the Twenty-Sixth International Florida Artificial Intelligence Research Society Conference 2013*, pp. 492–495
5. F. Aladwan, M. Alshraideh, M. Rasol, A genetic algorithm approach for breaking of simplified data encryption standard. Int. J. Secur. Its Appl. **9**(9), 295–304 (2015)
6. J. Barraza, L. Rodríguez, O. Castillo, P. Melin, F. Valdez, A new hybridization approach between the fireworks algorithm and grey wolf optimizer algorithm. J. Optim. Res. Artic. **2018** (18 p), Article ID 6495362
7. M.A. Sanchez, O. Castillo, J.R. Castro, P. Melin, Fuzzy granular gravitational clustering algorithm for multivariate data. Inf. Sci. **279**, 498–511 (2014)
8. D.H. Wolpert, W.G. Macready, No free lunch theorems for optimization. IEEE Trans. Evolut. Comput., pp. 67–82 (1997)
9. B.Y. Wu, On the intercluster distance of a tree metric. Theoret. Comput. Sci. **369**, 136–141 (2006)
10. J. Soto, P. Melin, Optimization of the fuzzy integrators in ensembles of ANFIS model for time series prediction: the case of Mackey-Glass, in *IFSA-EUSFLAT 2015*
11. L. Telescaa, M. Bernardib, C. Rovellib, Intra-cluster and inter-cluster time correlations in lightning sequences. Phys. A **356**, 655–661 (2005)
12. Y. Zheng, Q. Song, S.-Y. Chen, Multiobjective fireworks optimization for variable-rate fertilization in oil crop production. Appl. Soft Comput. **13**, 4253–4263 (2013)
13. N.R. Cázarez-Castro, L.T. Aguilar, O. Castillo, Designing Type-1 and Type-2 Fuzzy Logic Controllers via Fuzzy Lyapunov Synthesis for nonsmooth mechanical systems. Eng. Appl. AI **25**(5), 971–979 (2012)
14. O. Castillo P. Melin, Intelligent systems with interval type-2 fuzzy logic. Int. J. Innov. Comput. Inf. Control **4**(4), 771–783 (2008)
15. E. Rubio, O. Castillo, F. Valdez, P. Melin, C. I. González, G. Martinez: An extension of the fuzzy possibilistic clustering algorithm using Type-2 fuzzy logic techniques. Adv. Fuzzy Syst. **2017**, 7094046:1–7094046:23 (2017)
16. P Melin, O Castillo Modelling, simulation and control of non-linear dynamical systems: an intelligent approach using soft computing and fractal theory (CRC Press, 2001)
17. N.H. Abdulmajeed, M. Ayob, A firework algorithm for solving capacitated vehicle routing problem. Int. J. Adv. Comput. Technol. (IJACT) **6**(1), 79–86 (2014 Jan)
18. Y. Tan, *Fireworks algorithm* (Springer, Berlin Heidelberg, 2015), pp. 355–364
19. P. Melin, A. Mancilla, M. Lopez, O. Mendoza, A hybrid modular neural network architecture with fuzzy Sugeno Integration for time series forecasting. Appl. Soft Comput. **7**(4), 1217–1226 (2007)
20. P. Melin, O. Castillo, Intelligent control of complex electrochemical systems with a neuro-fuzzy-genetic approach. IEEE Trans. Ind. Electron. **48**(5), 951–955
21. H.A. Sturges, The choice of a class interval. J. Am. Stat. Assoc. **21**(153), pp. 65–66 (1926 Mar)
22. P. Melin, D. Sánchez, O. Castillo, Genetic optimization of modular neural networks with fuzzy response integration for human recognition. Inf. Sci. **197**, 1–19 (2012)
23. E. Ontiveros, P. Melin, O. Castillo, High order α-planes integration: a new approach to computational cost reduction of General Type-2 Fuzzy Systems. Eng. Appl. AI **74**, 186–197 (2018)

24. C.I. González, P. Melin, J.R. Castro, O. Mendoza, O. Castillo, An improved sobel edge detection method based on generalized type-2 fuzzy logic. Soft. Comput. **20**(2), 773–784 (2016)
25. J. Barraza, P. Melin, F. Valdez, C.I. González, Fuzzy fireworks algorithm based on a sparks dispersion measure. Algorithms **10**(3), 83 (2017)
26. J. Barraza, P. Melin, F. Valdez, C.I. González, O. Castillo, Iterative fireworks algorithm with fuzzy coefficients. FUZZ-IEEE, 1–6 (2017)
27. J. Barraza, P. Melin, F. Valdez, Fuzzy FWA with dynamic adaptation of parameters, in *IEEE CEC*, pp. 4053–4060 (2016)
28. J. Barraza, F. Valdez, P. Melin, C. Gonzalez, Fireworks Algorithm (FWA) with adaptation of parameters using fuzzy logic, in *Nature-Inspired Design of Hybrid Intelligent Systems 2017*; M.G. Simoes, K. Bose, J. Spiegel, Fuzzy logic based intelligent control of a variable speed cage machine wind generation system. IEEE Trans. Power Electron. **12**(1), pp. 87–95 (1997 Jan)
29. Y. Tan, S. Zheng, Dynamic search in fireworks algorithm, in *Evolutionary Computation (CEC 2014)*
30. L.A. Zadeh, Knowledge representation in fuzzy logic. IEEE Trans. Knowl. Data Eng. **I**(I), 89-0084 (1989 Mar)

Metaheuristics: Theory and Applications

Harmony Search with Dynamic Adaptation of Parameters for the Optimization of a Benchmark Set of Functions

Cinthia Peraza, Fevrier Valdez and Oscar Castillo

Abstract In this paper a fuzzy search algorithm harmony (FHS) is presented. The main difference between previous work is that this method uses a fuzzy system for dynamic parameter adaptation of the two main parameters throughout the iterations of the algorithm, which are: harmony memory accepting (HMR) and pitch adjustment (PArate), with the rules of the fuzzy system control the intensification and diversification of the search space is achieved. This method was applied to the mathematical functions provided by the CEC 2017, which are unimodal, multimodal, hybrid and composite functions to verify the efficiency of the proposed method. A comparison is presented to verify the results obtained with the original harmony search algorithm and the fuzzy harmony search algorithm.

Keywords Harmony search · Fuzzy logic · Dynamic parameter adaptation

1 Introduction

This article presents the optimization of the mathematical functions CEC 2017. In previous works other classical benchmark functions are considered using the FHS method [1–3]. The proposed method is an adaptation of the original harmony search algorithm (HS), which uses the improvisation process of jazz musicians [4]. This method has been used to solve variety optimization problems and hybrid methods such as in [5–8]. The main difference between the variants and the existing methods in the literature is that this method is based on the original harmony search algorithm and uses fuzzy logic to dynamically adjust the HMR and PArate parameters whose main function is to maintain control of the exploitation and exploration of the search

C. Peraza · F. Valdez · O. Castillo (✉)
Tijuana Institute of Technology, Tijuana, BC, Mexico
e-mail: ocastillo@tectijuana.mx

C. Peraza
e-mail: cinthia_sita@hotmail.com

F. Valdez
e-mail: fevrier@tectijuana.mx

© Springer Nature Switzerland AG 2020
O. Castillo and P. Melin (eds.), *Hybrid Intelligent Systems in Control, Pattern Recognition and Medicine*, Studies in Computational Intelligence 827, https://doi.org/10.1007/978-3-030-34135-0_8

space, applied to CEC 2017 mathematical functions. The main objective of the article is to present results of the proposed method applied to complex mathematical functions to be able to validate the effectiveness of the proposed method and verify that by making use of fuzzy logic in the adjustment of the parameters, significant improvements are achieved, we can find in the literature articles where this technique is applied as in: [9, 10]. Fuzzy logic is used in order to control the diversification and intensification, processes in the search space and enable finding the global optimum, avoiding stagnation and premature convergence, this is achieved based on knowledge of the rules. Fuzzy logic has also recently been used in existing metaheuristic methods because it uses linguistic variables and rules with which fuzzy models make decisions [6, 11–13], and have been used in areas of evolutionary algorithms with fuzzy logic [14–18], engineering problems [19–22] among others.

This article describes the implementation of the FHS method that can be applied for solving global optimization problems using two important metrics in the inputs that are iterations and diversity. The proposed method uses fuzzy logic to achieve an efficient adjustment of parameters and CEC 2017 benchmark functions are evaluated with 10, 30, 50 and 100 dimensions. Different methods have used these functions as reference [23, 24].

This article is organized as follows. The original harmony search algorithm is shown in Sect. 2. The proposed method with fuzzy logic is shown in Sect. 3. The methodology for parameter adaptation is presented in Sect. 4. The comparison between the original harmony search algorithm (HS) and FHS is shown in Sect. 5. Conclusions are shown in Sect. 6.

2   Harmony Search Algorithm

This algorithm is inspired by music and was created by ZW Geem in 2001 [6].

This algorithm uses three operators which are: harmony memory accepting (HMR), pitch adjustment (PArate) and randomization [25], and the range of these parameters is from 0 to 1.

HMR parameter represents the exploitation in a search space and its objective is to choose the best harmonies.

$$HMR \in [0, 1] \tag{1}$$

The pitch adjustment (PArate) is the second component, this parameter is given by Eq. (2).

$$x_{new} = x_{old} + b_p(2rand - 1) \tag{2}$$

Randomization is the third component, this parameter is given by Eq. (3).

$$P_a = P_{lower\,limit} + P_{range} * rand \tag{3}$$

Objective function f (x), x = (x₁, ..., xₙ) ᵀ
Initial generate harmonics (matrices of real numbers)
Define pitch adjustment rate (PArate) and limits of tone
Define acceptance rate of the harmony memory (HMR)
        while (t < Maximum number of iterations)
                Generate a new harmony and accept the best harmonies
                Setting the tone for new harmonies (solutions)
        if (rand> HMR)
                Choose an existing harmony randomly
        else if (rand> PArate)
        Setting the tone at random within a bandwidth   (2)
        else
                Generate a new harmony through a randomization (3)
        End if
                Accepting new harmonies (solutions) best
        End while
To find the best solutions.

Fig. 1 Pseudocode for harmony search

where "rand" is a generator of random numbers in the range of 0 and 1 (Search space).

2.1 Pseudocode for Harmony Search

The pseudocode of this algorithm is explained in detail in Fig. 1.

3 Proposed Method

In previous works the harmony search algorithm has been studied and applied to problems of optimization, benchmark mathematical functions, benchmark control problems, among others [26–30].

The main difference with the previous works is that are have added a second input to the fuzzy system, which is called diversity and this is calculated with the Euclidean distance.

The HMR and PArate parameters are adjusted dynamically by means of a fuzzy system to control the exploration and exploitation of the search space by fuzzy rules.

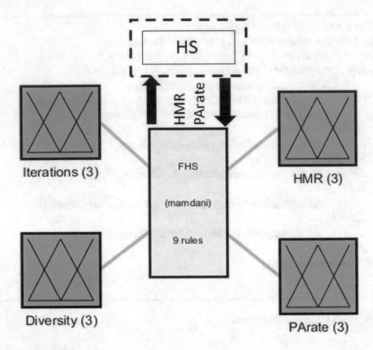

Fig. 2 Proposed fuzzy system

The main difference between the original and the fuzzy method is that the original leaves the parameters fixed throughout the iterations and the fuzzy will move those parameters in each iteration.

The FHS uses a fuzzy system to be responsible for dynamically changing the *PArate* and HMR parameters in the range from 0 to 1 in each iteration number as shown in Fig. 2 and expressed as follows.

4 Methodology for Parameter Adaptation

Based on the previous study of the HS algorithm, it was decided to adjust the HMR parameter in order to control the exploitation and the PArate parameter to achieve the control of the exploration. These parameters will be representing the outputs of the fuzzy system in a range of 0 to 1.

The inputs of the fuzzy system will be the iterations that are represented by Eq. (4) and a second input is diversity, that is defined by Eq. (5). The design of the proposed method can be find in Fig. 3.

$$\text{Iteration} = \frac{\text{Current Iteration}}{\text{Maximum of Iterations}} \tag{4}$$

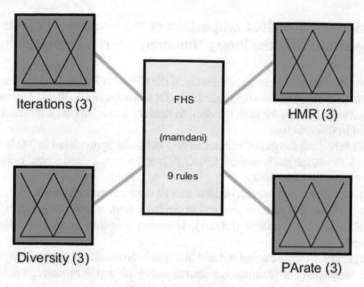

Fig. 3 Fuzzy system for dynamic parameter adaptation

$$\text{Diversity}\,(S(t)) = \frac{1}{n_S} \sum_{i=1}^{n_S} \sqrt{\sum_{j=1}^{n_x}(x_{ij}(t) - \bar{x}_j(t))^2} \qquad (5)$$

The design of the input and output variables can be appreciated in Fig. 3, these are granulated into three triangular membership functions. The linguistic values are Low, Medium and High.

The rules were created based on the function of each parameter and are defined in Fig. 4.

1. If (Iterations is Low) and (D is Low) then (HMR is High) (PArate is Low) (1)
2. If (Iterations is Low) and (D is Medium) then (HMR is Medium) (PArate is Medium) (1)
3. If (Iterations is Low) and (D is High) then (HMR is Medium) (PArate is Medium) (1)
4. If (Iterations is Medium) and (D is Low) then (HMR is Medium) (PArate is Medium) (1)
5. If (Iterations is Medium) and (D is Medium) then (HMR is Medium) (PArate is Medium) (1)
6. If (Iterations is Medium) and (D is High) then (HMR is Medium) (PArate is Medium) (1)
7. If (Iterations is High) and (D is Low) then (HMR is Medium) (PArate is High) (1)
8. If (Iterations is High) and (D is Medium) then (HMR is Medium) (PArate is Medium) (1)
9. If (Iterations is High) and (D is High) then (HMR is Low) (PArate is High) (1)

Fig. 4 Rules for the fuzzy system FHS

5 Simulation Results Comparison of the Harmony Search Algorithm with the Fuzzy Harmony Search Algorithm

Thirty benchmark mathematical functions of the CEC 2017 were used for the experimentation of the original and fuzzy methods. Dimensions of 10, 30, 50, 100 variables were used with 51 runs for each function to achieve a comparison with the methods presented in this article.

For all functions the global minimum is 0 as can be appreciated in Table 1.

Table 1 shows the mathematical functions, search domain and global optimum of each functions that was used.

The parameters used in each method can be appreciated in Table 2.

Table 2 shows the parameters used in each method, which are Harmony memory (HMS), pitch adjustment (*PArate*), Harmony memory accepting (*HMR*) and dimensions.

51 experiments were carried out and in Table 3 the results of the average obtained by each mathematical function are shown using 10 and 30 dimensions for each method.

Table 4 shows the results of the average obtained by each mathematical function are shown using 50 and 100 dimensions for each method.

6 Conclusions

In this paper a complete fuzzy system based on the HS algorithm is presented, with which the diversity of the solutions is achieved. In this case we applied a fuzzy system to be responsible to dynamically adjust the HMR and PArate parameters and it is applied to the CEC 2017 benchmark mathematical functions.

A comparison of two methods was performed between the Fuzzy HS and original harmony search algorithm, applied to 30 Benchmark mathematical functions using 10, 30, 50 and 100 dimensions.

It can be observed that by using the FHS method it is possible to obtain better results in most of the mathematical functions. As future work, we can deal with other types of applications, like in [31–36].

Table 1 Benchmark functions

	No.	Functions	$F_i^* = F_i(x^*)$
Unimodal functions	1	Shifted and Rotated Bent Cigar Function	100
	2	Shifted and Rotated Sum of different power Function	200
	3	Shifted and Rotated Zakharov Function	300
Simple multimodal functions	4	Shifted and Rotated Rosenbrock's Function	400
	5	Shifted and Rotated Rastrigin's Function	500
	6	Shifted and Rotated Expanded Scaffer's F6 Function	600
	7	Shifted and Rotated Lunacek Bi_Rastrigin Function	700
	8	Shifted and Rotated Non-Continous Rastrigin's Function	800
	9	Shifted and Rotated Levy Function	900
	10	Shifted and Rotated Schwefel's Function	1000
Hybrid functions	11	Hybrid function 1 (N = 3)	1100
	12	Hybrid function 1 (N = 3)	1200
	13	Hybrid function 2 (N = 3)	1300
	14	Hybrid function 3 (N = 3)	1400
	15	Hybrid function 4(N = 4)	1500
	16	Hybrid function 5 (N = 4)	1600
	17	Hybrid function 6 (N = 4)	1700
	18	Hybrid function 6 (N = 5)	1800
	19	Hybrid function 6 (N = 5)	1900
	20	Hybrid function 6 (N = 5)	2000
	21	Composition function 1 (N = 3)	2100
	22	Composition function 2 (N = 3)	2200
	23	Composition function 3 (N = 4)	2300
	24	Composition function 4 (N = 4)	2400
	25	Composition function 5 (N = 5)	2500
	26	Composition function 6 (N = 5)	2600
	27	Composition function 7 (N = 6)	2700
	28	Composition function 8 (N = 6)	2800
	29	Composition function 9 (N = 3)	2900
	30	Composition function 10 (N = 3)	3000

Table 2 Parameters used for test problems

Methods	HMS	PArate	HMR	Dimensions
Simple HS	40	0.75	0.95	10.30.50, 100
Fuzzy HS	40	Dynamic	Dynamic	10.30.50, 100

Table 3 Benchmark optimization average results for each method

Function	HS	FHS	HS	FHS
	10	10	30	30
1	**2.54E+03**	2.72E+03	**4.02E+03**	5.95E+03
2	1.96E−02	**0.00E+00**	**1.96E−02**	5.39E+13
3	1.18E−08	**0.00E+00**	**1.96E+01**	1.94E+01
4	3.67E+00	**2.12E+00**	**1.19E+02**	1.24E+02
5	8.04E+00	**4.23E+00**	**5.56E+01**	5.52E+01
6	2.79E−08	**0.00E+00**	1.29E−02	**8.94E−04**
7	1.88E+01	**1.26E+01**	1.01E+02	**1.01E+02**
8	9.15E+00	**4.37E+00**	6.30E+01	**6.09E+01**
9	0.00E+00	**0.00E+00**	3.75E+01	**3.56E+01**
10	3.03E+02	**1.86E+02**	**2.49E+03**	2.53E+03
11	4.76E+00	**3.08E+00**	**5.57E+01**	6.20E+01
12	2.12E+04	**1.54E+04**	2.57E+05	**2.46E+05**
13	7.32E+03	**1.25E+03**	1.11E+04	**9.88E+03**
14	2.68E+01	**8.28E+00**	8.04E+03	**5.64E+03**
15	3.39E+01	**1.73E+00**	**1.06E+04**	1.61E+04
16	1.77E+01	**3.06E−01**	**5.16E+02**	5.79E+02
17	1.38E+01	**4.78E+00**	1.58E+02	**1.55E+02**
18	4.55E+03	**7.43E+02**	**2.23E+05**	2.90E+05
19	2.39E+01	**1.56E+00**	2.00E+04	**1.89E+04**
20	9.63E+00	**5.31E+00**	**1.42E+02**	1.51E+02
21	**1.82E+02**	1.87E+02	**2.63E+02**	2.64E+02
22	9.64E+01	**9.65E+01**	1.93E+03	**1.91E+03**
23	3.08E+02	**3.06E+02**	4.03E+02	**3.99E+02**
24	3.25E+02	**3.20E+02**	**4.82E+02**	4.83E+02
25	4.15E+02	**4.07E+02**	4.03E+02	**3.97E+02**
26	3.48E+02	**3.19E+02**	1.62E+03	**1.44E+03**
27	3.98E+02	**3.93E+02**	5.25E+02	**5.25E+02**
28	4.22E+02	**3.94E+02**	4.22E+02	**4.20E+02**
29	2.62E+02	**2.37E+02**	5.95E+02	**5.80E+02**
30	5.59E+05	**2.06E+05**	**8.62E+03**	9.47E+03

Bold values indicates the best results

Table 4 Benchmark optimization average results for each method

Function	HS	FHS	HS	FHS
	50	50	100	100
1	1.62E+07	**7.74E+04**	7.84E+08	**3.81E+06**
2	**1.96E−02**	2.56E+36	5.25E+119	**2.38E+102**
3	1.16E+05	**1.22E+04**	5.06E+05	**1.64E+05**
4	3.35E+02	**2.49E+02**	8.73E+02	**5.63E+02**
5	1.50E+02	**1.32E+02**	5.98E+02	**3.53E+02**
6	3.99E+00	**2.60E−01**	2.17E+01	**8.89E+00**
7	3.23E+02	**2.44E+02**	1.20E+03	**8.57E+02**
8	1.49E+02	**1.33E+02**	5.86E+02	**3.52E+02**
9	1.46E+03	**5.43E+02**	1.48E+04	**7.01E+03**
10	9.01E+03	**5.51E+03**	2.69E+04	**1.57E+04**
11	3.65E+02	**2.21E+02**	2.80E+04	**1.54E+03**
12	1.65E+07	**3.23E+06**	3.68E+08	**6.35E+07**
13	**8.10E+03**	8.52E+03	3.75E+05	**4.84E+04**
14	2.08E+05	**7.89E+04**	4.99E+06	**2.04E+06**
15	4.98E+04	**1.07E+04**	1.96E+06	**1.88E+04**
16	1.51E+03	**1.31E+03**	5.12E+03	**3.40E+03**
17	1.25E+03	**1.03E+03**	3.57E+03	**2.91E+03**
18	3.29E+06	**1.52E+06**	1.10E+07	**4.56E+06**
19	1.29E+04	**1.11E+04**	1.39E+06	**6.63E+03**
20	9.51E+02	**6.00E+02**	4.40E+03	**3.09E+03**
21	3.65E+02	**3.32E+02**	8.39E+02	**6.14E+02**
22	9.79E+03	**6.04E+03**	2.84E+04	**1.65E+04**
23	6.21E+02	**5.62E+02**	9.97E+02	**8.28E+02**
24	8.44E+02	**6.58E+02**	1.71E+03	**1.28E+03**
25	6.62E+02	**6.14E+02**	1.70E+03	**1.22E+03**
26	2.82E+03	**2.59E+03**	9.32E+03	**7.31E+03**
27	7.94E+02	**6.61E+02**	1.12E+03	**8.88E+02**
28	5.87E+02	**5.30E+02**	1.55E+03	**9.10E+02**
29	1.20E+03	**8.08E+02**	4.55E+03	**3.26E+03**
30	3.44E+06	**2.15E+06**	3.18E+05	**3.11E+04**

Bold values indicates the best results

Acknowledgements We would like to express our thanks to CONACYT and Tijuana Institute of Technology for the facilities and resources granted for the development of this research.

References

1. F. Olivas, F. Valdez, O. Castillo, P. Melin, Theory and background, in *Dynamic Parameter Adaptation for Meta-Heuristic Optimization Algorithms Through Type-2 Fuzzy Logic* (pp. 3–10). (Springer International Publishing, Cham, 2018)
2. L. Amador-Angulo, O. Castillo, A new fuzzy bee colony optimization with dynamic adaptation of parameters using interval type-2 fuzzy logic for tuning fuzzy controllers. Soft. Comput. **22**(2), 571–594 (2018)
3. C. Caraveo, F. Valdez, O. Castillo, A new optimization meta-heuristic algorithm based on self-defense mechanism of the plants with three reproduction operators. Soft Comput. (Apr. 2018)
4. C.-M. Wang, Y.-F. Huang, Self-adaptive harmony search algorithm for optimization. Expert Syst. Appl. **37**(4), 2826–2837 (2010)
5. K.S. Lee, Z.W. Geem, A new meta-heuristic algorithm for continuous engineering optimization: harmony search theory and practice. Comput. Methods Appl. Mech. Eng. **194**(36–38), 3902–3933 (2005)
6. P. Ochoa, O. Castillo, J. Soria, Interval Type-2 fuzzy logic dynamic mutation and crossover parameter adaptation in a fuzzy differential evolution method, in *Intuitionistic Fuzziness and Other Intelligent Theories and Their Applications*, vol. 757, ed. by M. Hadjiski, K.T. Atanassov (Springer International Publishing, Cham, 2019), pp. 81–94
7. D. Zou, L. Gao, Y. Ge, P. Wu, A novel global harmony search algorithm for chemical equation balancing, in *2010 International Conference On Computer Design and Applications*, Qinhuangdao, China, pp. V2-1–V2-5 (2010)
8. R. Eberhart, J. Kennedy, A new optimizer using particle swarm theory, in *MHS'95. Proceedings of the Sixth International Symposium on Micro Machine and Human Science*, Nagoya, Japan, pp. 39–43 (1995)
9. E. Bernal, O. Castillo, J. Soria, F. Valdez, Imperialist competitive algorithm with dynamic parameter adaptation using fuzzy logic applied to the optimization of mathematical functions. Algorithms **10**(1), 18 (2017)
10. E. Bernal, O. Castillo, J. Soria, F. Valdez, Galactic swarm optimization with adaptation of parameters using fuzzy logic for the optimization of mathematical functions, in *Fuzzy Logic Augmentation of Neural and Optimization Algorithms: Theoretical Aspects and Real Applications*, vol. 749, ed. by O. Castillo, P. Melin, J. Kacprzyk (Springer International Publishing, Cham, 2018), pp. 131–140
11. Z.W. Geem, K.-B. Sim, Parameter-setting-free harmony search algorithm. Appl. Math. Comput. **217**(8), 3881–3889 (2010)
12. P. Ochoa, O. Castillo, J. Soria, Differential evolution algorithm using a dynamic crossover parameter with fuzzy logic applied for the CEC 2015 benchmark functions, in *Fuzzy Information Processing*, vol. 831, ed. by G.A. Barreto, R. Coelho (Springer International Publishing, Cham, 2018), pp. 580–591
13. O. Castillo, P. Ochoa, J. Soria, Differential evolution with fuzzy logic for dynamic adaptation of parameters in mathematical function optimization, in *Imprecision and Uncertainty in Information Representation and Processing*, vol. 332, ed. by P. Angelov, S. Sotirov (Springer International Publishing, Cham, 2016), pp. 361–374
14. M.H. Mashinchi, M.A. Orgun, M. Mashinchi, W. Pedrycz, A tabu-harmony search-based approach to fuzzy linear regression. IEEE Trans. Fuzzy Syst. **19**(3), 432–448 (2011)
15. O. Castillo, C. Soto, F. Valdez, A review of fuzzy and mathematic methods for dynamic parameter adaptation in the firefly algorithm, in *Advances in Data Analysis with Computational Intelligence Methods*, vol. 738, ed. by A.E. Gawęda, J. Kacprzyk, L. Rutkowski, G.G. Yen (Springer International Publishing, Cham, 2018), pp. 311–321

16. B. González, P. Melin, F. Valdez, G. Prado-Arechiga, Ensemble neural network optimization using a gravitational search algorithm with interval type-1 and type-2 fuzzy parameter adaptation in pattern recognition applications, in *Fuzzy Logic Augmentation of Neural and Optimization Algorithms: Theoretical Aspects and Real Applications*, vol. 749, ed. by O. Castillo, P. Melin, J. Kacprzyk (Springer International Publishing, Cham, 2018), pp. 17–27
17. J. Barraza, L. Rodríguez, O. Castillo, P. Melin, F. Valdez, A new hybridization approach between the fireworks algorithm and grey wolf optimizer algorithm. J. Optim. **2018**, 1–18 (2018)
18. M.L. Lagunes, O. Castillo, J. Soria, M. Garcia, F. Valdez, Optimization of granulation for fuzzy controllers of autonomous mobile robots using the firefly algorithm. Granul. Comput. (July 2018)
19. M. Mahdavi, M. Fesanghary, E. Damangir, An improved harmony search algorithm for solving optimization problems. Appl. Math. Comput. **188**(2), 1567–1579 (2007)
20. Y.Y. Moon, Z.W. Geem, G.-T. Han, Vanishing point detection for self-driving car using harmony search algorithm. Swarm Evol. Comput. **41**, 111–119 (2018)
21. Y.-H. Kim, Y. Yoon, Z.W. Geem, A comparison study of harmony search and genetic algorithm for the max-cut problem. Swarm Evol. Comput. (Feb 2018)
22. Z.W. Geem, S.Y. Chung, J.-H. Kim, Improved optimization for wastewater treatment and reuse system using computational intelligence. Complexity **2018**, 1–8 (2018)
23. A.W. Mohamed, A.A. Hadi, A.M. Fattouh, K.M. Jambi, LSHADE with semi-parameter adaptation hybrid with CMA-ES for solving CEC 2017 benchmark problems, in *2017 IEEE Congress on Evolutionary Computation (CEC)*, Donostia, San Sebastián, Spain, pp. 145–152 (2017)
24. J. Brest, M. S. Maucec, B. Boskovic, Single objective real-parameter optimization: Algorithm jSO, in *2017 IEEE Congress on Evolutionary Computation (CEC)*, Donostia, San Sebastián, Spain, pp. 1311–1318 (2017)
25. D. Manjarres et al., A survey on applications of the harmony search algorithm. Eng. Appl. Artif. Intell. **26**(8), 1818–1831 (2013)
26. Cinthia Peraza, Fevrier Valdez, Patricia Melin, Optimization of intelligent controllers using a type-1 and interval type-2 fuzzy harmony search algorithm. Algorithms **10**(3), 82 (2017)
27. C. Peraza, F. Valdez, M. Garcia, P. Melin, O. Castillo, A new fuzzy harmony search algorithm using fuzzy logic for dynamic parameter adaptation. Algorithms **9**(4), 69 (2016)
28. C. Peraza, F. Valdez, J.R. Castro, O. Castillo, Fuzzy dynamic parameter adaptation in the harmony search algorithm for the optimization of the ball and beam controller. Adv. Oper. Res. **2018**, 1–16 (2018)
29. C. Peraza, F. Valdez, O. Castillo, Study on the use of type-1 and interval type-2 fuzzy systems applied to benchmark functions using the fuzzy harmony search algorithm, in *Fuzzy logic in intelligent system design*, vol. 648, ed. by P. Melin, O. Castillo, J. Kacprzyk, M. Reformat, W. Melek (Springer International Publishing, Cham, 2018), pp. 94–103
30. C. Peraza, F. Valdez, O. Castillo, Improved method based on type-2 fuzzy logic for the adaptive harmony search algorithm, in *Fuzzy Logic Augmentation of Neural and Optimization Algorithms: Theoretical Aspects and Real Applications*, vol. 749, ed. by O. Castillo, P. Melin, J. Kacprzyk (Springer International Publishing, Cham, 2018), pp. 29–37
31. C. Leal Ramírez, O. Castillo, P. Melin, A. Rodríguez Díaz, Simulation of the bird age-structured population growth based on an interval type-2 fuzzy cellular structure. Inf. Sci. **181**(3), 519–535 (2011)
32. N.R. Cázarez-Castro, L.T. Aguilar, O. Castillo, Designing type-1 and type-2 fuzzy logic controllers via fuzzy lyapunov synthesis for nonsmooth mechanical systems. Eng. Appl. of AI **25**(5), 971–979 (2012)
33. O. Castillo, P. Melin, Intelligent systems with interval type-2 fuzzy logic. Int. J. Innov. Comput. Inf. Control **4**(4), 771–783 (2008)

34. G.M. Mendez, O. Castillo, Interval type-2 TSK fuzzy logic systems using hybrid learning algorithm, in *The 14th IEEE International Conference on Fuzzy Systems FUZZ'05*, 230–235 (2005)
35. P. Melin, O. Castillo, Intelligent control of complex electrochemical systems with a neuro-fuzzy-genetic approach. IEEE Trans. Ind. Electr. **48**(5), 951–955
36. E. Rubio, O. Castillo, F. Valdez, P. Melin, C.I. González, G. Martinez, An extension of the fuzzy possibilistic clustering algorithm using type-2 fuzzy logic techniques. Adv. Fuzzy Syst., 7094046:1–7094046:23 (2017)

Type-2 Fuzzy Logic for Dynamic Parameter Adaptation in the Imperialist Competitive Algorithm

Emer Bernal, Oscar Castillo, José Soria and Fevrier Valdez

Abstract In this paper we propose the utilization of type-2 fuzzy systems for the dynamic adjustment of parameters in the imperialist competitive algorithm (ICA), a type-1 fuzzy system was used as a basis, with decades as the input variable and the beta parameter as the output variable, then it was extended to interval type-2 fuzzy systems, and three variants with triangular, Gaussian and trapezoidal membership functions were performed. The imperialist competitive algorithm is based on the concept of imperialism, where the strongest countries absorb the weakest and make then their colonies. To measure the performance of the proposed method 10 mathematical functions with different number of decades are used and finally, a comparison was made between our variants and the results obtained with the type-1 fuzzy system to observe their behavior in the face of optimization problems.

Keywords Imperialist competitive algorithm · Mathematical functions · Fuzzy system · Type-2 fuzzy system

1 Introduction

The dynamic adaptation of parameters utilizing fuzzy logic, in particular type-1 and type-2 fuzzy systems, in metaheuristic algorithms, has gained great importance in recent years because it has shown significant improvements in the performance of various metaheuristics algorithms found in the literature [1–5].

Fuzzy logic and fuzzy sets were originally proposed by Zadeh, and this originated the creation of fuzzy systems with different applications such as in metaheuristic algorithms and modeling and control systems [1, 2, 6]. Initially type-1 fuzzy systems, that represent the imprecision with numerical values in a range of [0–1], came to replace traditional sets when it is difficult to establish or find an exact value in some type of measurement [7]. In addition, when the problem or situation contains a higher

E. Bernal · O. Castillo (✉) · J. Soria · F. Valdez
Tijuana Institute of Technology, Tijuana, BC, Mexico
e-mail: ocastillo@tectijuana.mx

© Springer Nature Switzerland AG 2020
O. Castillo and P. Melin (eds.), *Hybrid Intelligent Systems in Control,
Pattern Recognition and Medicine*, Studies in Computational Intelligence 827,
https://doi.org/10.1007/978-3-030-34135-0_9

degree of uncertainty, type-2 fuzzy systems can be used because they work better with high levels of uncertainty or lack of information [7, 8].

We propose dynamic adjustment of the parameters in the imperialist competitive algorithm using fuzzy logic. Three distinct fuzzy systems were designed to dynamically adjust the beta parameter and to measure the performance of the proposed fuzzy imperialist competitive algorithm (FICAT2) utilizing interval type-2 fuzzy systems versus fuzzy imperialist competitive algorithm (FICA) using type-1 fuzzy systems.

The rest of the paper is conformed as follows. Section 2 provides a review of imperialist competitive algorithm. Section 3 presents the methodology used in the proposed approach. Section 4 presents the mathematical benchmark functions. Section 5 summarizes the simulation of results obtained and finally the conclusions.

2   Imperialist Competitive Algorithm

Atashpaz-Gargari and Lucas proposed the imperialist competitive algorithm (ICA) in 2007 [9], this algorithm is inspired by imperialism, where all the most powerful countries aspire to make a colony the less powerful countries and thus absorb them. In the field of metaheuristic algorithms, the imperialist competitive algorithm takes as a basis the social political progress unlike other metaheuristics or evolutionary algorithms that are based on bio-inspired phenomena [6, 10].

In the imperialist competitive algorithm, we start with a randomly generated population, where all individuals are called countries. The best positioned countries are considered the imperialist countries and the remaining countries are the colonies. All the colonies are divided among the imperialist countries according to their power, in this case their fitness function [9, 11, 12].

Once the initial population is generated and divided into imperialist countries and colonies, the colonies begin to move towards their imperialist country (known as assimilation process). The colonies move X distance toward their imperialist, where the movement X is a random number that is generated by a uniform distribution within the interval $(0, \beta d)$, where β is a number in the range of 1–2 and d is the distance between the colony and the imperialist [3, 13] (Fig. 1).

$$x \sim U(0, \beta d) \tag{1}$$

The power of an empire is calculated based on the power of the imperialist country and the power of the colonies that are part of the empire. Each empire tries to take possession of other empires, which leads to an imperialist competition that diminishes the power of the weaker empires and increases the power of the strongest empires, causing the weak empires to collapse and their colonies become part of the most powerful empires [9, 14].

After a period of time, all the empires will collapse periodically except the strongest empire, in this way all the colonies will belong and be controlled by a single imperialist country ending the imperialist competition [15, 16].

Fig. 1 Motion of a colony toward its imperialist

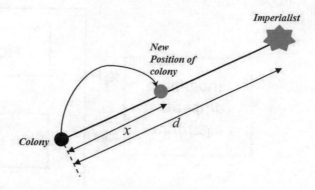

3　Proposed Methodology

In this paper we propose a method based on interval type-2 fuzzy systems for the dynamic adjustment of parameters in the imperialist competitive algorithm in specific beta parameters. Initially, a type-1 fuzzy system presented in [1], where the decades represent the input variable and the beta parameter the output variable with triangular membership functions labeled as "low", "medium" and "high" as shown in Fig. 2 [1].

To use the parameter of decades as input variable, we use a percentage, so when the algorithm starts the decades will be "low" and as time passes will be considered "high". The following mathematical expression is used to model the concept of the input variable decades [1, 17].

$$Decades = \frac{Current\ Decade}{Total\ Number\ of\ Decades} \tag{2}$$

The proposed approach based on interval type-2 (IT2) fuzzy systems in the competitive imperialist algorithm for the dynamic adjustment of the beta parameter is illustrated in Fig. 3.

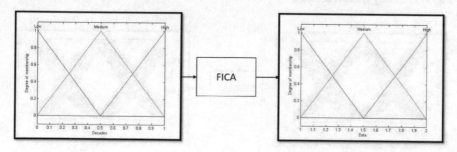

Fig. 2 Type 1 fuzzy system for Beta

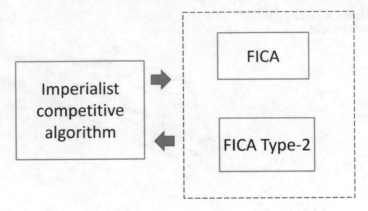

Fig. 3 General proposal dynamic parameter adaptation

The first IT2 fuzzy system proposed can be found in Fig. 4, which was designed with triangular membership functions that are represented as follows [18].

$$\tilde{\mu}(x) = \left[\underline{\mu}(x), \bar{\mu}(x)\right] = itritype2(x, [a_1, b_1, c_1, a_2, b_2, c_2]),$$

$$where \, a_1 < a_2, b_1 < b_2, c_1 < c_2 \tag{3}$$

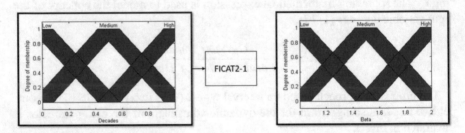

Fig. 4 IT2 fuzzy system FICAT2-1

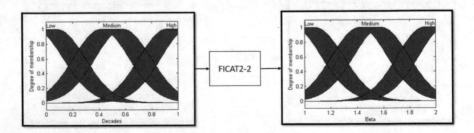

Fig. 5 IT2 fuzzy system FICAT2-2

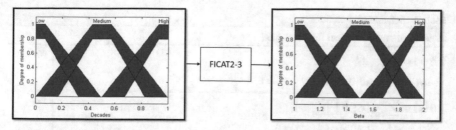

Fig. 6 IT2 fuzzy system FICAT2-3

The second IT2 fuzzy system proposed is shown in Fig. 5, which was designed with Gaussian membership functions that are represented as follows [18].

$$\tilde{\mu}(x) = \left[\underline{\mu}(x), \bar{\mu}(x)\right] = igaussmtype2(x, [\sigma, m_1, m_2]),$$

$$where\, m_1 < m_2 \tag{4}$$

Finally the third IT2 fuzzy system proposed can be found in Fig. 6, which was designed with trapezoidal membership functions that are represented as follows [18].

$$\tilde{\mu}(x) = \left[\underline{\mu}(x), \bar{\mu}(x)\right]$$

$$= itrapatype2(x, [a_1, b_1, c_1, d_1, a_2, b_2, c_2, d_2]),$$

$$where\, a_1 < a_2, b_1 < b_2, c_1 < c_2, d_1 < d_2 \tag{5}$$

The rules of all IT2 fuzzy systems are the same as those used in the type-1 fuzzy system [1].

4 Benchmark Functions

In this section we present the mathematical functions used to measure the performance of the imperialist competitive algorithm and our proposal using interval type-2 (IT2) fuzzy systems to the adaptation of parameters in the imperialist competitive algorithm.

In the area of metaheuristic algorithms it is common to use benchmark functions to measure their performance, and in this work we use 10 mathematical functions that are presented below with their mathematical expression, search space and their optimal value [1, 19] (Table 1).

Table 1 Benchmark functions

Benchmark mathematical functions [19]	
$f_1(x) = \sum_{1=1}^{n} x_i^2$ $with\ x_j \in [-5.12, 5.12]\ and\ f(x^*) = 0$	(6)
$f_2(x) = \sum_{i=1}^{n} |x_i| + \prod_{i=1}^{n} |x_i|$ $with\ x_j \in [-10, 10]\ and\ f(x^*) = 0$	(7)
$f_3(x) = \sum_{i=1}^{n} \left(\sum_{j-1}^{i} x_j \right)^2$ $with\ x_j \in [-100, 100]\ and\ f(x^*) = 0$	(8)
$f_4(x) = max_i\{|x_i|, 1 \le i \le n\}$ $with\ x_j \in [-100, 100]\ and\ f(x^*) = 0$	(9)
$f_5(x) = \sum_{i=1}^{n-1} \left[100\left(x_{i+1} - x_i^2\right)^2 + (x_i - 1)^2 \right]$ $with\ x_j \in [-30, 30]\ and\ f(x^*) = 0$	(10)
$f_6(x) = \sum_{i=1}^{n} ([x_i + 0.5])^2$ $with\ x_j \in [-100, 100]\ and\ f(x^*) = 0$	(11)
$f_7(x) = \sum_{i=1}^{n} ix_i^4 + random[0, 1]$ $with\ x_j \in [-1.28, 1.28]\ and\ f(x^*) = 0$	(12)
$f_8(x) = \sum_{i-1}^{n} -x_i \sin\left(\sqrt{|x_i|}\right)$ $with\ x_j \in [-500, 500]\ and\ f(x^*) = -418.9829x5$	(13)
$f_9(x) = \sum_{i=1}^{n} \left[x_i^2 - 10\cos(2\pi x_i) + 10 \right]$ $with\ x_j \in [-5.12, 5.12]\ and\ f(x^*) = 0$	(14)
$f_{10}(x) = -20\exp\left(-0.2\sqrt{\frac{1}{n} \sum_{i=1}^{n} x_i^2}\right) - \exp\left(\frac{1}{n} \sum_{i=1}^{n} \cos(2\pi x_i)\right) + 20 + e$ $with\ x_j \in [-32, 32]\ and\ f(x^*) = 0$	(15)

5　Experimental Results

Imperialist competitive algorithm (ICA) and our proposal based on interval type 2 (IT2) fuzzy system for the adjustment of the beta parameter will be implemented for 10 mathematical functions in all cases for 30 dimensions, the results obtained by the competitive imperialist algorithm and the proposed method are shown in tables by number of decades. The parameters that were used to perform the experiments are as follows [1]: (Table 2).

Table 2 Parameters for ICA and FICAT2 [1]

Parameters	Value
No. dimensions	30
No. countries	200
No. imperialists	10
Revolution rate	0.2
Xi	0.02
Beta	Dynamic

Table 3 Simulation of results for 1000 decades

1000 decades					
Function	ICA	FICA Beta	FICAT2-1	FICAT2-2	FICAT2-3
f1	1.12E−23	8.79E−25	1.8645E−24	1.15E−24	**2.9585E−25**
f2	**3.39E−56**	4.41E−52	4.461E−52	4.10567E−52	5.6336E−51
f3	5822.35897	5909.17803	5576.79494	7663.438814	**4647.69011**
f4	8.60023687	**0.83147595**	0.95756351	1.046168329	1.17481712
f5	24.1293542	171.579304	292.409739	**18.64982835**	22.2757672
f6	6.59E−17	**1.06E−22**	2.7874E−21	9.94226E−22	2.5728E−22
f7	**9.76E−41**	11.6622472	11.820379	11.90426893	11.825276
f8	2535.76878	2195.79881	2270.00064	2217.73828	**2018.75216**
f9	123.807243	**101.011796**	104.006802	106.2300716	103.533745
f10	13.0151646	12.90137	**12.3078055**	14.27351109	13.3222873

Table 3 shows the average obtained by implementing the imperialist competitive algorithm and our proposal using IT2 fuzzy systems to dynamically adjust the parameters of the ICA algorithm, the bold results represent the best obtained for each of the benchmark functions for 1000 decades.

Table 4 shows the average obtained by implementing the imperialist competitive algorithm and our proposal using IT2 fuzzy systems to dynamically adjust the parameters of the ICA algorithm, the bold results represent the best obtained for each of the benchmark functions for 2000 decades.

Table 5 shows the average obtained by implementing the imperialist competitive algorithm and our proposal using IT2 fuzzy systems to dynamically adjust the parameters of the ICA algorithm, the bold results represent the best obtained for each of the benchmark functions for 3000 decades.

Table 4 Simulation of results for 2000 decades

2000 decades					
Function	ICA	FICA Beta	FICAT2-1	FICAT2-2	FICAT2-3
f1	9.26E−51	3.24E−51	5.1683E−50	**1.8721E−52**	4.3847E−51
f2	**2.22E−141**	4.42E−116	1.819E−114	9.083E−117	4.307E−115
f3	7928.65015	**3638.93304**	3794.41715	4876.59037	4404.91186
f4	0.61591754	**0.0102226**	0.01062303	0.01594462	0.03324432
f5	11.799352	8.03929545	474.115769	**7.14853337**	11.5097085
f6	1.06E−17	2.47E−20	1.9507E−16	2.97E−20	**9.0064E−23**
f7	**9.95E−92**	11.2356722	11.3735362	11.2075171	11.0072419
f8	2349.08987	1889.56348	1837.32693	1934.45719	**1750.6182**
f9	114.795875	70.1576847	57.4281423	58.5647031	**36.0353448**
f10	12.8467107	12.9715618	13.9097643	13.2210342	**11.5814051**

Table 5 Simulation of results for 3000 decades

3000 decades					
Function	ICA	FICA Beta	FICAT2-1	FICAT2-2	FICAT2-3
f1	1.18E−29	1.51E−75	**4.059E−79**	1.4847E−77	2.1053E−67
f2	**7.50E−231**	3.19E−183	7.317E−182	7.688E−183	1.098E−178
f3	6863.925722	7563.274419	5910.23245	6980.23264	**4796.69413**
f4	0.010497636	**0.00013876**	0.00027069	0.00014173	0.00083529
f5	13.04971328	9.935683436	10.0435337	**7.82405887**	8.42344034
f6	5.40E−19	1.86E−19	1.0484E−20	**1.81E−27**	3.5166E−15
f7	**9.87E−144**	10.95189055	11.0305008	11.1722702	10.9066825
f8	2345.035647	**1804.99772**	1965.09724	2020.07303	1935.32963
f9	114.289348	68.03826234	67.041847	62.6639841	**43.2350318**
f10	15.34792029	**11.982714**	13.6105129	13.0193872	12.240195

6 Conclusions

In this paper, several type 1 and interval type 2 fuzzy systems were performed for the dynamic adjustment of the parameters in the imperialist competitive algorithm. A comparison was made of the results obtained that were divided by number of decades.

We can conclude that improvements have been obtained on average since in some cases our proposal manages to overcome the imperialist competitive algorithm and the method using type-1 fuzzy systems according to Tables 3, 4 and 5.

The use of interval type-2 fuzzy systems can be an alternative to achieve significant improvements by dynamically adapting the parameters in metaheuristic algorithms [20–24]. Also, we can consider other areas of application like in [25–30].

Acknowledgements We want to show our gratitude to CONACYT and Tijuana institute of technology for the resources provided for the development of our research.

References

1. E. Bernal, O. Castillo, J. Soria, A fuzzy logic approach for dynamic adaptation of parameters in galactic swarm optimization, in *Annual Conference of the North American Fuzzy Information Processing Society (NAFIPS)* (IEEE, 2017), pp. 1–6
2. F. Valdez, P. Melin, O. Castillo, An improved evolutionary method with fuzzy logic for combining particle swarm optimization and genetic algorithms. Appl. Soft Comput. **11**(2), 2625–2632 (2011)
3. E. Bernal, O. Castillo, J. Soria, F. Valdez, Imperialist competitive algorithm with dynamic parameter adaptation using fuzzy logic applied to the optimization of mathematical functions. Algorithms **10**(1), 18 (2017)
4. B. González, F. Valdez, P. Melin, A gravitational search algorithm using type-2 fuzzy logic for parameter adaptation, in *Nature-Inspired Design of Hybrid Intelligent Systems*, vol. 667 (Springer, Cham, 2017), pp. 127–138
5. P. Ochoa, O. Castillo, J. Soria, Differential evolution using fuzzy logic and a comparative study with other metaheuristics, in *Nature-Inspired Design of Hybrid Intelligent Systems*, vol. 667 (Springer, Cham, 2017), pp. 257–268
6. M.J. Mahmoodabadi, H. Jahanshahi, Multi objective optimized fuzzy-PID controllers for fourth order nonlinear systems. Int. J. Eng. Sci. Technol. **19**(2), 1084–1098 (2016)
7. L. Zadeh, Fuzzy logic=computing with words. IEEE Trans. Fuzzy Syst. **4**(2), 103–111 (1996)
8. E. Ontiveros-robles, P. Melin, O. Castillo, Comparative analysis of noise robustness of type 2 fuzzy logic controllers. Kybernetika **54**(1), 175–201 (2018)
9. E. Atashpaz-gargari, C. Lucas, Imperialist competitive algorithm: an algorithm for optimization inspired by imperialistic competition. Evol. Comput., 4661–4667 (2007)
10. E. Atashpaz-gargari, F. Hashemzadeh, R. Rajabioun, C. Lucas, Colonial competitive algorithm: a novel approach for PID controller design in mimo distillation column process. Int. J. Intell. Comput. Cybern. **1**(3), 337–355 (2008)
11. E. Bernal, O. Castillo, J. Soria, Imperialist competitive algorithm applied to the optimization of mathematical functions: a parameter variation study, in *Design of Intelligent Systems Based on Fuzzy Logic, Neural Networks and Nature-Inspired Optimization*, vol. 601 (Springer International Publishing, 2015), pp. 219–232
12. H. Duan, L.Z. Huang, Imperialist competitive algorithm optimized artificial neural networks for UCAV global path planning. Neurocomputing **125**, 166–171 (2013)
13. E. Atashpaz-gargari, C. Lucas, Imperialist competitive algorithm for minimum bit error rate beamforming. Int. J. Bio Inspired Comput. **1**(1–2), 125–133 (2009)
14. E. Bernal, O. Castillo, J. Soria, Fuzzy logic for dynamic adaptation in the imperialist competitive algorithm, in *IEEE Symposium Series on Computational Intelligence (SSCI)* (IEEE, 2017), pp. 1–7
15. E. Bernal, O. Castillo, J. Soria, Imperialist competitive algorithm with dynamic parameter adaptation applied to the optimization of mathematical functions, in *Nature Inspired Design of Hybrid Intelligent Systems*, vol. 667 (Springer International Publishing, 2017), pp. 329–341
16. M. Mitchell, *An Introduction to Genetic Algorithms* (Mit Press, Massachusetts, 1999)

17. P. Melin, F. Olivas, O. Castillo, F. Valdez, J. Soria, M. Valdez, Optimal design of fuzzy clas-sification systems using PSO with dynamic parameter adaptation through fuzzy logic. Expert Syst. Appl. **40**(8), 3196–3206 (2012)
18. J.R. Castro, O. Castillo, L.G. Martinez, Interval type-2 fuzzy logic toolbox. Eng. Lett. **15**(1), 89–98 (2007)
19. S. Mirjalili, S.M. Mirjalili, A. Lewis, Grey wolf optimizer. Adv. Eng. Softw. **69**, 46–61 (2014)
20. C. Leal Ramírez, O. Castillo, P. Melin, A. Rodríguez Díaz, Simulation of the bird age-structured population growth based on an interval type-2 fuzzy cellular structure. Inf. Sci. **181**(3), 519–535 (2011)
21. N.R. Cázarez-Castro, L.T. Aguilar, O. Castillo, Designing type-1 and type-2 fuzzy logic con-trollers via fuzzy Lyapunov synthesis for nonsmooth mechanical systems. Eng. Appl. AI **25**(5), 971–979 (2012)
22. O. Castillo, P. Melin, Intelligent systems with interval type-2 fuzzy logic. Int. J. Innov. Comput. Inf. Control **4**(4), 771–783 (2008)
23. G.M. Mendez, O. Castillo, Interval type-2 TSK fuzzy logic systems using hybrid learning algorithm, in *The 14th IEEE International Conference on Fuzzy Systems*. FUZZ'05, (2005), pp. 230–235
24. E. Rubio, O. Castillo, F. Valdez, P. Melin, C.I. González, G. Martinez: An extension of the fuzzy possibilistic clustering algorithm using type-2 fuzzy logic techniques. Adv. Fuzzy Syst., 7094046:1-7094046:23 (2017)
25. P. Melin, O. Castillo, Intelligent control of complex electrochemical systems with a neuro-fuzzy-genetic approach. IEEE Trans. Ind. Electron. **48**(5), 951–955
26. L. Aguilar, P. Melin, O. Castillo, Intelligent control of a stepping motor drive using a hybrid neuro-fuzzy ANFIS approach. Appl. Soft Comput. **3**(3), 209–219 (2003)
27. P. Melin, O. Castillo, Adaptive intelligent control of aircraft systems with a hybrid approach combining neural networks, fuzzy logic and fractal theory. Appl. Soft Comput. **3**(4), 353–362 (2003)
28. P. Melin, J. Amezcua, F. Valdez, O. Castillo, A new neural network model based on the LVQ algorithm for multi-class classification of arrhythmias. Inf. Sci. **279**, 483–497 (2014)
29. P. Melin, O. Castillo, Modelling, Simulation and Control of Non-Linear Dynamical Systems: An Intelligent Approach Using Soft Computing and Fractal Theory (CRC Press, 2001)
30. P. Melin, D. Sánchez, O. Castillo, Genetic optimization of modular neural networks with fuzzy response integration for human recognition. Inf. Sci. **197**, 1–19 (2012)

Fuzzy Flower Pollination Algorithm to Solve Control Problems

Hector Carreon, Fevrier Valdez and Oscar Castillo

Abstract Pollination is an essential process for the proper functioning of ecosystems and the production of food, through the transfer of pollen. Knowing these mechanisms, the Flower Pollination Algorithm (FPA) was developed in 2012 by Yang (Nature-inspired optimization algorithms. Elsevier, London-New York, pp 155–173, 2014 [1]). The Fuzzy Flower Pollination Algorithm (FFPA) was tested on two optimization problems: (1) Optimization of 8 mathematical functions: Sphere, Ackley, Rastrigin, Zakharov, Griewank, Sum of Different Powers, Michalewicz and Rosenbrock, for 30 and 100 dimensions. (2) Optimization of the fuzzy controller. For the water tank plant. The FFPA method obtained excellent results when compared with other bioinspired algorithms such as BCO and PSO, knowing that the FPA is relatively new in the field of collective intelligence, it opens a very promising area of research.

Keywords Flower pollination algorithm · Fuzzy flower pollination algorithm · Fuzzy logic · Bio-inspired algorithm · Pollination

1 Introduction

When using heuristics as a noun refers to the discipline, art or science of discovery, when used as an adjective it refers to something more concrete such as strategies, rules or syllogisms and conclusions, in this research it is used without a doubt as a noun.

We define metaheuristics as an art, technique or practical or informal procedure to solve problems.

Metaheuristic algorithms, from the Greek "meta" (beyond) and "heuristic" (find) [2], are methods developed to solve very general problems more quickly than with traditional methods or to find an approximate solution or close to the optimum when classical methods fail to find the exact solution [3].

H. Carreon · F. Valdez (✉) · O. Castillo
Tijuana Institute of Technology, Tijuana, BC, Mexico
e-mail: fevrier@tectijuana.mx

© Springer Nature Switzerland AG 2020
O. Castillo and P. Melin (eds.), *Hybrid Intelligent Systems in Control,*
Pattern Recognition and Medicine, Studies in Computational Intelligence 827,
https://doi.org/10.1007/978-3-030-34135-0_10

119

In this sense, in contrast to what happens with deterministic algorithms, as a rule metaheuristics do not guarantee that an optimal global solution is found for some types of problems. Actually, the goal of heuristic techniques is to produce a solution that is good enough for the problem in question within a reasonable period of time. That is, this solution may not be the best and simply be an approximation, but its value lies in the fact that the search time is not prohibitively long.

In this article we focus on the application of the FPA algorithm with a fuzzy approach to the water tank plant, to minimize the error, obtaining very good results.

There are other investigations where the FPA algorithm has been used, but had not previously been applied to solve control problems.

In the following sections we describe the FFPA algorithm, its inspiration, structure and application in solving control problems through experiments and comparing ourselves with other methods obtaining better results.

2   Flower Pollination Algorithm (FPA)

From the point of view of the biological evolution, the objective of the pollination of the flowers is the survival of the fittest and the optimal reproduction of the plants in terms of numbers.

In nature, the main purpose of flowers is reproduction through pollination, which is the transfer of pollen, made by pollinators such as insects, birds, other animals or the wind. Some types of flowers have special pollinators for successful pollination [4].

2.1   Pollination

Pollination, understood as the transfer of pollen from the male part of a flower to the female part of the same or another flower, is an essential process for maintaining the viability and genetic diversity of flowering plants, as well as improving the quality and quantity of seeds and fruits, as well as the characteristics of the offspring [5].

2.2   Types of Pollination

Self-pollination or Direct Pollination, Fig. 1a, also called autogamy, occurs if the pollination is done in the same flower, is when the pollen of a flower pollinates the same flower or flowers of the same plant, and 10% of the usually pollination is achieved through direct pollination [6].

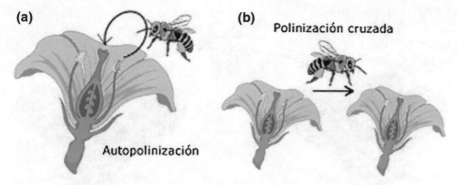

Fig. 1 Types of pollination

Crossed or Indirect Pollination, Fig. 1b, also called allogamy, occurs if pollination is carried out between flowers of different individuals, and usually 90% of the pollination is achieved through cross-pollination [6].

Pollination can also be abiotic, as can be seen in Fig. 2, by transporting pollen by wind or water, or biotic using animals as vectors in transport, as can be seen in Fig. 3 [7].

Fig. 2 Abiotic pollination

Fig. 3 Biotic pollination

2.3 Levy Flight

The French mathematician Paul Levy (1886–1971) introduced Levy Flights in 1937; Levy's flight is a statistical description of the movement that extends beyond the more traditional Brownian movement discovered more than a hundred years ago [8, 9].

Levy flights are a random walk whose length of passage it taken from Levy's distribution, often in terms of a power law formula $L(s) \sim |s|^{-1-\beta}$ where $0 < \beta \leq 2$. The Levy distribution is a distribution of the sum of N random variables of identical and independent distribution whose Fourier transform takes the following form [10], represented in Eq. (1):

$$F_N(k) = \exp\left[-N|k|^\beta\right] \tag{1}$$

2.4 Algorithm Rules

The FPA is designed with 4 basic rules that are described below:

1. Biotic and cross-pollination, occurs when pollen is transported from one plant to another, pollinators move according to the Levy distribution, this process is also known as global pollination, as can be seen in Fig. 4.
2. Self-pollination occurs when the plant fertilizes itself, also known as local pollination, as shown in Fig. 5.
3. Flower constancy involves the association between pollinators and flowers.
4. Probability of change p E [0 1] controls global and local pollination, the value of p is generally 0.8 [4].

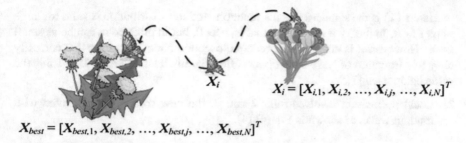

$$X_i = [X_{i,1}, X_{i,2}, ..., X_{i,j}, ..., X_{i,N}]^T$$

$$X_{best} = [X_{best,1}, X_{best,2}, ..., X_{best,j}, ..., X_{best,N}]^T$$

Fig. 4 Global or biotic pollination

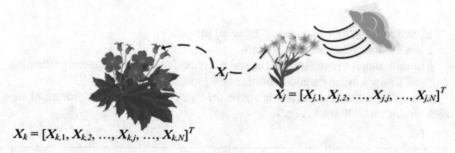

$$X_j = [X_{j,1}, X_{j,2}, ..., X_{j,j}, ..., X_{j,N}]^T$$

$$X_k = [X_{k,1}, X_{k,2}, ..., X_{k,j}, ..., X_{k,N}]^T$$

Fig. 5 Local or abiotic pollination

There are two important steps in the FPA algorithm:

1. The global pollination that involves rules 1 and 3 that are used to find the solution of the next stage x_i^{t+1}, using the values from the previous stage (step t) defined as x_i^t. The equation of global pollination is shown in (2) [4].

$$x_i^{t+1} = x_i^t + \gamma L(\lambda)(g^* - x_i^t) \tag{2}$$

Parameter definition:
i represents the i-th pollen.
g^* is the best current solution.
γ it is a scale factor, which is used to control the size of the step.
$L(\lambda)$ is the pollination force, basically it is a step size.
Levy's flight is used to mimic the movement of insects over long distances, where $L > 0$, Eq (3).

$$L \sim \frac{\lambda \Gamma(\lambda)\sin(\pi\lambda/2)}{\pi} \frac{1}{s^{1+\lambda}}, (s \gg s_0 > 0) \tag{3}$$

Here $\Gamma(\lambda)$ is the standard gamma function, and this distribution is valid for large steps $s > 0$. In theory it is required that $|s_0| \gg 0$, but in practice s_0 can be as small as 0.1 However, it is not trivial to generate pseudorandom step sizes that correctly obey this Equation of Levy's distribution Eq. (3). In all simulations, $\lambda = 1.5$ is the value being used [7].

2. Local pollination involves rules 2 and 3, the new solution is generated with random walks as shown in Eq. (4) [4].

$$x_i^{t+1} = x_i^t + \varepsilon\left(x_j^t - x_k^t\right) \qquad (4)$$

where:

x_j^t and x_k^t represent solutions of different plants.

ε is a random value between 0 and 1.

p (fourth rule) is a probability of change that is used to decide the type of pollination that will be used in the iteration optimization process.

The optimization pseudocode is shown in Fig. 6 and the logical process of the FPA algorithm is shown in Fig. 7.

Begin
Initialize a population of n flowers: n
Define d - dimensional objective function, $f(x)$
Find the best solution g*, *in the initial population*
Define switching probability $p \in [0,1]$
Do Until the iteration counter < maximum number of iterations
    For $i = 1:n$ (all population n)
        If *rand* < *p*
            Draw a (d-dimensional) step vector L (Levy distribution)
            Global Pollination: $x_i^{t+1} = x_i^t + \gamma L(\lambda)(g^* - x_i^t)$ (4.1)
        else
            Local Pollination: $x_i^{t+1} = x_i^t + \varepsilon\left(x_j^t - x_k^t\right)$ (4.3)
        end if
            Evaluate new solutions x_i^{t+1}
        if x_i^{t+1} are better x_i^t
        update $x_i^t = x_i^{t+1}$
        end if
    end for
        Find the current best solution g*
End Until
Output the best solution found
End

Fig. 6 Pseudocode of the FPA algorithm [4]

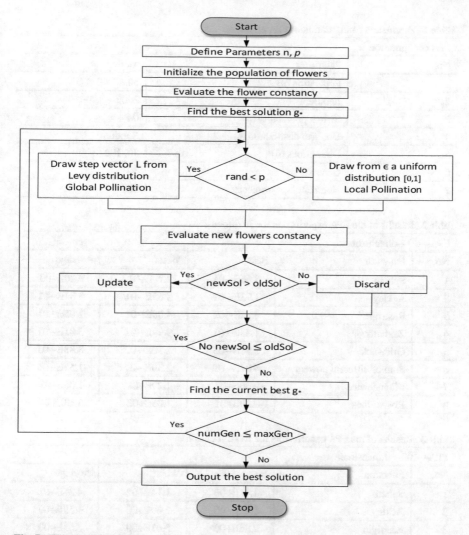

Fig. 7 FPA algorithm flowchart [11]

2.5 Analysis of the Performance of the FPA with Benchmark Functions

Experiments performed with the FPA for 30 and 100 dimensions with the mathematical functions: Sphere, Ackley, Rastrigin, Zakharov, Griewank, Sum of Different Powers, Michalewicz and Rosenbrock, are presented Table 1 shows the configuration and Tables 2 and 3 show the obtained results.

Table 1 shows the parameters with which all the experiments in this investigation were made.

Table 1 Parameters configuration of FPA algorithm

FPA configuration		
No.	Parameter	Value
1	Population (n)	30
2	Iterations	2000
3	Switch probability (p)	0.8
4	Uniform distribution (ϵ)	Random
5	Dimensions (d)	30, 100
6	Generations	30
7	Number of experiments	480

Table 2 Results of the FPA experiments for 30 dimensions

FPA—30—dimensions				
No.	Function	Best	Worst	Average
1	Sphere	3.66E−01	2.78E+01	9.20E+00
2	Ackley	1.93E−01	8.65E−01	4.84E−01
3	Rastrigin	6.88E−02	4.00E−01	2.05E−01
4	Zakharov	1.82E−01	9.99E−01	3.81E−01
5	Griewank	2.38E−03	2.90E−02	8.88E−03
6	Sum of different powers	4.07E−09	2.30E−04	1.55E−05
7	Michalewicz	−9.06E+00	−1.29E+01	−1.08E+01
8	Rosenbrock	3.20E+01	1.05E+02	5.13E+01

Table 3 Results of the FPA experiments for 100 dimensions

FPA—100—dimensions				
No.	Function	Best	Worst	Average
1	Sphere	1.41E+00	1.14E+02	4.37E+01
2	Ackley	1.27E+00	2.00E+00	1.59E+00
3	Rastrigin	2.26E+00	5.63E+00	3.53E+00
4	Zakharov	1.85E+00	4.45E+00	3.16E+00
5	Griewank	2.42E−02	7.28E−02	5.10E−02
6	Sum of different powers	8.10E−08	3.41E−03	1.83E−04
7	Michalewicz	−1.46E+01	−2.40E+01	−1.81E+01
8	Rosenbrock	3.37E+02	6.89E+02	4.44E+02

Table 2 shows the results of the best, worst and average global minimum of each of the eight mathematical functions, for 30 dimensions.

Table 3 shows the results of the best, worst and average global minimum of each of the eight mathematical functions, for 100 dimensions.

3 Fuzzy Logic

The term Fuzzy Logic refers to a method that exploits a numerical representation of common sense rules that are approximate in nature and modeled by interpolation, a control law or a process, the approximate reasoning is outside the scope of classical logic, because It is concerned with the forms of reasoning that formulate precise analyzes [12].

Sensory information is interpreted by the human brain incompletely and inaccurately. To treat linguistic information, it is through fuzzy set theory that provides a systematic calculation and also performs numerical calculations with linguistic labels stipulated by membership functions. In a fuzzy inference system (FIS), the selection of 'IF-THEN' rules that model the human experience in an application are key [13].

In 1965 the term fuzzy logic was introduced by Lofti A. Zadeh that emerged with the development of the theory of fuzzy sets that provides a method to treat linguistic terms [14], a fuzzy system includes:

- Fuzzification it is the process of converting a crisp input value into a diffuse value that is done by using the information in the knowledge base.
- The knowledge base is expressed through a series of IF-THEN rules based on an expert. The fuzzy inference mechanism performs operations through the knowledge base and aims to convert fuzzy inputs into fuzzy outputs.
- Defuzzification is the process of obtaining a unique number of the output of the diffuse set, it is used to transfer diffuse inference results to a crisp output.

3.1 Membership Functions

Triangular MF represented in Fig. 8, is a function with 3 parameters $\{a, b, c\}$ defined as [15]:

$$triangular(\mu; a, b, c) = \begin{cases} 0 & \mu \leq a \\ \frac{\mu-a}{b-a} & a \leq x \leq b \\ \frac{c-x}{c-b} & b \leq x \leq c \\ 0 & c \leq x \end{cases} \tag{5}$$

Fig. 8 Triangular
membership function

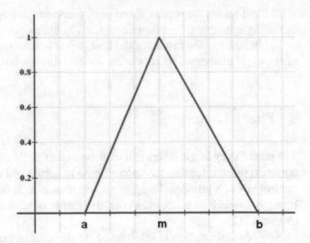

The parameters $\{a, b, c\}$ (*with* $a < b < c$) determine the coordinates in x of the three corners of the fundamental membership function.

Trapezoidal MF (function pi π) represented in Fig. 9, is a function with 4 parameters $\{a, b, c, d\}$, defined as [15]:

$$\pi(\mu; a, b, c, d) = \begin{cases} 0 & \mu \leq a \\ \frac{\mu-a}{b-a} & a \leq \mu \leq b \\ 1 & b \leq \mu \leq c \\ \frac{d-\mu}{d-c} & c \leq \mu \leq d \\ 0 & d \leq \mu \end{cases} \quad (6)$$

The parameters $\{a, b, c, d\}$ (*with* $a < b < c < d$) determine the coordinates in x of the four corners of the trapezoidal membership function.

Fig. 9 Trapezoidal
membership function

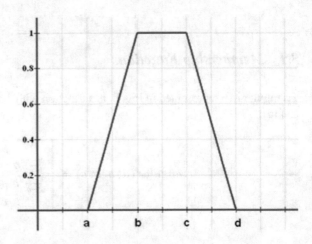

Fig. 10 Gaussian
membership function

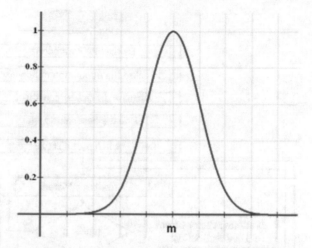

MF Gaussian represented in Fig. 10, is defined by the parameters c and σ, the parameter c determines the center and the parameter σ determines the width [15]:

$$gaussian(\mu; c, \sigma) = e^{-\frac{1}{2}\left(\frac{\mu-c}{\sigma}\right)^2} \tag{7}$$

4 Fuzzy Flower Pollination Algorithm (FFPA)

The FFPA algorithm (Fuzzy Flower Pollination Algorithm) is a bio-inspired algorithm that uses a fuzzy system or several to adapt its parameters; in this case, two are used to update the parameters p (change probability) and ε (epsilon) of the FPA algorithm, shown in Fig. 11.

Development of the proposal:

Experiments were performed with the FPA and the eight mathematical functions, for 30 and 100 dimensions.

Experiments were also performed with the FFPA updating the parameters p and ε, and the eight mathematical functions, for 30 and 100 dimensions.

Comparative performance analyzes were conducted among the FPA, FFPA and other methods described below.

The FFPA algorithm was used to optimize the fuzzy control of the water tank plant. The fuzzy system of the water tank optimized by means of the FPA algorithm.

In this proposal, we used several fuzzy system configurations (16 variants), which are described below.

With each of the variants experiments were performed with eight mathematical functions that are described each one below.

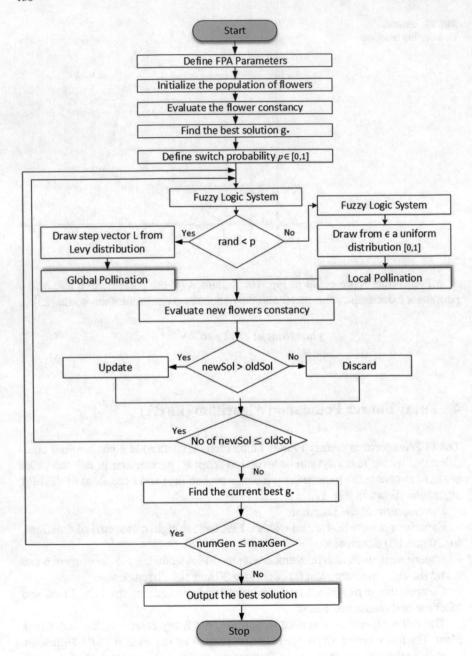

Fig. 11 Flowchart of the FFPA for fuzzy control

Figure 11 shows the flowchart of the original FPA with the two fuzzy systems for the parameters p and ε (epsilon).

4.1 FFPA Variants

Table 4 shows the eight variants used of the FFPA to develop the experiments, the parameters used are p and ε, probability of change and the change in the error respectively.

4.2 Combination of Variants

Table 5 shows the 16 combinations that were made with the eight variants of Table 4, with these combinations the experiments were performed whose results are shown later in this chapter.

4.3 Analysis of the Performance of the FFPA (Applying Fuzzy Logic), with Benchmark Functions

Experiments were performed with the FFPA for 30 and 100 dimensions with the mathematical functions: Sphere, Ackley, Rastrigin, Zakharov, Griewank, Sum of Different Powers, Michalewicz and Rosenbrock. Table 6 shows the configuration and Tables 7 and 8 shows the results obtained.

Table 6 shows the parameters with which the experiments in this investigation were made.

Table 7 shows the results of the best, worst and average global minimum of each of the eight mathematical functions, for 30 dimensions.

Table 8 shows the results of the best, worst and average global minimum of each of the eight mathematical functions, for 100 dimensions.

4.4 Comparative Analysis of FPA Versus FFPA

After showing the experiments performed with the original FPA and the FFPA, a comparisons made with both methods is presented below.

Table 9 shows the best results of the FPA and FFPA experiments with 8 mathematical functions and 30 dimensions.

Table 4 FFPA variants

F I S	Parameters		Input	MF		Output	MF		Rules
	P	E		No.	Type		No.	Type	
FFPA1	FPA1		1	3	Gauss	1	3	Gauss	3
FFPA2	FPA2		1	5	Gauss	1	5	Gauss	5
FFPA3		FPA3	1	3	Gauss	1	3	Gauss	3
FFPA4		FPA4	1	5	Gauss	1	5	Gauss	5
FFPA5	FPA5		2	3	Gauss	1	3	Gauss	9
FFPA6	FPA6		2	5	Gauss	1	5	Gauss	25
FFPA7		FPA7	2	3	Gauss	1	3	Gauss	9
FFPA8		FPA8	2	5	Gauss	1	5	Gauss	25

Table 5 16 combinations of the eight variants of the FFPA

FFPA	p	ε
1	FPA1	FPA3
2	FPA2	FPA3
3	FPA1	FPA4
4	FPA2	FPA4
5	FPA5	FPA7
6	FPA5	FPA8
7	FPA6	FPA7
8	FPA6	FPA8
9	FPA1	FPA7
10	FPA2	FPA8
11	FPA5	FPA3
12	FPA6	FPA4
13	FPA1	FPA8
14	FPA2	FPA7
15	FPA5	FPA4
16	FPA6	FPA3

Table 6 Parameters configuration of FFPA algorithm

FFPA configuration		
No.	Parameter	Value
1	Population (n)	30
2	Iterations	2000
3	Switch probability (p)	Fuzzy
4	Uniform distribution (ϵ)	Fuzzy
5	Dimensions (d)	30, 100
6	Generations	30
7	Number of experiments	960

Table 10 shows the best results of the FPA and FFPA experiments with 8 mathematical functions and 100 dimensions.

Table 11 shows a summary with the results of the 16 variants with the global minimum, with 30 dimensions, the variant with the best result was the FFPA3 function and with worse result was FFPA4. In Fig. 12 the function Rosenbrock shows the convergence of the experiments done with the 16 variants of the FFPA to the global minimum.

Table 12 shows a summary with the results of the 16 variants with the global minimum, with 100 dimensions, the variant with the best result was the function FFPA5 and with worse result was FFPA3. In Fig. 13 the Griewank function shows

Table 7 Results of the FFPA experiments for 30 dimensions

FFPA—30—dimensions

No.	Variant	Function	Best	Worst	Average
1	FFPA13	Sphere	1.65E−04	2.08E−02	4.35E−03
2	FFPA7	Ackley	1.69E−02	2.21E+00	7.62E−01
3	FFPA14	Rastrigin	2.63E−02	1.07E+02	9.79E+00
4	FFPA13	Zakharov	2.79E−03	1.38E−01	3.89E−02
5	FFPA13	Griewank	2.53E−06	6.31E−04	1.41E−04
6	FFPA9	Sum of different powers	5.01E−17	2.51E−11	1.12E−12
7	FFPA8	Michalewicz	−9.52E+00	−1.31E+01	−1.18E+01
8	FFPA3	Rosenbrock	2.37E+01	4.69E+01	3.23E+01

Table 8 Results of the FFPA experiments for 100 dimensions

FFPA—100—dimensions

No.	Variant	Function	Best	Worst	Average
1	FFPA14	Sphere	3.57E−01	1.49E+00	7.52E−01
2	FFPA13	Ackley	3.96E−01	7.43E−01	5.89E−01
3	FFPA1	Rastrigin	5.89E−01	1.03E+00	7.91E−01
4	FFPA9	Zakharov	1.11E+00	4.10E+00	2.08E+00
5	FFPA5	Griewank	2.66E−03	6.53E−03	4.02E−03
6	FFPA9	Sum of different powers	6.63E−14	1.09E−08	7.83E−10
7	FFPA13	Michalewicz	−1.40E+01	−2.35E+01	−1.91E+01
8	FFPA14	Rosenbrock	1.27E+02	2.17E+02	1.68E+02

Table 9 Comparison of the best results of the FPA and FFPA experiments

FPA versus FFPA—30—dimensions

No.	Function	Best FPA	Best FFPA
1	Sphere	3.66E−01	1.65E−04
2	Ackley	1.93E−01	1.69E−02
3	Rastrigin	6.88E−02	2.63E−02
4	Zakharov	1.82E−01	2.79E−03
5	Griewank	2.38E−03	2.53E−06
6	Sum of different powers	4.07E−09	5.01E−17
7	Michalewicz	−9.06E+00	−9.52E+00
8	Rosenbrock	3.20E+01	2.37E+01

Table 10 Comparison of the best results of the FPA and FFPA experiments

FPA versus FFPA—100—dimensions			
No.	Function	Best FPA	Best FFPA
1	Sphere	1.41E+00	3.57E−01
2	Ackley	1.27E+00	3.96E−01
3	Rastrigin	2.26E+00	5.89E−01
4	Zakharov	1.85E+00	1.11E+00
5	Griewank	2.42E−02	2.66E−03
6	Sum of different powers	8.10E−08	6.63E−14
7	Michalewicz	−1.46E+01	−1.40E+01
8	Rosenbrock	3.37E+02	1.27E+02

Table 11 Summary of the variants with 30 dimensions

FPA	Global min
FFPA1	2.57E+01
FFPA2	2.70E+01
FFPA3	**2.37E+01**
FFPA4	*2.79E+01*
FFPA5	2.56E+01
FFPA6	2.41E+01
FFPA7	2.61E+01
FFPA8	2.48E+01
FFPA9	2.63E+01
FFPA10	2.65E+01
FFPA11	2.64E+01
FFPA12	2.70E+01
FFPA13	2.48E+01
FFPA14	2.73E+01
FFPA15	2.53E+01
FFPA16	2.73E+01

the convergence of the experiments done with the 16 variants of the FFPA to the global minimum.

4.5 Comparison of FPA and FFPA with Other Methods for 30 Dimensions

Table 13 shows the results of the best, worst and average with the methods GA (Genetic Algorithm) [16], DE (Differential Evolution) [17] and PSO taken from [18],

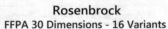

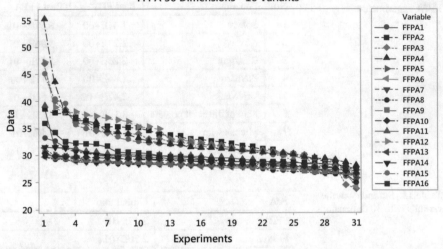

Fig. 12 Convergence of the Rosenbrock function with the 16 variants of the FFPA

Table 12 Summary of the variants with 100 dimensions

FPA	Global min
FFPA1	8.20E−03
FFPA2	1.06E−02
FFPA3	*1.65E−02*
FFPA4	1.17E−02
FFPA5	**2.66E−03**
FFPA6	3.12E−03
FFPA7	3.00E−03
FFPA8	3.38E−03
FFPA9	6.24E−03
FFPA10	4.85E−03
FFPA11	5.69E−03
FFPA12	8.36E−03
FFPA13	5.38E−03
FFPA14	6.47E−03
FFPA15	1.15E−02
FFPA16	6.76E−03

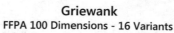

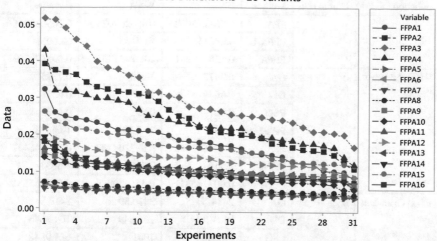

Fig. 13 Convergence of the Griewank function with the 16 variants of the FFPA

Table 13 Comparison of FPA and FFPA with other methods for 30 dimensions

Function	Global opt	Method	Best	Worst	Average
F1(x) sphere	0	GA	0.0008	0.1978	0.0535
		DE	0.004	0.3654	0.0932
		PSO	0.0099	0.0719	0.0237
		FPA-L	0.0302	0.3213	0.1255
		FPA	0.366134784	27.82367305	9.199649305
		FFPA-L	0.0073	0.0224	0.0152
		FFPA	0.000164833	0.123954607	0.001100593
F2(x) Ackley	0	GA	0.0169	3.6098	0.6513
		DE	0.5545	4.5842	2.2593
		PSO	0.8398	2.3545	1.7527
		FPA-L	2.6333	4.4619	3.6088
		FPA	0.193309572	0.864812093	0.483790943
		FFPA-L	0.001	4.5938	2.6244
		FFPA	0.016870845	2.468933377	0.077729571
F3(x) Rastrigin	0	GA	0.0563	19.1326	4.6809
		DE	5.0154	12.3525	9.7992
		PSO	46.1047	116.193	79.3988
		FPA-L	64.1695	95.6657	85.2551

(continued)

Table 13 (continued)

Function	Global opt	Method	Best	Worst	Average
		FPA	0.068775902	0.400207024	0.205039916
		FFPA-L	45.8711	85.7563	63.6051
		FFPA	0.026275201	106.7951431	0.095479703
F4(x) Zakharov	0	GA	0.2972	1.6587	0.7395
		DE	0.0649	1.9861	0.6328
		PSO	0.0613	0.692	0.2835
		FPA-L	3.28738	23.2394	13.4698
		FPA	0.181843546	0.999208282	0.380948447
		FFPA-L	1.558	7.5484	4.8033
		FFPA	0.002785331	0.555450839	0.002785331
F5(x) Griewank	0	GA	2414.34	2420.59	2416.512
		DE	91.0157	206.649	164.3646
		PSO	361.002	1081	621.9848
		FPA-L	1.157	2.1337	1.6363
		FPA	0.002381614	0.02901503	0.008881239
		FFPA-L	0.8348	0.8348	1.0556
		FFPA	2.53268E−06	0.004606034	5.99613E−05
F6(x) Sum of different powers	0	GA	1.00E−06	5.41E−04	1.03E−04
		DE	8.94E−11	1.97E−06	3.12E−07
		PSO	1.65E−16	2.45E−14	5.24E−15
		FPA-L	3.72E−15	4.84E−12	8.64E−13
		FPA	4.07E−09	2.30E−04	1.55E−05
		FFPA-L	1.23E−15	9.28E−13	1.46E−13
		FFPA	5.01E−17	1.27E−05	1.20E−13
F7(x) Michalewicz	−28.98	GA	−24.9448	−22.3162	−23.9686
		DE	−23.7439	−17.0515	−20.0908
		PSO	−22.9425	−14.9033	19.0958
		FPA-L	−15.7955	−13.7761	−14.6617
		FPA	−12.8905296	−9.061366703	−10.83657064
		FFPA-L	−18.7786	−15.6142	−16.7907
		FFPA	−9.52389972	−16.58966204	−11.46948907

FPA-L and FFPA-L taken from [19], FPA and FFPA, for 7 mathematical functions and 30 dimensions, and we obtained better results in 6 of 7 functions with the proposed method.

4.6 Comparison of FPA and FFPA with Other Methods for 100 Dimensions

Table 14 shows the results of the best, worst and average with the GA and PSO (Particle Swarm Optimization) methods taken from [18], FPA-L and FFPA-L taken from [19], FPA and FFPA, for 8 mathematical functions and 100 dimensions, and we obtained better results in 7 of 8 functions.

4.7 Comparative Summary Against Other 30 Dimensional Methods

Table 15 is a comparative summary of the results of GA, DE, PSO methods taken from [18], and FPA-L and FFPA-L, taken from [19].

4.8 Comparative Summary Against Other 100 Dimensional Methods

Table 16 is a comparative summary of the results of methods GA, PSO, taken from [18], FPA-L and FFPA-L, taken from [19].

4.9 Hypothesis Test Z

Table 17 is a comparative table of the hypothesis test for 30 samples, FPA and FFPA (16 variants), and the results favor the FFPA in 13 of 16 mathematical functions.

5 Fuzzy Controller

The inference engine is the core of a fuzzy controller (and any fuzzy system) operation. Its operation is divided into 3 steps as shown in Fig. 14.

Fuzzification—actual inputs are fuzzified and fuzzy inputs are obtained.

1. Fuzzy processing—processing fuzzy inputs according to the rules set and producing fuzzy outputs.
2. Defuzzification—producing a crisp real value for a fuzzy output.

Table 14 Comparison of FPA and FFPA with other methods for 100 dimensions

Function	Global opt	Method	Best	Worst	Average
F1(x) sphere	0	GA	32.5424	53.6624	47.1804
		PSO	13.7761	14.5472	−14.3083
		FPA-L	5.961	15.0828	9.2921
		FPA	1.408878094	113.8408057	43.73101603
		FFPA-L	0.0668	4.1207	2.0055
		FFPA	0.356624568	7.046194417	0.751806571
F2(x) Ackley	0	GA	5.0154	12.3525	9.7992
		PSO	1.9744	4.2044	2.1508
		FPA-L	3.6392	6.8716	5.0354
		FPA	1.267168363	2.000186611	1.592057602
		FFPA-L	2.7539	5.0853	3.962
		FFPA	0.396368897	2.268066577	1.476324027
F3(x) Rastrigin	0	GA	271.081	317.289	296.7937
		PSO	45.8711	85.7563	66.0546
		FPA-L	225.0241	421.9527	317.289
		FPA	2.260669401	5.625308941	3.526953238
		FFPA-L	226.9238	391.0345	321.9301
		FFPA	0.588983904	157.1162256	0.791248063
F4(x) Zakharov	0	GA	1089.3483	1100.4062	1096.3845
		PSO	361.002	621.9848	583.8823
		FPA-L	302.9564	880.3748	543.8928
		FPA	1.850332473	4.45476385	3.16389122
		FFPA-L	323.6544	1030.5314	553.6868
		FFPA	1.114344246	7.233919374	1.346607005
F5(x) Griewank	0	GA	43.4633	55.9846	47.6626
		PSO	164.3646	206.649	361.002
		FPA-L	7.4012	20.738	13.317
		FPA	0.024221432	0.07279436	0.050962251
		FFPA-L	6.8286	22.1806	13.4716
		FFPA	0.002660399	0.051748819	0.004001742
F6(x) sum of different powers	0	GA	93.6477	240.5874	205.0356
		PSO	472.5683	501.4673	480.3562
		FPA-L	3.52E−17	7.78E−10	1.81E−11
		FPA	8.10E−08	3.41E−03	1.83E−04
		FFPA-L	5.95E−17	8.49E−11	2.85E−12
		FFPA	6.63E−14	1.35E−02	4.70E−10

(continued)

Table 14 (continued)

Function	Global opt	Method	Best	Worst	Average
F7(x) Michalewicz	−96.6	GA	−23.5645	−10.5634	−20.1547
		PSO	−44.3251	15.4451	−40.8757
		FPA-L	−40.3673	−29.131	−33.4871
		FPA	−23.99781919	−14.58863176	−18.1426139
		FFPA-L	−64.0829	−27.7103	−53.7921
		FFPA	−13.95058343	−26.06566596	−17.85246755
F8(x) Rosenbrock	0	GA	20473.7437	25041.6674	21377.8401
		PSO	62488.8044	84673.3572	71974.6825
		FPA-L	18140.031	63,521.2689	33,108.9655
		FPA	372.4,182,473	726.8,505,802	476.1,743,174
		FFPA-L	15,662.2387	52,351.933	32,232.2923
		FFPA	126.6,149,718	732.5440282	168.013269

In a real control system, the controller output should be used to control a real object or process (robot, plant, etc.). Therefore, we need to know a crisp value for every output signal. Defuzzification produces these output values based on the membership functions [20, 21].

Figure 15 shows the flow diagram of the original FPA algorithm, which optimizes the membership functions of the fuzzy system used by the water tank plant.

Figure 16 shows the model of the water tank, which we use in this work for optimization of the parameters (MF) of the fuzzy system.

Figure 17 shows the fuzzy system architecture of the water tank plant, with the inputs, output, as well as the desired reference and controller output.

In the model of the water tank of the Matlab version R2017b (9.3.0.713579), the error is of 0.1260, calculated with the Equation of the Mean Square Error (MSE) (5), and the objective of this work is achieving a lower error.

The Mean Square Error Equation is shown below:

$$MSE = \frac{1}{N} \sum_{K=1}^{N} \left[y(k) - \hat{y}(k) \right]^2 \tag{5}$$

$y(k)$ = Current result in the instant k.
$\hat{y}(k)$ = Outlook the value in the instant k.
N = Total number of samples considered.

Table 15 Comparative summary of the FPA and FFPA with other 30 dimensional methods

F(x)	GA	DE	PSO	FPA-L	FPA	FFPA-L	FFPA
1	0.0535	0.0932	0.0237	0.1255	9.199649305	0.0152	0.001100593
2	0.6513	2.2593	1.7527	3.6088	0.483790943	2.6244	0.077729571
3	4.6809	9.7992	79.3988	85.2551	0.205039916	63.6051	0.095479703
4	0.7395	0.6328	0.2835	13.4698	0.380948447	4.8033	0.002285331
5	2416.512	164.3646	621.9848	1.6363	0.008881239	1.0556	5.99613E−05
6	0.000103	0.000000312	5.24E−15	8.64E−13	1.54617E−05	1.4594E−13	1.19606E−13
7	−23.9686	−20.0908	19.0958	−14.6617	−10.83657064	−16.7907	−11.46948907

Table 16 Comparative summary of the FPA and FFPA with other 100 dimensional methods

F(x)	GA	PSO	FPA-L	FPA	FFPA-L	FFPA
1	47.1804	−14.3083	9.2921	43.73101603	2.0055	0.751806571
2	9.7992	2.1508	5.0354	1.592057602	3.962	1.476324027
3	296.7937	66.0546	317.289	3.526953238	321.9301	0.791248063
4	1096.3845	583.8823	543.8928	3.16389122	553.6868	1.346607005
5	47.6626	361.002	13.317	0.050962251	13.4716	0.004001742
6	205.0356	480.3562	1.8078E−11	0.000182627	2.853E−12	4.70205E−10
7	−20.1547	−40.8757	−33.4871	−18.1426139	−53.7921	−17.85246755
8	21377.8401	71974.6825	33108.9655	476.1743174	32232.2923	168.013269

6 Experimental Results

The parameters of the fuzzy system for the water plant were optimized in order to reduce the error of 0.1260, and in Table 18 the experiments carried out with different fuzzy system architectures are presented.

Table 19 presents the parameter configuration of Fig. 18; the optimized fuzzy system architecture is shown, with three Gaussian MFs in each of the two inputs and with five triangular membership functions in the output, (they were not modified from the original design), in the last box the simulation with the optimized MFs and the error result is shown.

The parameters that were modified in the FPA are dimensions = 15 and iterations = 1000.

Table 20 presents the parameter configuration of Fig. 19, the best optimized fuzzy system architecture is shown, with two trapezoidal and 1 triangular MFs in each of the two inputs and with 5 triangular membership functions in the output (they were not modified from the original design), and in the last box the simulation with the optimized MFs is shown.

The result of the error = 0.013381, the parameters that were modified in the FPA are: for dimensions = 5 and iterations = 1000.

The FPA was used to optimize the fuzzy controller that uses the water tank plant; the results are shown in Table 21. The best result using only Gaussian MF, with 5 dimensions and 1000 generations was with an error of 0.075692. The best result in general was using 2 Trapezoidal MF, and 1 Triangular with an error of 0.013381.

6.1 Statistical Test

Table 22 shows the statistical test of Trapezoidal 5 × 1000 (dimensions × generations) and Gaussian 5 × 1000 methods.

Table 17 Hypothesis test Z, for 30 and 100 dimensions FPA and FFPA

Var.	Function	Dimension	Average	Std Dev	Difference	Z	Z crit. val.	Evidence
FFPA13	Sphere	30	**0.00435**	0.00500	-9.195	-8.034	-1.645	Yes
FPA			9.19965	6.26900				
FFPA14		100	**0.75181**	0.28283	-42.979	-7.328	-1.645	Yes
FPA			43.73102	32.12249				
FFPA7	Ackley	30	0.76202	0.98256	0.278	1.559	-1.645	*No*
FPA			0.48379	0.14642				
FFPA13		100	**0.58886**	0.07735	-1.003	-27.823	-1.645	Yes
FPA			1.59206	0.18224				
FFPA14	Rastrigin	30	9.79098	25.22391	9.586	2.116	-1.645	*No*
FPA			**0.20504**	0.08263				
FFPA1		100	**0.79125**	0.10602	-2.736	-19.196	-1.645	Yes
FPA			3.52695	0.77357				
FFPA13	Zakharov	30	**0.03889**	0.03669	-0.342	-9.886	-1.645	Yes
FPA			0.38097	0.18606				
FFPA9		100	**2.08244**	0.60863	-1.082	-6.436	-1.645	Yes
FPA			3.16400	0.69902				
FFPA13	Griewank	30	**0.0001413**	0.0001317	-0.009	-9.628	-1.645	Yes
FPA			0.0088812	0.0049705				
FFPA5		100	0.00402	0.00089	-0.047	-18.975	-1.645	Yes
FPA			0.05096	0.01352				
FFPA9	Sum D P	30	**1.117E−12**	4.546E−12	−1.546E−05	-1.865	-1.645	Yes

(continued)

Table 17 (continued)

Var.	Function	Dimension	Average	Std Dev	Difference	Z	Z crit. val.	Evidence
FPA			1.546E−05	4.539E−05				
FFPA9		100	7.829E−10	2.168E−09	−1.825E−04	−1.608	−1.645	*No*
FPA			**1.825E−04**	6.218E−04				
FFPA8	Michalewicz	30	**−11.84849**	0.82699	−1.009	−4.524	−1.645	Yes
FPA			−10.83967	0.91098				
FFPA13		100	**−19.06384**	2.05746	−0.924	−1.886	−1.645	Yes
FPA			−18.14000	1.76178				
FFPA3	Rosenbrock	30	**3.23E+01**	4.37E+00	−19.028	−6.866	−1.645	Yes
FPA			5.13E+01	1.46E+01				
FFPA14		100	**1.680E+02**	2.089E+01	−276.388	−17.481	−1.645	Yes
FPA			4.444E+02	8.412E+01				

Fig. 14 Operation of a
fuzzy controller

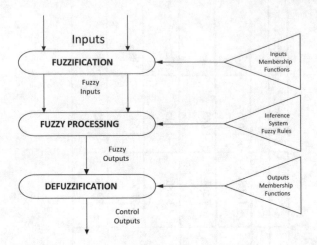

It was found with a level of significance of 5% that there exists significant evidence, stating us that the average of the trapezoidal 5 × 1000 method is less than that of 5 × 1000 Gaussian [22], see Table 22.

Table 23 shows the values of Z (critical) and Z (observed).

6.2 Statistical Comparison Between FPA and Other Methods Such as BCO and PSO

Table 24 shows the data of the means and standard deviation of the FPA, BCO (Bee Colony Optimization) [23], PSO [24] and Michigan Approach methods [25].

Figure 20 shows the statistical results of the aforementioned methods and the method that had the best performance was the FPA.

Table 25 shows the results of the simulations made with the FFPA and BCO methods [17] in the water tank plant; because of the experiments, the FFPA method obtained a better performance.

7 Conclusions

The FPA algorithm is a relatively new metaheuristic algorithm inspired by the pollination process in flowers (2012), which has proven to be an efficient optimization tool, in this work it was applied to the fuzzy controller of the water tank plant, fulfilling its purpose by the results obtained and previously presented.

The experiments were performed with 8 mathematical functions: Sphere, Ackley, Rastrigin, Zakharov, Griewank, Sum of different powers, Michalewicz and Rosenbrock for 30 and 100 dimensions, using only the FPA algorithm. Experiments

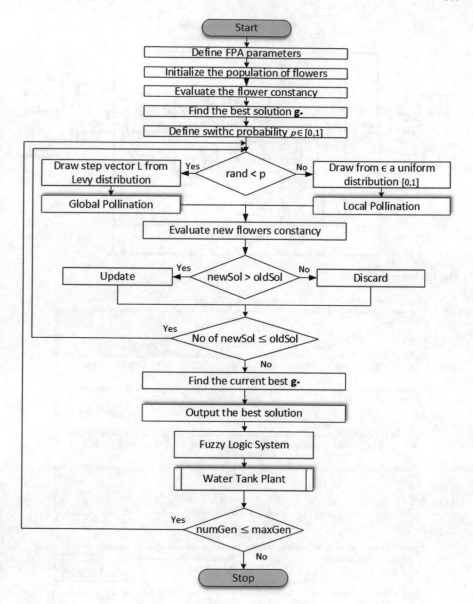

Fig. 15 Flowchart FFPA—water tank

were also performed with the same mathematical functions, the FPA and two fuzzy systems, and statistical comparisons were made between these experiments.

Statistical comparisons were also made between the FFPA and other methods such as GA, DE, PSO and FFPA-L that used the same mathematical functions, obtaining very good results.

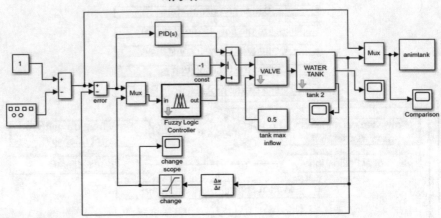

Fig. 16 Water tank plant

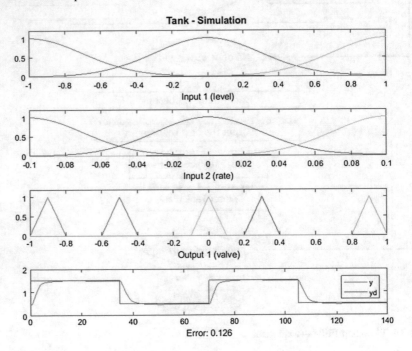

Fig. 17 Behavior of the controller in the water tank plant

Table 18 Experiments carried out in the optimization of the fuzzy controller

FIS		Inputs	MF		Outputs	MF		No. rules
No	Name		No.	Type		No.	Type	
1	fpaTankDinGauSNR	2	3	Gauss	1	5	Tri	5
2	fpaTankDinGauSNR	2	3	Gauss	1	5	Tri	5
3	fpaTankDinGauSR	2	3	Gauss	1	5	Tri	5
4	fpaTankDinGauSR	2	3	Gauss	1	5	Tri	5
5	fpaTankDinGauSR	2	3	Gauss	1	5	Tri	5
6	fpaTankDinGauSR	2	3	Gauss	1	5	Tri	5
7	fpaTankDinTraTriR	2	3	Trap-Tri	1	5	Tri	5
8	fpaTankDinTraTriR	2	3	Trap-Tri	1	5	Tri	5
9	fpaTankDinTraTriR	2	3	Trap-Tri	1	5	Tri	5
10	fpaTankDinTraTriR	2	3	Trap-Tri	1	5	Tri	5
11	fpaTankDinTraTriR	2	3	Trap-Tri	1	5	Tri	5
12	fpaTankDinTraTriR	2	3	Trap-Tri	1	5	Tri	5
13	fpaTankDinTraTriRR	2	3	Trap-Tri	1	5	Tri	5
14	fpaTankDinTraTriR	2	3	Trap-Tri	1	5	Tri	5
15	fpaTankDinTraTriRR	2	3	Trap-Tri	1	5	Tri	5
16	fpaTankDinTraTriRR	2	3	Trap-Tri	1	5	Tri	5

Table 19 Parameters configuration

FPA—parameters configuration						
FIS	Generations	Dimensions	Population	Iterations	p	ϵ
6	100	15	30	1000	0.8	Random

Statistical comparisons were made among other methods such as BCO, PSO, Michigan Approach, etc., obtaining good results.

For all of the above, it can be concluded that the FPA is efficient, fast and robust in the search for the optimal parameters of the membership functions of the fuzzy system applied to control systems.

As future work, experiments can be carried out with the FFPA (Fuzzy Logic Pollination Algorithm) algorithm with other plants, with neural networks, hybridization with other algorithms, etc. In addition we could consider type-2 fuzzy logic [26–28] or other application areas like in [29–34].

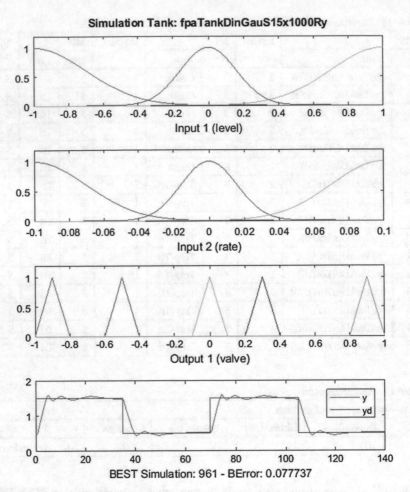

Fig. 18 Simulation of the fuzzy system in the controller of the water tank plant

Table 20 Parameters configuration

FPA—parameters configuration						
FIS	Generations	Dimensions	Population	Iterations	p	ϵ
16	100	5	30	1000	0.8	Random

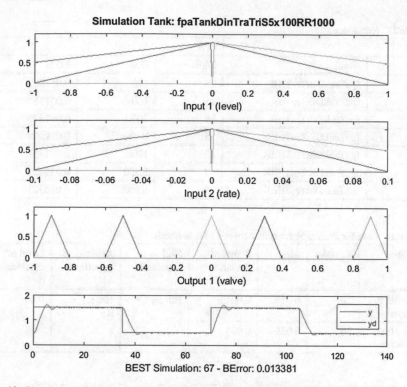

Fig. 19 Simulation of the fuzzy system in the controller of the water tank plant

Table 21 Results of the experiments in the optimization of the fuzzy controller

FIS		FPA		
No	Nombre	Dim	Gen	MSE
1	fpaTankDinGauSNR	5	100	0.085990
2	fpaTankDinGauSNR	5	200	0.084581
3	fpaTankDinGauSR	5	1000	*0.075692*
4	fpaTankDinGauSR	15	100	0.088328
5	fpaTankDinGauSR	15	200	0.090815
6	fpaTankDinGauSR	15	1000	0.077737
7	fpaTankDinTraTriR	5	100	0.067146
8	fpaTankDinTraTriR	5	200	0.046063
9	fpaTankDinTraTriR	5	1000	0.038354

(continued)

Table 21 (continued)

FIS		FPA		
No	Nombre	Dim	Gen	MSE
10	fpaTankDinTraTriR	15	100	0.056902
11	fpaTankDinTraTriR	15	200	0.057562
12	fpaTankDinTraTriR	15	1000	0.057576
13	fpaTankDinTraTriRR	5	R100	0.019908
14	fpaTankDinTraTriRR	15	1000	0.017757
15	fpaTankDinTraTriRR	5	R500	0.015100
16	fpaTankDinTraTriRR	5	R1000	**0.013381**

Table 22 Data for statistical analysis between FPA methods

Method	Samples	Mean μ	Standard deviation	Null hypothesis	Alternative hypothesis	Level of significance α
5×1000 Trapezoidal	30	0.015	0.001	H0: $\mu_1 \geq \mu_2$	Ha: $\mu_1 < \mu_2$	0.05
5×1000 Gaussian	30	0.089	0.004			

Table 23 Result of the statistical test

Z	Value
Critical +Z	-1.645
Observed Z	-102.994

Table 24 Comparative table between different methods

Method	Samples	Mean	Standard deviation
FFPA	30	**0.0154**	0.00139
BCO	30	0.0380	0.00345
PSO	30	0.0510	0.12800
Michigan approach	30	0.2006	0.07530

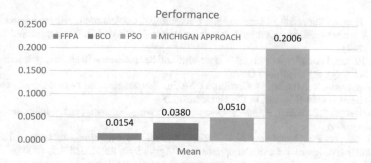

Fig. 20 Performance result between the different methods

Table 25 Results of the mean square error

Method	MSE
FFPA	**0.013381**
BCO	0.030400

References

1. X.-S. Yang, Flower pollination algorithms, in *Nature-Inspired Optimization Algorithms* (Elsevier, London-New York, 2014), pp. 155–173
2. J.A. Moreno Pérez, *Metaheurísticas: Concepto y Propiedades* (Universidad de la Laguna, Spain, Tenerife, 2004)
3. M. Macías, I. Tutor, J.F. Jiménez, A. Tutor, Andrés, S. Pérez, Estudio Comparativo de Técnicas de Optimización para la Actualización de Modelos de Elementos Finitos, 2016, Universidad de Sevilla, Spain, Sevilla
4. X. Yang, M. Karamanoglu, X. He, X. Yang, M. Karamanoglu, X. He, Flower pollination algorithm: a novel approach for multiobjective optimization. Eng. Optim. **00**, 1–16 (2014)
5. M. Garc, L. Alberto, La polinización en los sistemas de producción agrícola: revisión sistemática de la literatura Pollination in agricultural systems: a systematic literature review (IDESIA, Chile, 2016), pp. 53–68
6. D.P. Abrol, *Pollination Biology: Biodiversity Conservation and Agricultural Production* (Springer, Dordrecht Heidelberg, London, New York, 2012)
7. X.-S. Yang, Flower pollination algorithm for global optimization, in *Unconventional Computation and Natural Computation Lecture Notes Computer Science*, vol. 7445 (Springer, Heidelberg Dordrecht London, New York, 2012), pp. 240–249
8. A.F. Kamaruzaman, A.M. Zain, S.M. Yusuf, A. Udin, Levy flight algorithm for optimization problems—a literature review. Appl. Mech. Mater. **421**, 496–501 (2013)
9. X.-S. Yang, Random walks and optimization, in *Nature-Inspired Optimization Algorithms* (Elsevier, London, 2014), pp. 45–65
10. X.-S. Yang, *Engineering Optimization An Introduction with Metaheuristic Applications* (Wiley, New Jersey, 2010)
11. S.D. Madasu, M.L.S. Sai Kumar, A.K. Singh, A flower pollination algorithm based automatic generation control of interconnected power system. Ain Shams Eng. J. (2015)
12. R. Berenji, Hamid, in *An Introduction to Fuzzy Logic Applications in Intelligent Systems* (The Kluwer International Series in Engineering and Computer Science, Springer Science + Business Media New York, 1992)

13. J. R. Jang, C. Sun, *Neuro Fuzzy and Soft Computing A Computational Approach to Learning and Machine intelligence*, ed. by J. Jyh-Shing Roger (Prentice-Hall, Inc., Upper Saddle River, NJ, 1997)
14. L.A. Zadeh, The concept of a linguistic variable and its application to approximate reasoning-I. Inf. Sci. (Ny). **8**(3), 199–249 (1975)
15. R. Sepulveda, R; Montiel, O. Castillo, O. Melin, *Fundamentos de Logica Difusa*, vol. 2002 (Ediciones ILCSA, Tijuana, B. C., México, 2002)
16. R.L. Haupt, S.E. Haupt, *Practical Genetic Algorithms*, 2nd edn. (Wiley Interscience, New Jersey, 2004)
17. D.G. Mayer, B.P. Kinghorn, A.A. Archer, Differential evolution—an easy and efficient evolutionary algorithm for model optimisation. Agric. Syst. **83**(3), 315–328 (2005)
18. S.P. Lim, H. Haron, Performance comparison of genetic algorithm, differential evolution and particle swarm optimization towards benchmark functions, in *2013 IEEE Conference Open System*, no. (Research Gate, Sarawak Malaysia, 2013), pp. 41–46
19. L. Valenzuela, F. Valdez, P. Melin, Nature-inspired design of hybrid intelligent systems, flower pollination algorithm with fuzzy approach for solving optimization problems, in *Studies in Computational Intelligence*, vol. 667 (Springer International Publishing, Switzerland, 2017)
20. L. Reznik, L. Reznik, Fuzzy controllers, in *Victoria University of Technology Melbourne, Australia* (Butterworth Heinemann Newnes, Oxford, 1997), p. 287
21. L.-X. Wang, A course in fuzzy systems and control, in *Design*, (Prentice-Hall International, Inc., 1997), p. 448
22. R. Larson, B. Farber, *Elementary Statistics Fifth Edition* (Prentice-Hall, 2013)
23. C.P. Lim, L.C. Jain, S. Dehuri, *Studies in Computational Intelligence: Innovations in Swarm Intelligence*, vol. 248 (Springer, Berlin Heidelberg, 2009)
24. M. Couceiro, P. Ghamisi, *Fractional Order Darwinian Particle Swarm Optimization Applications and Evaluation of an Evolutionary Algorithm* (Springer Briefs in Applied Sciences and Technology, Springer Cham Heidelberg New York Dordrecht London, 2016)
25. C. Caraveo, F. Valdez, O. Castillo, Optimization of fuzzy controller design using a new bee colony algorithm with fuzzy dynamic parameter adaptation. Appl. Soft Comput. J. **43**, 131–142 (2016)
26. C.I. González, P. Melin, J.R. Castro, Olivia Mendoza, O. Castillo, An improved sobel edge detection method based on generalized type-2 fuzzy logic. Soft. Comput. **20**(2), 773–784 (2016)
27. C.I. González, P. Melin, J.R. Castro, O. Castillo, O. Mendoza, Optimization of interval type-2 fuzzy systems for image edge detection. Appl. Soft Comput. **47**, 631–643 (2016)
28. E. Ontiveros, P. Melin, O. Castillo, High order α-planes integration: a new approach to computational cost reduction of general type-2 fuzzy systems. Eng. Appl. AI **74**, 186–197 (2018)
29. P. Melin, A. Mancilla, M. Lopez, O. Mendoza, A hybrid modular neural network architecture with fuzzy Sugeno integration for time series forecasting. Appl. Soft Comput. **7**(4), 1217–1226 (2007)
30. P. Melin, O. Castillo, *Modelling, Simulation and Control of Non-Linear Dynamical Systems: An Intelligent Approach Using Soft Computing and Fractal Theory* (CRC Press, 2001)
31. P. Melin, D. Sánchez, O. Castillo, Genetic optimization of modular neural networks with fuzzy response integration for human recognition. Inf. Sci. **197**, 1–19 (2012)
32. P. Melin, I. Miramontes, G. Prado-Arechiga, A hybrid model based on modular neural networks and fuzzy systems for classification of blood pressure and hypertension risk diagnosis. Expert Syst. Appl. **107**, 146–164 (2018)
33. P. Melin, D. Sánchez, Multi-objective optimization for modular granular neural networks applied to pattern recognition. Inf. Sci. **460–461**, 594–610 (2018)
34. D. Sánchez, P. Melin, O. Castillo, Optimization of modular granular neural networks using a firefly algorithm for human recognition. Eng. Appl. AI **64**, 172–186 (2017)

Constrained Real-Parameter Optimization Using the Firefly Algorithm and the Grey Wolf Optimizer

Luis Rodríguez, Oscar Castillo, Mario García and José Soria

Abstract The main goal of this paper is to present the performance of two popular algorithms, the first is the Firefly Algorithm (FA) and the second one is the Grey Wolf Optimizer (GWO) algorithm for complex problems. In this case the problems that we are presenting are of the CEC 2017 Competition on Constrained Real-Parameter Optimization in order to realize a brief analysis, study and comparison between the FA and GWO algorithms respectively.

Keywords Grey wolf optimizer · Firefly algorithm · Constraints · Complex problems · Study · Optimization

1 Introduction

In the last decades technology and computer science have improved at a very fast level and now it is more common that the computers realize arithmetic operations in a few seconds in comparisons with the last years. Also they now have the ability to solve problems when the complexity and cost computational is higher.

The main goal of computational intelligence is to use methodologies and approaches that are bio-inspired and the most popular techniques are the metaheuristics, fuzzy logic and artificial neuronal networks that basically imitate behaviors based on the nature of humans or phenomena in the real world.

Computer Science has specifically areas to solve optimization problems when the main goal is to maximize or minimize time, costs, money or others problems that exist in the real world and that the humans always are searching new methods for improve the results of these problems.

In addition we can note that researchers develop new metaheuristics in order to improve the performance and the results in complex problems. Basically studied new techniques to find good results, and these techniques are based on the behavior in

L. Rodríguez · O. Castillo (✉) · M. García · J. Soria
Tijuana Institute of Technology, Tijuana, BC, Mexico
e-mail: ocastillo@tectijuana.mx

© Springer Nature Switzerland AG 2020
O. Castillo and P. Melin (eds.), *Hybrid Intelligent Systems in Control,*
Pattern Recognition and Medicine, Studies in Computational Intelligence 827,
https://doi.org/10.1007/978-3-030-34135-0_11

the nature (bio-inspired). Metaheuristics are very popular in the last years because they are: simple, flexible, stochastic and avoid the local optimal.

Finally in this research we are presenting the FA and GWO algorithms that are popular algorithms in the optimization area. In addition we can find many papers in the literature that solve different problems of optimization using these algorithms. In this work we are presenting the set of ten first problems that was used in the CEC 2017 competition for constrained functions.

This paper is organized as follows: Sect. 2 shows a brief explanation of the Firefly Algorithm, Sect. 3 describes the Grey Wolf Optimizer algorithm, in Sect. 4 we are presenting the simulation results and finally in Sect. 5 we present some conclusions.

2 Firefly Algorithm

According to the inspiration of the metaheuristics these are classified as evolutionary [1], Based on physics [2] and Swarm Intelligence [3]. In addition the No free lunch theorem [4] for optimization proved that there is no meta-heuristic appropriately suited for solving all optimization problems. For example, a particular meta-heuristic can give very promising results for a set of problems, but the same algorithm can show poor performance in an another set of different problems.

The Firefly algorithm [5] was created in 2008 by Xin-She Yang and was inspired by the flashing patterns and behavior of fireflies. This algorithm has basically three rules that we can find in more detail in the original paper [6]. In this algorithm we can find the two main equations, the first is the attractiveness or the light intensity of each firefly and is represented by the following equation:

$$\beta = \beta_0 e^{-\gamma r^2} \tag{1}$$

where β_0 and γ are parameters that could be optimized according to the problem and r represents the distance among two fireflies, and with this equation the attractiveness of each firefly is presented. Finally the movement of a firefly is calculated with the following equation:

$$x_i^{t+1} = x_i^t + \beta_0 e^{-\gamma r_{ij}^2}\left(x_j^t - x_i^t\right) + \alpha_t\, \epsilon_i^t \tag{2}$$

where x_i^{t+1} is the new position of the firefly, x_i^t is the current position of the firefly, where we are adding the attractiveness that exists between the other best solution (x_j^t) and the current solution (x_i^t) and finally adding the randomness parameter that is α_t being the randomization parameter, and ϵ_i^t is a vector of random numbers drawn from a Gaussian distribution or uniform distribution at time t. Finally in Fig. 1 we are presenting a general flowchart of the Firefly Algorithm.

Fig. 1 General flowchart of
the Firefly Algorithm

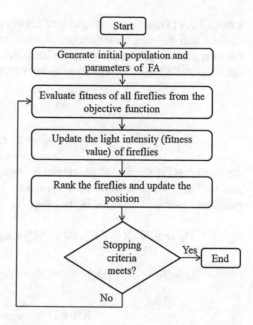

3 Grey Wolf Optimizer Algorithm

The second metaheuristic that we are presenting in this work is the Grey Wolf Optimizer (GWO) [7] algorithm was proposed by Seyedali Mirjalili in 2014. Thus algorithm is based on the behavior of the Canis lupus (Grey Wolf) and the justification is because in the literature and specifically in the Swarm Intelligence area there was not a technique that mimics the behavior of this species.

The original author of the GWO algorithm designed the mathematical model based on two main features of the grey wolf that are the following: the hunting process and the hierarchy in the members of the pack that C. Muro presented in his work [8] also this researcher show that the hunting mechanism of the grey wolves has 3 main phases and are the following: follow and approach the prey, after, encircling, and harassing the prey until it stops moving and finally attack the prey.

Finally we can find the hierarchy pyramid [9] of the pack as a leadership strategy, in other words the best wolf is called alpha (α) and is considered as the best solution in the algorithm, the second best solution is called beta (β) and the third best solution is delta (δ) respectively. Finally, the rest of the candidate solutions are consider as the omega (x) wolves.

In order to mathematically model the two main inspirations described above, we present the following equations:

$$D = \left\| \mathbf{C} \cdot \mathbf{X}_p(t) - \mathbf{X}(t) \right\| \tag{3}$$

$$X(t + 1) = \mathbf{X}_p(t) - \mathbf{AD} \tag{4}$$

where: Eq. (1) represents the distance between the best solution with a randomness method and the current individual that we are analyzing. Equation (2) represents the next position of the current individual based on the distance of Eq. (1). Finally, coefficients "A" and "C" are represented by the following equations:

$$A = 2a \cdot r_1 - a \tag{5}$$

$$C = 2 \cdot r_2 \tag{6}$$

where for the Eq. (5) the "a" parameter is linearly decreasing through of the iterations and "r_1" is a random value between 0 and 1. In addition for the Eq. (6) the "r_2" parameter is a random value in the range [0, 1].

$$D_\alpha = \|C_1 \cdot X_\alpha - X\|, D_\beta = \|C_2 \cdot X_\beta - X\|, D_\delta = \|C_3 \cdot X_\delta - X\| \tag{7}$$

$$X_1 = X_\alpha - A_1 \cdot (D_\alpha), \ X_2 = X_\beta - A_2 \cdot (D_\beta), \ X_3 = X_\delta - A_3 \cdot (D_\delta) \tag{8}$$

$$X(t + 1) = \frac{X_1 + X_2 + X_3}{3} \tag{9}$$

In addition Eqs. (7) and (8) are the same equations as Eqs. (3) and (4) respectively, but in this case the results are based on alpha, beta and delta respectively as we mention above. Finally, in order to calculate the next position of the current individual the original author proposed Eq. (9), which we can described as an average based on the results obtained in Eq. (8) that represent the leaders of the pack.

Finally we can find the general flow of the GWO algorithm as in Fig. 2, which describes all phases that we mentioned above.

4   Simulations Results

In this paper we are presenting a set of problems that contain the presence of constraints that alter the shape of the search space making it more difficult to solve in comparison with the conventional benchmark functions [10–12].

In this section we selected the first 10 problems of the CEC 2017 Competition on Constrained Real-Parameter Optimizer [13] which can be transformed into the following format:

$$\text{Minimize:} f(X), \ X = (x_1, x_2, \ldots, x_n) \quad and \ X \in S \tag{10}$$

$$\text{Subject to } \begin{array}{l} g_i(X) \leq 0, \ i = 1, \ldots, p \\ h_j(X) = 0, \ j = p + 1, \ldots, m \end{array} \tag{11}$$

Fig. 2 General flowchart of
the GWO algorithm

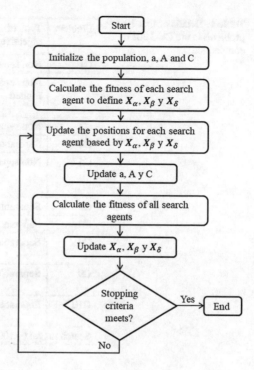

Usually equality constraints are transformed into inequalities of the form

$$|h_j(X)| - \varepsilon \le 0, \quad for \; j = p+1, \ldots, m \tag{12}$$

A solution X is regarded as feasible if $g_i(X) \le 0$, $for \; i = 1, \ldots, p$ and $|h_j(X)| - \varepsilon \le 0$, for $j = p+1, \ldots, m$. In this competition (CEC 2017) ε is set to 0.0001.

In Table 1 we can find a brief explanation of the first 10 constraints problems, where D is the number of decision variables, I is the number of inequality constraints and E is the number of equality constraints.

In this paper we can describe the basic performance evaluation criteria as the following: the maximum function evaluations are 20,000 * D, where D is the dimensions that has each optimization problem, so you are free to have an appropriate population size to suit your algorithm while not exceeding the number maximum of function evaluations.

In addition it is important to mention that in the following tables we are presenting the mean of the violation at one solution, these violations depend on the total numbers that we presented in Table 1, and the equation for calculate this mean violations is the following:

$$v = \frac{\left(\sum_{i=1}^{p} G_i(X) + \sum_{j=p+1}^{m} H_j(X) \right)}{m} \tag{13}$$

Table 1 Details of first 10 problems in the CEC 2017 competition

Problem	Type of objective	Number of constraints	
		E	I
C1	Non separable	0	1; separable
C2	Non separable, rotated	0	1; non separable, rotated
C3	Non separable	1; separable	1; separable
C4	Separable	0	2; separable
C5	Non separable	0	2; non separable, rotated
C6	Separable	6; separable	0
C7	Separable	2; separable	0
C8	Separable	2; non separable	0
C9	Separable	2; non separable	0
C10	Separable	2; non separable	0

Search range: $(-100, 100)^{\wedge}D$

where

$$G_i(X) = \begin{cases} g_i(X) & if \ g_i(X) > 0 \\ 0 & if \ g_i(X) \le 0 \end{cases} \tag{14}$$

$$H_j(X) = \begin{cases} |h_j(X)| & if \ |h_j(X)| - \varepsilon > 0 \\ 0 & if \ |h_j(X)| - \varepsilon \le 0 \end{cases} \tag{15}$$

For the simulation results, we are presenting the averages and standard deviations as the result of 31 independent executions for each problem of the CEC 17 competition with 30 individuals in both algorithms.

In the FA metaheuristic we are presenting tests with values for the alpha parameter in decrement with respect to the iterations from 0.8 to 0, the minimum value of beta was 0.2 and for the gamma value we considered a value of 1, and these parameters were used based on the paper of Lagunes et al. [14].

In Table 2 we can find the averages and standard deviations with FA optimizer when the problems have 10 dimensions, also we can note the averages and standard deviations of the violations that we described above. It is important to mention that the number of maximum iterations in the algorithm is 6666 for 10 dimensions.

In addition we can find in Table 3 the results of the FA metaheuristic when the problems have 50 dimensions and in this case the maximum number of iterations is

Table 2 Results of FA with 10 dimension in the CEC 2017 problems

FA—10 dimensions

Function	FA	STD	Violation	STD
F1	1.53E−04	4.75E−05	0	0
F2	1.73E−04	5.33E−05	0	0
F3	1.76E−04	4.50E−05	0.5	0
F4	7.5322	2.9612	0.2661	0.0624
F5	366.91	1158.22	0.1774	0.2432
F6	9.0410	3.6233	1	0
F7	−741.0755	54.7787	1	0
F8	−90.3668	0	1	0
F9	−90.6092	4.33E−14	0.5	0
F10	−59.7540	4.33E−14	1	0

Table 3 Results of FA with 50 dimension in the CEC 2017 problems

FA—50 dimensions

Function	FA	STD	Violation	STD
F1	0.0564	0.0142	0	0
F2	0.0563	0.0129	0	0
F3	0.0585	0.0133	0.5	0
F4	75.4605	13.3594	0.2903	0.0935
F5	83.0312	96.7234	0.0726	0.1606
F6	78.2756	21.2956	1	0
F7	−2852.58	201.07	1	0
F8	−90.36682	0.00	1	0
F9	−90.0593	5.78E−14	0.5	0
F10	−53.1833	4.33E−14	1	0

of 33,333 in order to respect the maximum number of evaluation functions allowed in the CEC 17 competition.

In addition in the GWO we are using the parameters that the author recommended in the original paper, basically the value for the parameter "a" is [0, 2] and for the "C" parameter the value is static and is 2 [15].

In Table 4 we can find the averages and standard deviations of GWO algorithm when the problems have 10 dimensions, also we can note the averages and standard deviations of the violations that we described above. Finally, in Table 5 we can find the results of the GWO algorithm when the problems have 50 dimensions and it is important mention that the maximum numbers of iterations are 6666 and 33,333 for 10 and 50 dimensions, respectively.

Table 4 Results of GWO with 10 dimension in the CEC 2017 problems

GWO—10 dimensions				
Function	GWO	STD	Violation	STD
F1	144.0604	259.6247	0	0
F2	103.4100	129.6063	0	0
F3	88.7982	118.2894	0.5	0
F4	12.4148	8.7777	0.3333	0.1217
F5	14.3888	43.0255	0.0215	0.1197
F6	18.0462	9.8407	0.8571	3.39E−16
F7	−726.2750	69.7004	0.6667	1.13E−16
F8	−90.3668	0	0.6667	1.13E−16
F9	−90.6092	4.33E−14	0.5	0
F10	−59.7540	4.33E−14	0.6667	1.13E−16

Table 5 Results of GWO with 50 dimension in the CEC 2017 problems

GWO—50 dimensions				
Function	GWO	STD	Violation	STD
F1	8657.98	2032.415	0	0
F2	8657.98	2032.42	0	0
F3	9734.57	2605.37	0.50	0
F4	297.0701	40.1308	0.2742	0.0751
F5	31,557.05	23,287.72	0	0
F6	513.4068	136.0203	1	0
F7	−1450.85	238.30	1	0
F8	−45.9427	40.9837	1	0
F9	−53.0457	41.4657	0.5	0
F10	−28.3182	2.2731	1	0

In this paper we are also presenting a hypothesis test [16], specifically the z-test in order to show statistically a comparison between the performance of the GWO and FA algorithm respectively. The main goal of hypothesis testing is demonstrate statistically which metaheuristic has better performance for these set of problem (constrained problems) and as a brief explanation, we can mention that if the z-value is less than −1.645 we can conclude that FA algorithm has better performance than the GWO algorithm.

Table 6 shows the results of the hypothesis test realized between the two algorithms that we presented above, we can find the Z-value just in the last column of the Table 5 and according of these results we can conclude that for 10 dimensions in the constrained problems the Firefly Algorithm has better performance that the GWO algorithm in 5 of the 7 first problems and for the last 3 problems we can

Table 6 Comparison between FA and GWO algorithms with 10 dimensions

10 dimensions					
Function	GWO	STD	FA	STD	Z-value
F1	144.0604	259.6247	**1.53E−04**	**4.75E−05**	−3.0894
F2	103.4100	129.6063	**1.73E−04**	**5.33E−05**	−4.4424
F3	88.7982	118.2894	**1.76E−04**	**4.50E−05**	−4.1796
F4	12.4148	8.7777	**7.5322**	**2.9612**	−2.9346
F5	**14.3888**	**43.0255**	366.91	1158.22	1.6935
F6	18.0462	9.8407	**9.0410**	**3.6233**	−4.7813
F7	**−726.2750**	**69.7004**	−741.0755	54.7787	−0.9296
F8	−90.3668	0	−90.3668	0	–
F9	−90.6092	4.33E−14	−90.6092	4.33E−14	0
F10	−59.7540	4.33E−14	−59.7540	4.33E−14	0

The meaning of bold is "Best results" in the table

note an interesting behavior, in function number 8 we can note that there is no standard deviations, so, we cannot realize the hypothesis test and finally, in F9 and F10 functions the results are exactly the same.

In addition we are presenting the results of hypothesis test in a graphical way, so in Fig. 3 we can find the results of the hypothesis tests of the first 7 problems that were analyzed in the CEC 17 constrained problems competition. According to these results, we can mention that the red bars represent problems where the Firefly Algorithm has a poor performance with respect to the Grey Wolf Optimizer Algorithm and the green bars are the results where the FA method has better performance than the

Fig. 3 Hypothesis tests in a graphical way with 10 dimensions

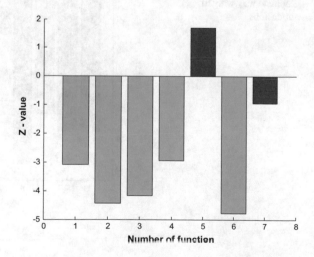

GWO method. Finally we can conclude that for low dimensions the FA algorithm has better performance that the GWO according of the statistical data presented above.

Finally in Table 7 and Fig. 4 we can find the results of hypothesis testing among the FA and GWO algorithms respectively with 50 dimensions for the constrained problems in the CEC 17 competition and according with these results we can conclude that the Firefly Algorithm is better in all constrained functions that were analyzed in this paper.

Table 7 Comparison between FA and GWO algorithms with 50 dimensions

50 dimensions					
Function	GWO	STD	FA	STD	Z-value
F1	8657.98	2032.415	**0.0564**	**0.0142**	−23.7182
F2	8657.98	2032.42	**0.0563**	**0.0129**	−23.7182
F3	9734.57	2605.37	**0.0585**	**0.0133**	−20.8030
F4	297.0701	40.1308	**75.4605**	**13.3594**	−29.1722
F5	31,557.05	23,287.72	**83.0312**	**96.7234**	−7.5249
F6	513.4068	136.0203	**78.2756**	**21.2956**	−17.5970
F7	−1450.85	238.30	**−2852.58**	**201.07**	−25.0305
F8	−45.9427	40.9837	**−90.36682**	**0.00**	−6.0352
F9	−53.0457	41.4657	**−90.0593**	**5.78E−14**	−4.9700
F10	−28.3182	2.2731	**−53.1833**	**4.33E−14**	−60.9058

The meaning of bold is "Best results" in the table

Fig. 4 Hypothesis tests in a graphical way with 10 dimensions

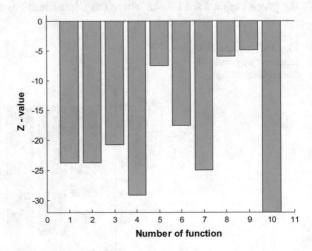

5   Conclusions

In this work we presented a comparative study between two popular algorithms that were the Firefly Algorithm (FA) and Grey Wolf Optimizer (GWO) and these algorithms were test with a set of 10 problems of the CEC 2017 Competition on Constrained Real-Parameter Optimizer in order to study their performance in complex problems.

In addition we presented a hypothesis test in order to demonstrate statistically which algorithm has better performance with the complex problems that we presented in this research (constrained benchmark functions), and for these problems we presented test with 10 and 50 dimensions respectively and we presented these general results in tables with information according with the guidelines that the competition describes in the original paper.

In addition we can conclude that for 10 dimensions the FA algorithm has better performance in approximately 70% of the problems analyzed and for 50 dimensions the FA shows that has better performance in all problems that we presented in this paper, so we can conclude that for these set of problems (constrained) the Firefly algorithm has better performance than the GWO algorithm according with the hypothesis test that we presented above.

Finally, it is important mention that today is more common that the computers solved mathematical operations in a few seconds and this is to main reason and inspiration of computer science to created new complex problems in order to test the performance of the metaheuristics when the problems simulate the real world problems and its complexity.

As a future work it is important improve the results in the experiments of the constrained problems of the CEC 2017 competition using other metaheuristics, for example GSA [17], ICA [18], FWA [19, 20] and include fuzzy logic [21] in order to dynamically adjust the parameters in the other algorithms [22, 23]. In addition, instead of using type-1 fuzzy logic we could consider type-2 fuzzy as in [24–32], or consider other areas of application as in [33–38].

References

1. H.R. Maier, Z. Kapelan, Evolutionary algorithms and other metaheuritics in water resources: Current status, research challenges and future directions. Environ. Model Softw. **62**, 271–299 (2014)
2. U. Can, Alatas B: physics based metaheuristic algorithms for global optimization. Am. J. Inf. Sci. Comput. Eng. **1**, 94–106 (2015)
3. X. Yang, M. Karamanoglu, Swarm intelligence and bio-inspired computation: an overview. Swarm Intell. Bio-Inspired Comput., 3–23 (2013)
4. D.H. Wolpert, W.G. Macready, No free lunch theorems for optimization. Evolut. Comput. IEEE Trans. **1**, 67–82 (1997)
5. X.-S. Yang, *Firefly Algorithm, Lévy Flights and Global Optimization* arXiv:1003.1464v1 (2010)

6. X.-S. Yang, *Firefly Algorithm: Recent Advances and Applications* arXiv:1308.3898v1 (2013)
7. S. Mirjalili, M. Mirjalili, A. Lewis, Grey wolf optimizer. Adv. Eng. Softw. **69**, 46–61 (2014)
8. C. Muro, R. Escobedo, L. Spector, R. Coppinger, Wolf-pack (Canis lupus) hunting strategies emerge from simple rules in computational simulations. Behav. Process. **88**, 192–197 (2011)
9. L. Rodríguez, O. Castillo, M. Valdez, J. Soria, A comparative study of dynamic adaptation of parameters in the GWO algorithm using type-1 and interval type-2 fuzzy logic. Fuzzy Logic Augmentation Neural Optim. Algorithms: Theor. Aspects Real Appl., 3–17 (2018)
10. J. Digalakis, K. Margaritis, On benchmarking functions for genetic algorithms. Int. J. Comput. Math. **77**, 481–506 (2001)
11. M. Molga, C. Smutnicki, Test functions for optimization needs. Test functions for optimization needs (2005)
12. X.-S. Yang, Test problems in optimization. arXiv, preprint arXiv: 1008.0549 (2010)
13. W. Guohua, R. Mallipeddi, P.N. Suganthan, *Problem Definitions and Evaluation Criteria for the CEC 2017 Competition on Constrained Real-Parameter Optimization* (2017)
14. M. Lagunes, O. Castillo J. Soria, Optimization of membership functions parameters for fuzzy controller of an autonomous mobile robot using the firefly algorithm, in *Fuzzy Logic Augmentation of Neural and Optimization Algorithms* (2018), pp 199–206
15. L. Rodriguez, O. Castillo, J. Soria, P. Melin, F. Valdez, C. Gonzalez, G. Martinez, J. Soto, A fuzzy hierarchical operator in the grey wolf optimizer algorithm. Appl. Soft Comput. **57**, 315–328 (2017)
16. R. Larson, B. Farber, *Elementary Statistics Picturing the World* (Pearson Education Inc. 2003), pp. 428–433
17. B. Gonzalez, P. Melin, F. Valdez, G. Prado-Arechiga, Ensemble neural network optimization using a gravitational search algorithm with interval type-1 and type-2 fuzzy parameter adaptation in pattern recognition applications, in *Fuzzy Logic Augmentation of Neural and Optimization Algorithms: Theoretical Aspects and Real Applications* (2018), pp 17–27
18. E. Bernal, O. Castillo, J. Soria, Imperialist competitive algorithm with dynamic parameter adaptation applied to the optimization of mathematical functions. Nat.-Inspired Des. Hybrid Int. Syst. (2017), pp 329–341
19. J. Barraza, P. Melin, F. Valdez, C.I. Gonzalez, *Fuzzy Fireworks Algorithm Based on a Sparks Dispersion Measure, Algorithms*, vol. 10 (2017)
20. J. Barraza, P. Melin, F. Valdez, C. Gonzalez, Fuzzy FWA with dynamic adaptation of parameters, in *IEEE CEC* (2016), pp. 4053–4060
21. L. Rodríguez, O. Castillo, M. García, J. Soria, A comparative study of dynamic adaptation of parameters in the GWO algorithm using type-1 and interval type-2 fuzzy logic, in *Fuzzy Logic Augmentation of Neural and Optimization Algorithms: Theoretical Aspects and Real Applications* (2018), pp 3–16
22. C. Caraveo, A. Fevrier O. Castillo, Optimization mathematical functions for multiple variables using the algorithm of self-defense of the plants. Nat.-Inspired Des. Hybrid Intell. Syst., 631–640 (2017)
23. M. Guerrero, O. Castillo, M. Garcia, Cuckoo search algorithm via Lévy flight with dynamic adaptation of parameter using fuzzy logic for benchmark mathematical functions, in *Design of Intelligent Systems Based on Fuzzy Logic, Neural Networks and Nature-Inspired Optimization. Studies in Computational Intelligence* (2016), pp 555–571
24. C. Leal Ramírez, O. Castillo, P. Melin, A. Rodríguez Díaz, Simulation of the bird age-structured population growth based on an interval type-2 fuzzy cellular structure. Inf. Sci. **181**(3), 519–535 (2011)
25. N.R. Cázarez-Castro, L.T. Aguilar, O. Castillo, Designing type-1 and type-2 fuzzy logic controllers via fuzzy Lyapunov synthesis for nonsmooth mechanical systems. Eng. Appl. AI **25**(5), 971–979 (2012)
26. E. Rubio, O. Castillo, F. Valdez, P. Melin, C.I. González, G. Martinez: An extension of the fuzzy possibilistic clustering algorithm using type-2 fuzzy logic techniques. Adv. Fuzzy Syst., 7094046:1-7094046:23 (2017)

27. O. Castillo, P. Melin, Intelligent systems with interval type-2 fuzzy logic. Int. J. Innov. Comput. Inf. Control **4**(4), 771–783 (2008)
28. G.M. Mendez, O. Castillo, Interval type-2 TSK fuzzy logic systems using hybrid learning algorithm, fuzzy systems, in *The 14th IEEE International Conference on FUZZ'05* (2005), pp. 230–235
29. P. Melin, C.I. González, J.R. Castro, O. Mendoza, O. Castillo, Edge-detection method for image processing based on generalized Type-2 fuzzy logic. IEEE Trans. Fuzzy Syst. **22**(6), 1515–1525 (2014)
30. C.I. González, P. Melin, J.R. Castro, O. Castillo, O. Mendoza, Optimization of interval type-2 fuzzy systems for image edge detection. Appl. Soft Comput. **47**, 631–643 (2016)
31. C.I. González, P. Melin, J.R. Castro, O. Mendoza, O. Castillo, An improved Sobel edge detection method based on generalized type-2 fuzzy logic. Soft. Comput. **20**(2), 773–784 (2016)
32. E. Ontiveros, P. Melin, O. Castillo, High order α-planes integration: a new approach to computational cost reduction of general type-2 fuzzy systems. Eng. Appl. AI **74**, 186–197 (2018)
33. P. Melin, O. Castillo, Intelligent control of complex electrochemical systems with a neuro-fuzzy-genetic approach. IEEE Trans. Ind. Electron. **48**(5), 951–955
34. L. Aguilar, P. Melin, O. Castillo, Intelligent control of a stepping motor drive using a hybrid neuro-fuzzy ANFIS approach. Appl. Soft Comput. **3**(3), 209–219 (2003)
35. P. Melin, O. Castillo, Adaptive intelligent control of aircraft systems with a hybrid approach combining neural networks, fuzzy logic and fractal theory. Appl. Soft Comput. **3**(4), 353–362 (2003)
36. P. Melin, J. Amezcua, F. Valdez, O. Castillo, A new neural network model based on the LVQ algorithm for multi-class classification of arrhythmias. Inf. Sci. **279**, 483–497 (2014)
37. P. Melin, O. Castillo, *Modelling, Simulation and Control of Non-Linear Dynamical Systems: An Intelligent Approach Using Soft Computing and Fractal Theory* (CRC Press, 2001)
38. P. Melin, D. Sánchez, O. Castillo, Genetic optimization of modular neural networks with fuzzy response integration for human recognition. Inf. Sci. **197**, 1–19 (2012)

The Differential Evolution Algorithm with a Fuzzy Logic Approach for Dynamic Parameter Adjustment Using Benchmark Functions

Patricia Ochoa, Oscar Castillo and José Soria

Abstract In this paper the main idea is to state that the use of fuzzy logic helps in the improvement of results in different optimization problems. For this particular paper we propose using the methodology of combining fuzzy logic with the differential evolution algorithm to perform experiments with a set of functions of the CEC2015, since these functions are more complicated than traditional benchmark functions.

Keywords Differential evolution · Fuzzy differential evolution · Mutation · Fuzzy logic · Dynamic parameters

1 Introduction

The importance of fuzzy logic in recent times has become more widely accepted, we in this work we will perform a dynamic adaptation of parameters with the of differential evolution algorithm [1–7]. Regarding related works that use fuzzy logic to optimize the performance of metaheuristic algorithms we can find [8–12]. In complement, related papers related to the competence of CEC2015 were reviewed, particularly those related to the Special Session & Competition on Real-Parameter Single Objective Optimization at CEC-2015 to mention some of them [13–22]. We describe the Fuzzy Differential Evolution (FDE) method employing a fuzzy system to achieve dynamical adjustment of the F parameter in the original algorithm and simulation results are presented to verify the effectiveness of the method using the CEC 2015 Benchmark Functions.

The paper is composed in the following form: Sect. 2 describes the concept of the Differential Evolution algorithm. Section 3 presents the proposed methods. Section 4 presents the CEC 2015 Benchmark Functions, Sect. 5 shows the experiments with the original DE and the FDE algorithms and Sect. 6 finally outline the Conclusions.

P. Ochoa · O. Castillo (✉) · J. Soria
Tijuana Institute of Technology, Tijuana, BC, Mexico
e-mail: ocastillo@tectijuana.mx

© Springer Nature Switzerland AG 2020
O. Castillo and P. Melin (eds.), *Hybrid Intelligent Systems in Control,
Pattern Recognition and Medicine*, Studies in Computational Intelligence 827,
https://doi.org/10.1007/978-3-030-34135-0_12

2 The Differential Evolution Algorithm

There are different algorithms in the literature, but the algorithm of differential evolution is one of the simplest and easiest to understand. Its structure is basically formed by four steps [23, 24]: initialization, mutation, crossing and selection.

It was proposed by Storn and Price with the aim of solving complex problems and that is why in this work we decided to work with this algorithm. This is the mathematical form of the Differential Evolution algorithm [12, 25]:

Population structure

$$P_{x,g} = (x_{i,g}), i = 0, 1, \ldots, Np, g = 0, 1, \ldots, g_{max} \tag{1}$$

$$x_{i,g} = (x_{j,i,g}), j = 0, 1, \ldots, D\text{-}1 \tag{2}$$

$$P_{v,g} = (v_{i,g}), i = 0, 1, \ldots, Np\text{-}1, g = 0, 1, \ldots, g_{max} \tag{3}$$

$$v_{i,g} = (v_{j,I,g}), j = 0, 1, \ldots, D\text{-}1 \tag{4}$$

$$P_{v,g} = (u_{i,g}), i = 0, 1, \ldots, Np\text{-}1, g = 0, 1, \ldots, g_{max} \tag{5}$$

$$u_{i,g} = (u_{j,I,g}), j = 0, 1, \ldots, D\text{-}1 \tag{6}$$

Initialization

$$x_{j,i,0} = rand_j(0, 1) \cdot (b_{j,U} - b_{j,L}) + b_{j,L} \tag{7}$$

Mutation

$$v_{i,g} = x_{r0,g} + F \cdot (x_{r1,g} - x_{r2,g}) \tag{8}$$

Crossover

$$U_{i,g} = (u_{j,i,g}) = \left\{ \begin{array}{l} v_{j,i,g} \ if(rand_j(0, 1) \leq Cr \ or \ j = j_{rand}) \\ x_{j,i,g} \ otherwise. \end{array} \right\} \tag{9}$$

Selection

$$X_{i,g+1} = \left\{ \begin{array}{l} U_{i,g} \ if \ f\left(U_{i,g}\right) \leq f\left(X_{i,g}\right) \\ X_{i,g} \ otherwise. \end{array} \right\} \tag{10}$$

3 Proposed Method

For this paper we use the methodology that we have been working on, in which the original algorithm is combined with fuzzy logic to dynamically adapt some of the parameters of this algorithm. For this work the parameter that is dynamically changed is F (mutation) [25]. The diagram of Fig. 1 represents the way in which a fuzzy system is integrated to be able to dynamically move the parameter F (mutation), during the execution of the algorithm.

Fig. 1 DE algorithm with a dynamic F parameter

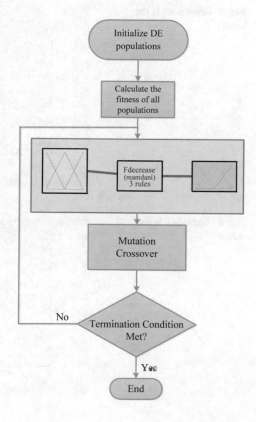

Fig. 2 Fuzzy system with F
in decrease

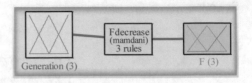

The way in which the fuzzy system is defined is presented below:

- The fuzzy system is of Mamdani type.
- Generation is the input and variable F is the output.

Figure 2 presents the structure of the fuzzy system.

The linguistic variables of the fuzzy system are granulated into three membership functions, which are called *Low, Medium and High.*

Figures 3 and 4 show the membership functions for the input and the output linguistic variables.

The rules of the fuzzy system to dynamically modify the F parameter (mutation) are defined in decreasing form and Fig. 5 presents the if-then rules.

Fig. 3 Generation is the
input

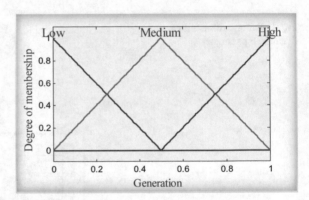

Fig. 4 Variable F is the
output

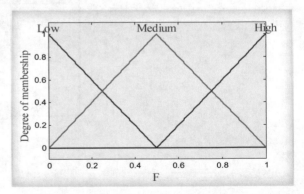

> 1.- If (Generations is Low) the (F is High)
> 2.- If (Generations is Medium) them (F is Medium)
> 3.- If (Generations is High) then (F is Low)

Fig. 5 If-then fuzzy rules

4 CEC 2015 Benchmark Functions

Below is a brief description of the set of CEC 2015 Benchmark Functions to perform the experiments and thereby be able to determine the efficiency of the proposed method [26, 27].

Below a brief description of the set of CEC 2015 Benchmark Functions is presented. Table 1 shows the set of 15 CEC 2015 Benchmark Functions used in this paper [16].

Table 2 presents the mathematical definitions of the Basic Functions, which are used to compose the whole set of CEC 2015 Benchmark Functions [28].

Table 1 Summary of the CEC'15 learning-based benchmark suite

	No.	Functions	$F_i^* = F_i(x^*)$
Unimodal functions	1	Rotated high conditioned elliptic function	100
	2	Rotated Cigar function	200
Simple multimodal functions	3	Shifted and rotated Ackley's function	300
	4	Shifted and rotated Rastrigin's function	400
	5	Shifted and rotated Schwefel's function	500

Table 2 Definitions of the basic functions

No.	Functions	Equations
1	High conditioned elliptic function	$f_1(x) = \sum_{i=1}^{D} (10^6)^{\frac{i-1}{D-1}} x_i^2$
2	Cigar function	$f_2(x) = x_1^2 + 10^6 \sum_{i=2}^{D} x_i^2$
3	Discus function	$f_3(x) = 10^6 x_1^2 + \sum_{i=2}^{D} x_i^2$
4	Rosenbrock's function	$f_4(x) = \sum_{i=1}^{D-1} \left(100 (x_i^2 - x_{i+1})^2 + (x_i - 1)^2 \right)$
5	Ackley's function	$f_5(x) = -20 \exp\left(-0.2 \sqrt{\frac{1}{n} \sum_{i=1}^{D} x_i^2} \right) - \exp\left(\frac{1}{D} \sum_{i=1}^{D} \cos(2\pi x_i) \right) + 20 + e$

5 Experiments

The way in which the experiments for the whole CEC 2015 Benchmark Functions were performed is by following the rules of the competition, which are explained below.

All test functions are for minimization problems defined as follows:

$$Min \ f(x), x = [x_1, x_2, \ldots, x_D]^T \tag{11}$$

The CEC 2015 Benchmark Functions are comprised of 15 minimization problems and the experiments for all functions are performed with different dimensions and for this competition are of D = 10, 30, 50, 100. The experiments were performed following the instructions indicated in the document associated to the competition [14]. The main points are:

For each of the dimensions, 51 runs are executed and we will use for the study a dimension $D = 10$ and the algorithms perform 100,000 evaluations of the fitness function.

Based on these guidelines the experiments were performed first with the original DE and subsequently with the proposed FDE to perform a comparison of results. Table 3 presents the parameters of the original algorithm and Table 4 shows the parameters using the proposed FDE where the F parameter is dynamic.

Table 3 Parameters of the original algorithm

Parameters
D = 10, 30, 50 and 100
$X_{min} = -100$
$X_{max} = 100$
CR = 0.5
F = 0.6
NP = 100
runs = 51

Table 4 Parameters of the proposed method

Parameters
D = 10, 30, 50 and 100
$X_{min} = -100$
$X_{max} = 100$
CR = 0.5
F = dynamic
NP = 100
runs = 51

Tables 5 and 6 show the results obtained with the Differential Evolution algorithm, the tables show the best, worst, median, mean and standard deviations of values for 51 runs.

For these experiments the parameters are constant during the execution of the algorithm.

Tables 7 and 8 show the results obtained with the FDE, the tables show the best, worst, median, mean and standard deviation of the simulation values for 51 runs.

For these experiments the F parameter (mutation) is dynamic during the execution of the algorithm.

Table 5 Results with the DE for $D = 10$

Experiments with the differential evolution algorithm					
$D = 10$					
	Best	Worst	Median	Mean	Std
f1	3.06E+07	7.17E+08	2.97E+08	2.97E+08	2.97E+08
f2	6.52E+09	2.94E+10	1.63E+10	1.69E+10	1.69E+10
f3	3.21E+02	3.21E+02	3.21E+02	3.21E+02	3.21E+02
f4	4.80E+02	5.30E+02	5.30E+02	5.28E+02	5.28E+02
f5	2.42E+03	2.97E+03	2.93E+03	2.97E+03	2.97E+03

Table 6 Results with the DE for $D = 30$

Experiments with the differential evolution algorithm					
$D = 30$					
	Best	Worst	Median	Mean	Std
f1	1.62E+09	6.45E+09	4.17E+09	4.14E+09	1.09E+09
f2	7.25E+10	1.92E+11	1.44E+11	1.43E+11	2.92E+10
f3	3.21E+02	3.21E+03	3.21E+02	3.21E+02	3.21E+02
f4	9.31E+02	1.13E+03	1.06E+03	1.05E+02	4.21E+01
f5	8.62E+03	1.09E+04	1.00E+04	9.99E+03	4.54E+02

Table 7 Results with the FDE for $D = 10$

Experiments with the fuzzy differential evolution algorithm					
$D = 10$					
	Best	Worst	Median	Mean	Std
f1	8.40E−03	6.35E−01	2.53E−02	4.38E−02	9.39E−02
f2	1.06E+00	6.35E−01	1.63E+00	1.80E+00	1.06E+00
f3	0.00E+00	2.10E+01	0.00E+00	6.59E+00	9.84E+00
f4	0.00E+00	1.63E+02	0.00E+00	4.00E+01	6.04E+01
f5	0.00E+00	2.86E+03	0.00E+00	7.62E+02	1.15E+03

Table 8 Results with the FDE for D = 30

Experiments with the fuzzy differential evolution algorithm

D = 30

	Best	Worst	Median	Mean	Std
f1	1.84E−02	3.23E−01	4.21E−02	4.77E−02	4.09E−02
f2	1.04E+00	7.47E+00	1.51E+00	1.63E+00	8.60E−01
f3	2.11E+01	2.15E+01	2.13E+01	2.13E+01	8.26E−02
f4	5.37E+02	7.79E+02	6.55E+02	6.58E+02	5.06E+01
f5	8.06E+03	1.00E+04	9.45E+03	9.33E+03	4.59E+02

Table 9 Comparison between the DE and FDE

Comparison between the DE and FDE

	D = 10		D = 30	
	Best (DE)	Best (FDE)	Best (DE)	Best (FDE)
f1	3.06E+07	8.40E−03	1.62E+09	1.84E−02
f2	6.52E+09	1.06E+00	7.25E+10	1.04E+00
f3	3.21E+02	0.00E+00	3.21E+02	2.11E+01
f4	4.80E+02	0.00E+00	9.31E+02	5.37E+02
f5	2.42E+03	0.00E+00	8.62E+03	8.06E+03

Table 9 represents a comparison between the original algorithm and the proposed method, the best results obtained by each of the mathematical functions are compared.

Analyzing the comparative table of the best result, the proposed method is better for all functions for dimensions of 10 and 30.

6 Conclusions

In conclusion we can say that the proposed FDE algorithm achieved good results in the CEC 2015 Benchmark Functions in comparison with the original DE algorithm.

With these results we affirm that the use of fuzzy logic to dynamically change parameters in an algorithm helps to improve solutions. The set of functions used are relatively simple compared to the rest used in the competition of the CEC 2015, but of greater complexity than the traditional functions used in previous works.

In the near future, the experimentation with all the functions of CEC 2015 will be carried out in order to perform the statistical test, and thus be able to affirm the improvement of the proposed method. In addition, we could use type-2 fuzzy logic like in [29–32] or [33–36]. Also, other applications could be considered like described in [37, 38].

References

1. F. Olivas, F. Valdez, O. Castillo, P. Melin, Dynamic parameter adaptation in particle swarm optimization using interval type-2 fuzzy logic. Soft. Comput. **20**(3), 1057–1070 (2016)
2. C. Peraza, F. Valdez, O. Castillo, A harmony search algorithm comparison with genetic algorithms, in *Fuzzy Logic Augmentation of Nature-Inspired Optimization Metaheuristics* (Springer International Publishing, Berlin, 2015), pp. 105–123
3. C. Peraza, F. Valdez, O. Castillo, Fuzzy control of parameters to dynamically adapt the HS algorithm for optimization, in *2015 Annual Conference of the North American Fuzzy Information Processing Society (NAFIPS) Held Jointly with 2015 5th World Conference on Soft Computing (WConSC)* (IEEE, New York, August 2015), pp. 1–6
4. R. Storn, On the usage of differential evolution for function optimization, in *Fuzzy Information Processing Society, 1996. NAFIPS, 1996 Biennial Conference of the North American* (IEEE, New York, June 1996), pp. 519–523
5. R. Storn, K. Price, *Differential Evolution-A Simple and Efficient Adaptive Scheme for Global Optimization Over Continuous Spaces*, vol. 3 (ICSI, Berkeley, 1995)
6. F. Valdez, P. Melin, O. Castillo, Evolutionary method combining particle swarm optimisation and genetic algorithms using fuzzy logic for parameter adaptation and aggregation: the case neural network optimisation for face recognition. Int. J. Artif. Intel. Soft Comput. **2**(1–2), 77–102 (2010)
7. F. Valdez, P. Melin, O. Castillo, An improved evolutionary method with fuzzy logic for combining particle swarm optimization and genetic algorithms. Appl. Soft Comput. **11**(2), 2625–2632 (2011)
8. O. Castillo, H. Neyoy, J. Soria, P. Melin, F. Valdez, A new approach for dynamic fuzzy logic parameter tuning in Ant Colony Optimization and its application in fuzzy control of a mobile robot. Appl. Soft Comput. **28**, 150–159 (2015)
9. O. Castillo, P. Melin, Intelligent adaptive model-based control of robotic dynamic systems with a hybrid fuzzy-neural approach. Appl. Soft Comput. **3**(4), 363–378 (2003)
10. R. Martínez-Soto, O. Castillo, L.T. Aguilar, Type-1 and Type-2 fuzzy logic controller design using a Hybrid PSO–GA optimization method. Inf. Sci. **285**, 35–49 (2014)
11. P. Melin, O. Castillo, A review on type-2 fuzzy logic applications in clustering, classification and pattern recognition. Appl. Soft Comput. **21**, 568–577 (2014)
12. P. Ochoa, O. Castillo, J. Soria, A fuzzy differential evolution method with dynamic adaptation of parameters for the optimization of fuzzy controllers, in *2014 IEEE Conference on Norbert Wiener in the 21st Century (21CW)* (IEEE, New York, June 2014), pp. 1–6
13. A. Al-Dujaili, K. Subramanian, S. Suresh, HumanCog: a cognitive architecture for solving optimization problems, in *2015 IEEE Congress on Evolutionary Computation (CEC)* (IEEE, New York, May 2015), pp. 3220–3227
14. N. Awad, M.Z. Ali, R.G. Reynolds, A differential evolution algorithm with success-based parameter adaptation for CEC2015 learning-based optimization, in *2015 IEEE Congress on Evolutionary Computation (CEC)* (IEEE, New York, May 2015), pp. 1098–1105
15. D. Aydın, T. Sfffitzle, A configurable generalized artificial bee colony algorithm with local search strategies, in *2015 IEEE Congress on Evolutionary Computation (CEC)* (IEEE, New York, May 2015), pp. 1067–1074
16. Q. Chen, B. Liu, Q. Zhang, J.J. Liang, P.N. Suganthan, B.Y. Qu, Problem definition and evaluation criteria for CEC 2015 special session and competition on bound constrained single-objective computationally expensive numerical optimization. Computational Intelligence Laboratory, Zhengzhou University, China and Nanyang Technological University, Singapore, Technical report (2014)

17. S.M. Guo, J.S.H. Tsai, C.C. Yang, P.H. Hsu, A self-optimization approach for L-SHADE incorporated with eigenvector-based crossover and successful-parent-selecting framework on CEC 2015 benchmark set, in *2015 IEEE Congress on Evolutionary Computation (CEC)* (IEEE, New York, May 2015), pp. 1003–1010

18. R. Poláková, J. Tvrdík, P. Bujok, Cooperation of optimization algorithms: a simple hierarchical model, in *2015 IEEE Congress on Evolutionary Computation (CEC)* (IEEE, New York, May 2015), pp. 1046–1052

19. J.L. Rueda, I. Erlich, Testing MVMO on learning-based real-parameter single objective benchmark optimization problems, in *2015 IEEE Congress on Evolutionary Computation (CEC)* (IEEE, New York, May 2015), pp. 1025–1032

20. K.M. Sallam, R.A. Sarker, D.L. Essam, S.M. Elsayed, Neurodynamic differential evolution algorithm and solving CEC2015 competition problems, in *2015 IEEE Congress on Evolutionary Computation (CEC)* (IEEE, New York, May 2015), pp. 1033–1040

21. C. Yu, L.C. Kelley, Y. Tan, Dynamic search fireworks algorithm with covariance mutation for solving the CEC 2015 learning based competition problems, in *2015 IEEE Congress on Evolutionary Computation (CEC)* (IEEE, New York, May 2015), pp. 1106–1112

22. Y.J. Zheng, X.B. Wu, Tuning maturity model of ecogeography-based optimization on CEC 2015 single-objective optimization test problems, in *2015 IEEE Congress on Evolutionary Computation (CEC)* (IEEE, New York, May 2015), pp. 1018–1024

23. K.V. Price, R.M. Storn, J.A. Lampinen, *The differential evolution algorithm. Differential Evolution: A Practical Approach to Global Optimization* (2005), pp. 37–134

24. K. Price, R.M. Storn, J.A. Lampinen, *Differential Evolution: A Practical Approach to Global Optimization* (Springer Science & Business Media, Berlin, 2006)

25. P. Ochoa, O. Castillo, J. Soria, Differential evolution with dynamic adaptation of parameters for the optimization of fuzzy controllers, in *Recent Advances on Hybrid Approaches for Designing Intelligent Systems* (Springer International Publishing, Berlin, 2014), pp. 275–288

26. X. Li, Decomposition and cooperative coevolution techniques for large scale global optimization, in *Proceedings of the Companion Publication of the 2014 Annual Conference on Genetic and Evolutionary Computation* (ACM, New York, July 2014), pp. 819–838

27. R. Tanabe, A. Fukunaga, Success-history based parameter adaptation for differential evolution, in *2013 IEEE Congress on Evolutionary Computation* (IEEE, New York, June 2013), pp. 71–78

28. L. Chen, C. Peng, H.L. Liu, S. Xie, An improved covariance matrix leaning and searching preference algorithm for solving CEC 2015 benchmark problems, in *2015 IEEE Congress on Evolutionary Computation (CEC)* (IEEE, New York, May 2015), pp. 1041–1045

29. C. Leal Ramírez, O. Castillo, P. Melin, A. Rodríguez Díaz, Simulation of the bird age-structured population growth based on an interval type-2 fuzzy cellular structure. Inf. Sci. **181**(3), 519–535 (2011)

30. N.R. Cázarez-Castro, L.T. Aguilar, O. Castillo, Designing Type-1 and Type-2 fuzzy logic controllers via fuzzy Lyapunov synthesis for nonsmooth mechanical systems. Eng. Appl. Artif. Intell. **25**(5), 971–979 (2012)

31. O. Castillo, P. Melin, Intelligent systems with interval type-2 fuzzy logic. Int. J. Innovative Comput. Inf. Control **4**(4), 771–783 (2008)

32. G.M. Mendez, O. Castillo, Interval type-2 TSK fuzzy logic systems using hybrid learning algorithm, in *The 14th IEEE International Conference on Fuzzy Systems, 2005. FUZZ'05*, pp. 230–235

33. P. Melin, C.I. González, J.R. Castro, O. Mendoza, O. Castillo, Edge-detection method for image processing based on generalized type-2 fuzzy logic. IEEE Trans. Fuzzy Syst. **22**(6), 1515–1525 (2014)

34. C.I. González, P. Melin, J.R. Castro, O. Castillo, O. Mendoza, Optimization of interval type-2 fuzzy systems for image edge detection. Appl. Soft Comput. **47**, 631–643 (2016)

35. C.I. González, P. Melin, J.R. Castro, O. Mendoza, O. Castillo, An improved sobel edge detection method based on generalized type-2 fuzzy logic. Soft. Comput. **20**(2), 773–784 (2016)

36. E. Ontiveros, P. Melin, O. Castillo, High order α-planes integration: a new approach to computational cost reduction of general type-2 fuzzy systems. Eng. Appl. Artif. Intell. **74**, 186–197 (2018)
37. P. Melin, O. Castillo, Intelligent control of complex electrochemical systems with a neuro-fuzzy-genetic approach. IEEE Trans. Ind. Electron. **48**(5), 951–955 (2001)
38. P. Melin, O. Castillo, *Modelling, Simulation and Control of Non-linear Dynamical Systems: An Intelligent Approach Using Soft Computing and Fractal Theory* (CRC Press, Boca Raton, 2001)

A Comparison of ACO, GA and SA for Solving the TSP Problem

Fevrier Valdez, Francisco Moreno and Patricia Melin

Abstract The ACO algorithm is an optimization algorithm, recognized for being very efficient in problems of finding routes and planning paths in roads. In terms of the problem of the traveling salesman, ACO algorithm has been able to find optimal solutions to the problem, we want to make a comparison with the algorithms GA and SA, to determine which of these obtains better results.

Keywords TSP (Travelling Salesman Problem) · ACO (Ant Colony Optimization) · Bio-inspired algorithms · GA (Genetic Algorithm) · SA (Simulated Annealing)

1 Introduction

The ACO algorithm was presented by Marco Dorigo in [1, 2] as a new metaheuristic algorithm, which use or objective is to find the shortest route or path between an initial point and a final point based on ant's behavior.

The most important parameter of this metaheuristic is the pheromones that the ant deposits on the way to the destination, that will reinforce the shortest path so the other ant's can follow it or find in the best way the food. However, it depends on other parameters as α, β and the max number of iterations, the number of ants, evaporation rate and other ones that will be explained later on Sect. 2.1.

This paper is organized as follows: Sect. 2 provides a literature review of the ACO algorithm, GA, SA algorithm and TSP problem. Section 3 presents results and

F. Valdez (✉) · F. Moreno · P. Melin
Tijuana Institute of Technology, Calzada Tecnologico S/N, Fracc. Tomás Aquino,
22414 Tijuana, BC, Mexico
e-mail: fevrier@tectijuana.mx

F. Moreno
e-mail: franciscojaviermorenofavela@gmail.com

P. Melin
e-mail: pmelin@tectijuana.mx

discussions of the comparison of all presented algorithms solving the TSP. Section 4 gives the conclusions of the experiments made in this research.

2 Literature Review

This section presents the basic concepts about ACO algorithms and the state of the art with other optimization algorithms as GA, SA are included.

2.1 Ant Colony Optimization

One of the first behaviors studied by entomologists was the ability of ants to find the shortest path between their nest and a food source. From these studies and observations followed the first algorithmic models of the foraging behavior of ants, as developed by Dorigo [1]. Collectively, algorithms that were developed as a result of studies of ant foraging behavior are referred to as instances of the ant colony optimization metaheuristic (ACO-MH) [2, 3].

Deneubourg et al. [4] studied the foraging behavior of the Argentine ant species Iridomyrmex humilis in order to develop a formal model to describe its behavior. In this laboratory experiment, as illustrated in Fig. 1, the nest is separated from the food source by a bridge with two equally long branches. Initially, both branches were free of any pheromones. The selection of one of the two paths is defined by random fluctuations in path selection, causing higher amounts of pheromones on one of the two paths.

From this experiment, referred to as the binary bridge experiment (and illustrated in Fig. 1), a simple formal model was developed to characterize the path selection

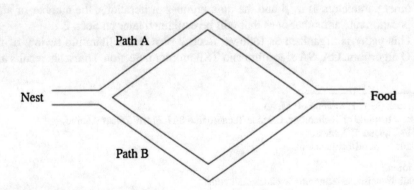

Fig. 1 Binary bridge experiment

process [5]. Let $n_A(t)$ and $n_B(t)$ denote the number of ants on paths A and B respectively at time step t. Pasteels et al. [5] found empirically that the probability of the next ant to choose path A at time step $t + 1$ is given as,

$$P_A(t + 1) = \frac{(c + n_A(t))^\alpha}{(c + n_A(t))^\alpha + (c + n_B)^\alpha} = 1 - P_B(t + 1) \tag{1}$$

where c quantifies the degree of attraction of an unexplored branch, and α is the bias to using pheromone deposits in the decision process. The larger the value of α, the higher the probability that the next ant will follow the path with a higher pheromone concentration even if that branch has only slightly more pheromone deposits. The larger the value of c, the more pheromone deposits are required to make the choice of path non-random. It was found empirically that $\alpha \approx 2$ and $c \approx 20$ provide the best fit to the experimentally observed behavior.

Dorigo and Gambardella described a qualified simulated ant colony for solving the traveling salesman problem (TSP). Ants of the simulated colony are able to generate one after another shorter feasible trip by using information gathered in the form of a pheromone trail dropped on the edges of the TSP graph. Experimental results prove that the artificial ant colony is able of producing good solutions to TSP. Also, other optimization methods as simulated annealing, neural networks and evolutionary computation, of the successful use of a natural metaphor to design an optimization algorithm [6].

2.2 Genetic Algorithm

Holland, from the University of Michigan initiated his work on genetic algorithms at the beginning of the 1960s. His first achievement was the publication of *Adaptation in Natural and Artificial System* [7] in 1975. He had two goals in mind: to improve the understanding of natural adaptation process, and to design artificial systems having properties similar to natural systems.

The traveling salesman problem (TSP) is implemented as an idea model for a wide-range class of problems. The TSP has become a target for the genetic algorithm (GA) researchers, because it is probably the significant problem in combinatorial optimization and many new ideas in combinatorial optimization have been tested on the TSP.

However, by using GA for solving TSPs, Tsujimura et al. [8] obtained a local optimal solution rather than a best estimated solution frequently. Thus, they obtained the best approximate solution of the TSP by using entropy-based GA [9].

2.3 Simulated Annealing Algorithm

SA algorithm is commonly said to be the oldest among the metaheuristics and surely one of the few algorithms that have explicit strategies to avoid local minima. The origins of SA are in statistical mechanics and it was first presented for combinatorial optimization problems. The main idea is to accept moves resulting in solutions of worse quality than the current solution in order to escape from local minima. The probability of accepting such a move is decreased during the search through parameter temperature. SA algorithm starts with an initial solution, and candidate solution is then generated (either randomly or using some specified rule) from the neighborhood of [10]. The Metropolis acceptance criterion [11] which models how a thermodynamic system moves from one state to another state in which the energy is being minimized, is used to decide whether to accept or not [12].

The SA algorithm can be described by Fig. 2.

2.4 Traveling Salesman Problem

The traveling salesman problem (TSP) is the first application to which an ACO algorithm was applied [1, 13]. It is an NP-hard combinatorial optimization problem [14], and is the most frequently used problem in ACO literature. This section shows how the ACO algorithm can be useful to solve problem types focused on permutations, particularly in the TSP.

The TSP is defined as to visit "n" cities, starting and ending with the same city, visiting each city once and making the tour with the lowest cost. The goal in the TSP is find a minimum length Hamiltonian circuit of the graph, where a Hamiltonian circuit is a closed path visiting each of n = |N| nodes of G exactly once [3]. Thus, an optimal solution to the TSP is a permutation π of the node indices 1, 2, ..., n such that the length $f(\pi)$ is minimal, where $f(\pi)$ is given by:

$$f(\pi) = \sum_{i=1}^{n-1} d_{\pi(i)\pi(i+1)} + d_{\pi(n)\pi(1)}$$

Two versions of the TSP are defined based on the characteristics of the distance matrix. If $d_{ij} = d_{ji}$ for all i, j = 1, ..., n π, then the problem is referred to as the symmetric TSP (STSP). If $d_{ij} = d_{ji}$, the distance matrix is asymmetric resulting in the Asymmetric TSP (ATSP) [15–17].

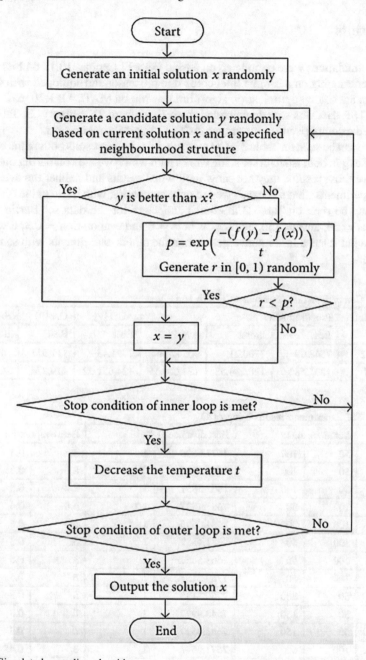

Fig. 2 Simulated annealing algorithm

3   Results

All the simulations were completed on a Mac OS x El Capitan 10.11 64 bits laptop computer, running on a 2.4 ghz Intel core2 duo processor and equipped with 4 gb of ram. The ant colony optimization algorithm was run on MATLAB R2016a.

The TSP data sets was downloaded from Heidelberg TSP library. In Table 1 is presented a comparison results among ACO, GA and SA algorithms.

As we can be seen on Table 3 ACO algorithm has better results both in the column best and avg in both tsp data set's, the works with which we are comparing ourselves only gave those results, they not show their worst results and neither the average of their experiments, that's because we are only comparing with those values.

As can be seen on Table 2 the best parameters for the data set Berlin 52 are iterations = 50, ants = 150, alpha = 1, beta = 6 and evaporation = 0.3, it was not included but it has a faster convergence than the others experiments with same best distance.

Table 1 Comparison results among ACO, GA and SA

TSP dataset	Proposed ACO			ACO [9]	GA [9]	SA [9]
	Best	Worst	Avg	Best	Best	Best
Berlin52	7548.99	7760.21	7654.5304	7721.43	11,551	10,586
Bier127	123,258.56	126,234.52	124,669.55	124,651.52	419,224	265,289

Table 2 TSP experiments results using ACO

No.	Iterations	ants	Best distance	alpha	Beta	evaporation
1	50	100	7677.6608	1	3	0.5
2	50	100	7663.8814	0.4	8	0.35
3	50	100	7663.5851	1	3	0.5
4	100	50	7681.4537	1.7	6.8	0.4
5	100	50	7684.0693	1	6	0.6
6	100	50	7681.4537	1	5	0.3
7	100	50	7663.5851	1	3	0.5
8	100	50	7760.2135	1	5	0.5
9	50	200	7760.2135	1	5	0.3
10	50	150	7548.9927	1	6	0.3
11	50	150	7681.4537	1	6	0.3
12	100	50	7571.9548	1	3	0.45
13	100	100	7548.9927	1	3	0.65
14	100	50	7548.9927	1	3	0.35
15	100	150	7681.4537	1	6	0.3
	Best	7548.99	Worst	7760.21	Avg.	7654.53

Table 3 TSP experiments results using ACO

No.	Iterations	ants	Best distance	alpha	Beta	evaporation
1	50	100	126,143.7473	1	5	0.3
2	50	100	125,963.0336	1	3	0.4
3	50	100	123,258.5556	1	5	0.4
4	50	100	126,234.5191	1	5	0.4
5	50	100	125,672.2016	1	7	0.4
6	50	100	123,654.7132	1	7	0.6
7	50	100	124,513.243	1	7	0.6
8	50	100	123,532.3992	1	9	0.7
9	50	100	124,233.0518	1	9	0.7
10	50	100	125,286.9117	1.5	10	0.6
11	50	100	125,121.3956	1	12	0.8
12	50	100	124,749.9212	1	13	0.7
13	50	100	125,162.4449	1	14	0.65
14	50	100	123,258.5556	1	5	0.4
15	50	100	123,258.5556	1	5	0.4
	Best	123,258.56	Worst	126,234.52	Avg.	124,669.55

We can see that ACO algorithm has a better performance than GA and SA in this particular kind of problem and it's the point we want to reach, so we compare all these algorithms to probe that.

As can be seen on Table 3 the best parameters for the data set Bier127 are iterations = 50, ants = 100, alpha = 1, beta = 5 and evaporation = 0.4, it was not included but it has a faster convergence than the others experiments with same best distance.

4 Conclusions

Given the previous results with the different algorithms we can see that ACO have a better performance against GA and SA algorithms solving this particular problem that is the TSP [3, 18–22], it found better solutions in a short period of time, now days it is necessary not only to complete a task in a good way but also in the minimum possible time, this can be applied on real time applications such as path planning where the response time is going to be an important fact to be considered, for more difficult problems the parameters of the ACO algorithm can be optimized by a genetic algorithm or a bio-inspired algorithm, to obtain optimum results in bigger complexity problems. In addition, we could consider applying the ideas presented in the paper in other areas of application like in [23 29]. In this paper we can conclude that the ACO algorithm is a good alternative solving optimization problems of routes, in this case, the algorithm was tested with the TSP problem achieving the best results that other optimization methods analyzed in this research work.

Acknowledgements The authors would like to express thank to the Consejo Nacional de Ciencia y Tecnología and Tecnológico Nacional de Mexico/Tijuana Institute of Technology for the facilities and resources granted for the development of this research.

References

1. M. Dorigo, *Optimization, Learning and Natural Algorithms*. (Ph.D. Thesis, Politecnico di Milano, Italian, 1992)
2. M. Dorigo, G.D. Caro, Ant colony optimization: a new meta-heuristic, in *Proceedings of the IEEE Congress on Evolutionary Computation*, vol. 2 (1999), pp. 1470–1477
3. M. Dorigo, L.M. Gambardella, Ant colony system: a cooperative learning approach to the traveling salesman problem. IEEE Trans. Evol. Comput. **1**(1), 53–66 (1997)
4. J.L. Deneubourg, S. Aron, S. Goss, J.M. Pasteels, The self-organizing exploratory pattern of the argentine ant. J. Insect Behav. **3**, 159–168 (1990)
5. J.M. Pasteels, J.L. Deneubourg, S. Goss, Self-organization mechanisms in ant societies (I): trail recruitment to newly discovered food sources. Experientia Suppl **76**, 579–581 (1989)
6. M. Dorigo, L.M. Gambardella, Ant colonies for the travelling salesman problem. Biosystems **43**(2), 73–81 (1997)
7. J.H. Holland, *Adaptation in Natural and Artificial Systems: An Introductory Analysis with Applications to Biology, Control, and Artificial Intelligence* (University of Michigan Press, Ann Arbor, MI, 1975)
8. Y. Tsujimura, M. Gen, Entropy-based genetic algorithm for solving TSP, in *1998 Second International Conference on Knowledge Based Intelligent Electronic Systems. Proceedings KES 98* (1998)
9. H.A. Mukhairez, A.Y.A. Maghari, Performance comparison of simulated annealing, GA and ACO applied TSP. Int. J. Intell. Comput. Res. (IJICR) **6**(4) (2015)
10. J.S.H. Zhan, Z.J. Lin, Y.W. Zhang, Zhong: List-based simulated annealing algorithm for traveling salesman problem. Comput. Intell. Neurosci. **2016**, Article ID 1712630, 12 p (2016)
11. N. Metropolis, A.W. Rosenbluth, M.N. Rosenbluth, A.H. Teller, E. Teller, Equation of state calculations by fast computing machines. J. Chem. Phys. **21**(6), 1087–1092 (1953)
12. L. Bo, M. Peisheng, Simulated annealing-based ant colony algorithm for traveling salesman problems. Nat. Sci. **11**, 26–30 (2009)
13. M. Dorigo, V. Maniezzo, A. Colorni, Ant system: optimization by a colony of cooperating agents. IEEE Trans. Syst. Man Cybern. Part B **26**(1), 29–41 (1996)
14. M.R. Garey, D.S. Johnson, *Computers and Intractability: A Guide to the Theory of NP-Completeness* (W.H. Freeman, San Francisco, 1979)
15. E.H.L. Aarts, J.K. Lenstra, The travelling salesman problem: a case study in local optimization, in *Local Search in Combinatorial Optimization* (1997)
16. R. Johnson, M.G. Pilcher, in *The Traveling Salesman Problem*, ed. by E.L. Lawler, J.K. Lenstra, A.H.G. Rinnooy Kan, D.B Shmoys, John Wiley (1988)
17. D.J. Rosenkrantz, R.E. Stearns, P.M. Lewis, An analysis of several heuristics for the traveling salesman problem. SIAM J. Comput. **6**, 563–581 (1977)
18. A. Acan, GAACO: A GA + ACO hybrid for faster and better search capability, in *Ant Algorithms* (2002), pp. 300–301
19. A. Colorni, M. Dorigo, V. Maniezzo, An investigation of some properties of an ant algorithm, in *Proceedings of Parallel Problem Solving from Nature Conference (PPSN 92)* (1992), pp. 509–520

20. B. Freisleben, P. Merz, New genetic local search operators for the traveling salesman problem, in *Proceedings of PPSN IVth International Conference on Parallel Problem Solving from Nature* (1996), pp. 890–899
21. P. Stodola, J. Mazal, M. Podhorec, Parameter tuning for the ant colony optimization algorithm used in ISR systems. Int. J. Appl. Math. Inform. **9** (2015)
22. T. Stutzle, M. Lopez, P. Pellegrini, M. Maur, M.M.D. Oca, M. Birattari, M. Dorigo, Parameter adaptation in ant colony optimization, Technical Report Series (2010)
23. B. Gonzalez, F. Valdez, P. Melin, A gravitational search algorithm using type-2 fuzzy logic for parameter adaptation, in *Nature-Inspired Design of Hybrid Intelligent Systems*, vol. 667 (Springer, Cham, 2017)
24. C.I. Gonzalez, P. Melin, J.R. Castro, O. Mendoza, O. Castillo, An improved sobel edge detection method based on generalized type-2 fuzzy logic. Soft. Comput. **20**(2), 773–784 (2016)
25. C.I. Gonzalez, P. Melin, J.R. Castro, O. Castillo, O. Mendoza, Optimization of interval type-2 fuzzy systems for image edge detection. Appl. Soft Comput. **47**, 631–643 (2016)
26. P. Melin, D. Sanchez, Multi-objective optimization for modular granular neural networks applied to pattern recognition. Inf. Sci. **460**, 594–610 (2018)
27. P. Ochoa, O. Castillo, J. Soria, Differential evolution using fuzzy logic and a comparative study with other metaheuristics, in *Nature-Inspired Design of Hybrid Intelligent Systems*, vol. 667 (Springer, Cham, 2017)
28. F. Olivas, F. Valdez, O. Castillo, C.I. González, G.E. Martinez, P. Melin, Ant colony optimization with dynamic parameter adaptation based on interval type-2 fuzzy logic systems. Appl. Soft Comput. **53**, 74–87 (2017). https://doi.org/10.1016/j.asoc.2016.12.015
29. D. Sanchez, P. Melin, O. Castillo, Optimization of modular granular neural networks using a firefly algorithm for human recognition. Eng. Appl. AI **64**, 172–186 (2017)

Medical Applications

Hybrid Model Based on Neural Networks and Fuzzy Logic for 2-Lead Cardiac Arrhythmia Classification

Eduardo Ramírez, Patricia Melin and German Prado-Arechiga

Abstract We present how to combine neural networks and fuzzy logic to form a hybrid model as a classification system using 2-lead for cardiac arrhythmias. This hybrid model is used samples of electrocardiograms contained in the MIT-BIH arrhythmia database. The samples of heartbeats are extracted and transformed from the electrocardiograms of this database. The hybrid model is trained and tested with 10 different classes of normal and cardiac arrhythmias heartbeats. The hybrid model used 2 leads included in the MIT-BIH arrhythmia database. The hybrid model used two basic module units, where each unit processing one lead. The basic module unit are composite by three classifiers. Finally, we combined the output results of the two basic module unit with a fuzzy system and we have achievement increase the global classification rate in the hybrid model proposed.

Keywords Fuzzy KNN algorithm · Neural network · Fuzzy system · 2-lead arrhythmia classification · Hybrid model

1 Introduction

Different solutions have been applied to resolve the interpretation of the different morphologies or electrocardiographic patterns that represent different diseases that are diagnosed by interpreting the electrocardiograms. Such as, Bayesian approach, linear discriminant systems, statistical and syntactic pattern recognition, self organizing maps (SOM), Markov models, learning vector quantization (LVQ) [1, 2], support vector machines (SVM) [3, 4], higher order statistics [5, 6], multilayer perceptron (MLP) [5, 7–14], expert systems [9, 15], type-1 and type-2 fuzzy systems [8, 16–22],

E. Ramírez · P. Melin (✉) · G. Prado-Arechiga
Tijuana Institute of Technology, Graduate Studies, Tijuana, BC, Mexico
e-mail: pmelin@tectijuana.mx

E. Ramírez
e-mail: eduarafl@hotmail.com

© Springer Nature Switzerland AG 2020
O. Castillo and P. Melin (eds.), *Hybrid Intelligent Systems in Control,
Pattern Recognition and Medicine*, Studies in Computational Intelligence 827,
https://doi.org/10.1007/978-3-030-34135-0_14

Fig. 1 Normal heartbeat
with intervals, segments and
fidutial points

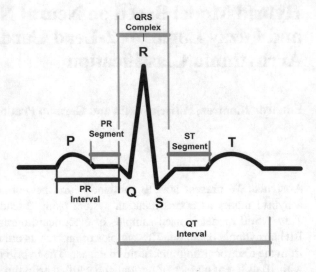

as well as hybrid systems [8, 9, 16, 23–30]. Some solutions are focus in the preprocessing process, for example: Discrete Wavelet Transform, LDA, PCA, ICA, and,
linear and nonlinear features [31–34].

The Electrocardiogram (ECG) provides a composite snapshot of the electrical
activity of the heart from several different projections. These projections focus on the
different regions of the heart. The electrical activity of the heart may give important
information about cardiac health and pathology [35].

The ECG is composite by electrical waves as a series. In Fig. 1, we illustrate the
normal heartbeat with intervals, segments and fidutial points.

Sinus rhythm is the only normal sustained rhythm. In young people the R-R
interval is reduced (i.e. the heart rate is increased) during inspiration, and this is
called sinus arrhythmia.

By cardiac arrhythmia can we define the alteration in the activity of the heart
rhythm, respect to the amplitude, duration or shape of the rhythm. An arrhythmia is
an abnormal heart rhythm. It may feel like fluttering or a brief pause [35–37].

The morphology of the QRS wave is governed by the sequence of ventricular
activation and the document vector force associated with each step. Normal ventricular depolarization can be simplified into two steps: depolarization of the septum
followed by depolarization of the ventricular free walls. Because the Purkinje fibers
are located just beneath the endocardium, activation of the ventricular walls spreads
from the endocardium to the epicardium. The left aspect of the septum is the first part
of the ventricles to depolarize. Normal septal depolarization occurs in a left-to-right
direction. This results in the small septal R wave in the right precordial lead V1 and
the small septal Q wave in V6. Depolarization of left and right ventricular free walls
normally occurs simultaneously. The right-to-left depolarization in the larger and
thicker left ventricle comprises the dominating vector force. The ECG interprets this
depolarization as a right-to-left force even though depolarization in the right ventricle

slightly opposes this force. This dominant vector accounts for the large S wave in V1 that transitions in V3/V4 to become the large R wave in the left precordial leads (V6) [35–37].

The P Wave. Atria are typically activated in a right-to-left direction as the electrical impulse spreads from the sinus node in the right atrium to the left atrium. The first half of the P wave therefore represents activation of the right atrium. In normal sinus rhythm, P waves should be upright in the inferior leads (reflecting the superior to inferior direction of the impulse from sinus to AV node). The P wave in V1 is upright or biphasic. *The QRS Wave* represents rapid ventricular depolarization and corresponds to phase 0 of the action potential. The QRS complex is widened by delay in the intra-ventricular conduction system and ventricular hypertrophy. *The T Wave* corresponds to rapid repolarization (phase 3 of the action potential). Repolarization of the epicardium is followed by repolarization of the endocardium. The axis of the T wave should parallel that of the QRS wave when depolarization is normal. *The U Wave* may by absent in the normal electrocardiogram. The source of the U wave is unclear but may represent His/Purkinje repolarization [35–37].

The PR Interval represents the time for an impulse to travel from the atria to the ventricles, including the time it takes to travel through the AV node and bundle of His. PR prolongation most often results from delayed conduction within the AV node. PR shortening classically occurs when an impulse travels form the atrium to the ventricle through an accessory pathway that bypasses the delay in conduction that occurs in the AV node. *The QT interval* represents ventricular depolarization and repolarization, corresponding to phases 0 to 3 of the action potential and ventricular systole. QT prolongation often results form delay in repolarization. *The R-R interval* corresponds to a complete cardiac cycle [35–37].

The rest of this work is organized as follows. In Sect. 2, the hybrid model, proposed method and problem description are presented. Background theory is presented in Sect. 3. In Sect. 4, we are presented the results of the experiments performed. Finally, the conclusions and future works are brief shown in Sect. 5. The last part of this chapter are listed the references and acknowledgements.

2 Problem Statement and Proposed Method

In this work cardiac arrhythmia classification of the Massachusetts Institute of Technology and Beth Israel Hospital (MIT-BIH) arrhythmia database achieved using neural networks and fuzzy logic. The purpose is combine different computational intelligence techniques to form a hybrid model as a classification method for cardiac arrhythmias.

A set of samples of heartbeat from the MIT-BIH database are extracted. The heartbeats are segmented and transformed in preprocessing stage. The samples of heartbeats are normal beat (N), left bundle branch block beat (LBBB), right bundle branch block beat (RDBB), premature ventricular contraction beat (PVC), fusion paced and normal beat (FPN), atrial premature beat (APB), aberrated atrial premature beat (AAPB), fusion of ventricular and normal beat (FVNB), ventricular escape beat

(VEB) and paced beat (PB). In Fig. 2, some examples of the MIT-BIH arrhythmia database are presented.

We extracted 1000 heartbeats, 100 samples of heartbeats per class. The hybrid model learns the 10 classes using samples of each class. We applied 10-fold cross validation to validate the hybrid model.

In this work, we used the MLII signal and (v1, v3 or v5) of MIT-BIH arrhythmia database, see Fig. 3, The records used for the classes were: N 113, 115, and 122, LBBB 109, 111, and 214, RBBB 118, 124, and 212, PVC 106, 200 and 208 FPN 217, APB 209 and 222, AAPB 201, 202, and 210, FVNB 208 and 214, VEB 207, and PB 217.

The hybrid model is composed for two basic module units, where each module represents one electrode signal or lead in the electrocardiogram signal lead being 2-lead cardiac arrhythmia classification. This hybrid model can be extended to 12-lead cardiac arrhythmia classification problem. We used an interval type-1 fuzzy inference system to determine the global classification for the hybrid model using the two basic module units. In Fig. 4, we are presented the architecture of the hybrid model.

In the preprocessing stage, we have taken the samples of the heartbeats of the electrocardiograms of the MIT-BIH arrhythmia database, the segmentation was performed manually and the transformation process consist in rearranging the values of voltage in the heartbeat signal, selected only 70 larger and 70 smaller voltage values, because the size of each heartbeat is different size [8, 25]. This means that each heartbeat is represented for a vector with 140 voltage values. This transformation process simplifies the signal of the heartbeat that will be classified for the hybrid model. The main focus for this work is the hybrid model as well as improving the global classification rate.

A basic module unit is used to resolve the classification for each signal lead. Each module unit is composed for three different classifiers are used based on the models: fuzzy KNN algorithm, multi layer perceptron with gradient descent and momentum (MLP-GDM), and multi layer perceptron with scaled conjugate gradient (MLP-SCG). We combined the output of the three classifiers used a Mamdani type-1 fuzzy system (type-1 FIS), see Fig. 5 to review the basic module unit for cardiac arrhythmia classification.

Every classifier receives the same set of vectors that corresponds to the selected heartbeats. Each classifier produces an output that represents the specific classification of the heartbeat. The outputs of the classifiers form the integration matrix. The integration matrix is the input of the both fuzzy inference systems above mentioned. Finally, the type-1 FIS throw out the final classification of the basic module unit for a ECG's signal electrode.

We combined the outputs of the classifiers in the basic module unit with a type-1 FIS and details of the results for the experiments are presented in the Sect. 3. The specifications for the type-1 fuzzy system are: Mamdani with 30 inputs; 10 inputs belonging for the first classifier fuzzy KNN algorithm, the next 10 inputs

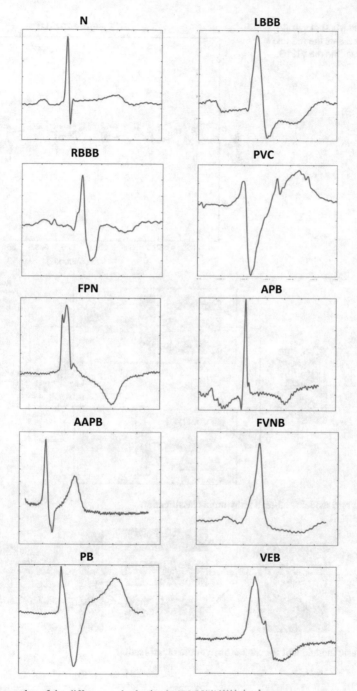

Fig. 2 Examples of the different arrhythmias in the MIT-BIH database

Fig. 3 The MLII electrode signal represents the red line and the blue line the v1, v3, or v5

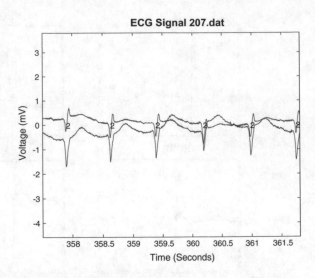

Fig. 4 Hybrid model for 2-lead arrhythmia classification

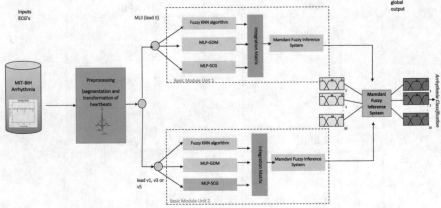

Fig. 5 Basic module unit for cardiac arrhythmia classification

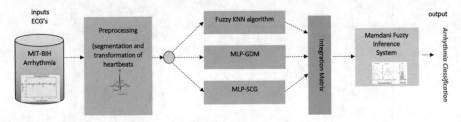

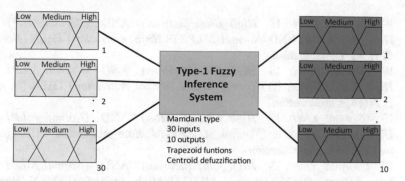

Fig. 6 Mamdani type-1 fuzzy inference system

correspond for the second classifier MLP-GDM, the last 10 inputs corresponding for the third classifier MLP-SCG, the inputs and outputs are trapezoid functions with fuzzy variables *Low, Medium,* and *High*. The defuzzification method is by centroid. We have 260 rules to combined the outputs of the classifiers regarding the membership of the fuzzy KNN algorithm and the activation of the both MLPs, in the next part of this section we present to review some examples of the fuzzy rules. We show the type-1 FIS in Fig. 6.

The parameters for inputs and outputs can be expressed as:

$$\mu_{Low}(x) = \begin{cases} 0, & x \leq -0.0529 \\ \frac{x+0.0529}{0.05033}, & -0.0529 \leq x \leq -0.00257 \\ 1, & -0.00257 \leq x \leq 0.2701 \\ \frac{0.322-x}{0.1149}, & 0.2071 \leq x \leq 0.322 \\ 0, & 0.322 \leq x \end{cases} \tag{1}$$

$$\mu_{Medium}(x) = \begin{cases} 0, & x \leq 0.323 \\ \frac{x-0.323}{0.134}, & 0.323 \leq x \leq 0.457 \\ 1, & 0.457 \leq x \leq 0.552 \\ \frac{0.705-x}{0.153}, & 0.552 \leq x \leq 0.705 \\ 0, & 0.705 \leq x \end{cases} \tag{2}$$

$$\mu_{High}(x) = \begin{cases} 0, & x \leq 0.702 \\ \frac{x-0.702}{0.103}, & 0.702 \leq x \leq 0.805 \\ 1, & 0.805 \leq x \leq 1.004 \\ \frac{0.705-x}{0.046}, & 1.004 \leq x \leq 1.05 \\ 0, & 1.05 \leq x \end{cases} \tag{3}$$

There are 260 rules in type-1 fuzzy system, 26 rules represent a specific basic rules for each class. The basic rules of the fuzzy system are listed as follows:

1. IF (*Normal_KNN* IS **High_Classification**) AND (*Normal_MLP1* IS **High_Activation**) AND (*Normal_MLP2* IS **High_Activation**) THEN (*Normal* IS **High_Classification**).
2. IF (*Normal_KNN* IS **High_Classification**) AND (*Normal_MLP1* IS **High_Activation**) AND (*Normal_MLP2* IS **Low_Activation**) THEN (*Normal* IS **High_Classification**).
3. IF (*Normal_KNN* IS **High_Classification**) AND (*Normal_MLP1* IS **High_Activation**) AND (*Normal_MLP2* IS **Medium_Activation**) THEN (*Normal* IS **High_Classification**).
4. IF (*Normal_KNN* IS **High_Classification**) AND (*Normal_MLP1* IS **Low_Activation**) AND (*Normal_MLP2* IS **High_Activation**) THEN (*Normal* IS **High_Classification**).
5. IF (*Normal_KNN* IS **High_Classification**) AND (*Normal_MLP1* IS **Medium_Activation**) AND (*Normal_MLP2* IS **High_Activation** THEN (*Normal* IS **High_Classification**).
6. IF (*Normal_KNN* IS **Low_Classification**) AND (*Normal_MLP1* IS **High_Activation**) AND (*Normal_MLP2* IS **High_Activation**) THEN (*Normal* IS **High_Classification**).
7. IF (*Normal_KNN* IS **Medium_Classification**) AND (*Normal_MLP1* IS **High_Activation**) AND (*Normal_MLP2* IS **High_Activation**) THEN (*Normal* IS **High_Classification**).
8. IF (*Normal_KNN* IS **High_Classification**) AND (*Normal_MLP1* IS **Medium_Activation**) AND (*Normal_MLP2* IS **Medium_Activation**) THEN (Normal IS **High_Classification**).
9. IF (*Normal_KNN* IS **High_Classification**) AND (*Normal_MLP1* IS **Medium_Activation**) AND (*Normal_MLP2* IS **Low_Activation**) THEN (Normal IS **High_Classfication**).
10. IF (*Normal_KNN* IS **High_Classification**) AND (*Normal_MLP1* IS **Low_Activation**) AND (*Normal_MLP2* IS **Low_Activation**) THEN (*Normal* IS **High_Classification**).
11. IF (*Normal_KNN* IS **High_Classification**) AND (*Normal_MLP1* IS **Low_Activation**) AND (*Normal_MLP2* IS **Medium_Activation**) THEN (*Normal* IS **High_Classification**).
12. IF (*Normal_KNN* IS **Medium_Classification**) AND (*Normal_MLP1* IS **High_Activation**) AND (*Normal_MLP2* IS **Medium_Activation**) THEN (*Normal* IS **High_Classification**).
13. IF (*Normal_KNN* IS **Medium_Classification**) AND (*Normal_MLP1* IS **Medium_Activation**) AND (*Normal_MLP2* IS **High_Activation**) THEN (*Normal* IS **High_Classification**).
14. IF (*Normal_KNN* IS **Medium_Classification**) AND (*Normal_MLP1* IS **High_Activation**) AND (*Normal_MLP2* IS **Low_Activation**) THEN (*Normal* IS **High_Classification**).
15. IF (*Normal_KNN* IS **Medium_Classification**) AND (*Normal_MLP1* IS **Low_Activation**) AND (*Normal_MLP2* IS **High_Activation**) THEN (*Normal* IS **High_Classification**).

16. IF (Normal_KNN IS *Low_Classification*) AND (*Normal_MLP1* IS *Medium_Activation*) AND (*Normal_MLP2* IS *High_Activation*) THEN (*Normal* IS *High_Classification*).

17. IF (*Normal_KNN* IS *Medium_Classification*) AND (*Normal_MLP1* IS *Medium_Activation*) AND (*Normal_MLP2* IS *Medium_Activation*) THEN (*Normal* IS *Medium_Classification*).

18. IF (*Normal_KNN* IS *Medium_Classification*) AND (*Normal_MLP1* IS *Medium_Activation*) AND (*Normal_MLP2* IS *Low_Activation*) THEN (*Normal* IS *Medium_Classification*).

19. IF (*Normal_KNN* IS *Medium_Classification*) AND (*Normal_MLP1* IS *Low_Activation*) AND (*Normal_MLP2* IS *Medium_Activation*) THEN (*Normal* IS *Medium_Classification*).

20. IF (*Normal_KNN* IS *Low_Classification*) AND (*Normal_MLP1* IS *Medium_Activation*) AND (*Normal_MLP2* IS *Medium_Activation*) THEN (*Normal* IS *Medium_Classification*).

21. IF (*Normal_KNN* IS *Low_Classification*) AND (*Normal_MLP1* IS *Low_Activation*) AND (*Normal_MLP2* IS *Medium_Activation*) THEN (*Normal* IS *Medium_Classification*).

22. IF (*Normal_KNN* IS *Medium_Classification*) AND (*Normal_MLP1* IS *Low_Activation*) AND (*Normal_MLP2* IS *Low_Activation*) THEN (*Normal* IS *Medium_Classification*).

23. IF (*Normal_KNN* IS *Low_Classification*) AND (*Normal_MLP1* IS *Medium_Activation*) AND (*Normal_MLP2* IS *Low_Activation*) THEN (*Normal* IS *Medium_Classification*).

24. IF (*Normal_KNN* IS *Low_Classifcation*) AND (*Normal_MLP1* IS *Low_Activation*) AND (*Normal_MLP2* IS *Low_Activation*) THEN (*Normal* IS *Low_Classification*).

25. IF (*Normal_KNN* IS *Low_Classifcation*) AND (*Normal_MLP1* IS *Low_Activation*) AND (*Normal_MLP2* IS *High_Activation*) THEN (*Normal* IS *High_Classification*).

26. IF (*Normal_KNN* IS *Low_Classifcation*) AND (*Normal_MLP1* IS *Low_Activation*) AND (*Normal_MLP2* IS *Medium_Activation*) THEN (*Normal* IS *Medium_Classification*).

In the fuzzy rules we take into consideration for a specific class the outputs of the three classifiers regarding the membership degrees from fuzzy KNN algorithm as first classifier and the activations of both neural networks (MLP-GDM and MLP-SCG) as second and third classifier respectively. We summarize the criteria used to develop the fuzzy rules: first, if the three classifiers obtain the same opinion respect to classify a specific sample then we keep the same fuzzy value; second, if two classifiers obtain the same linguistic value respect to classify a specific sample and the linguistic value of the different opinion classifier represent major value, then the consequent obtain the linguistic value of the different opinion classifier; third, If two classifiers obtain the same linguistic value respect to classify a specific sample and the linguistic value of the different opinion classifier represent lower linguistic value, then the consequent obtain the linguistic value of the two classifiers with the same opinion; if the three classifiers differs in opinion then the consequent obtain the higher linguistic value. We apply these three criteria to form the all fuzzy rules for the fuzzy system to combine all the output results of the classifiers in the basic module unit for the hybrid model.

In the following part of this chapter, we mentioned a brief description of some concepts of techniques used in this hybrid model.

3 Important Concept Review

In this section, we describe a brief resume of Fuzzy KNN algorithm, Multi Layer Perceptron.

3.1 Fuzzy KNN Algorithm

The Fuzzy KNN algorithm assigns class membership to a sample vector to a particular class. The advantage is that no arbitrary assignments are made by the algorithm. In addition, the vector's membership values should provide a level of assurance to accompany the resultant classification [16].

The basis of algorithm is to assign membership as a function of the vector's distance from its K nearest neighbors and those neighbors' memberships in the possible classes [16].

3.2 Multi Layer Perceptron with GDM and SCG

An artificial neural network (ANN) is a mathematical and computational model that simulates the abstract structure and functional aspects of biological neural networks. The basic computational elements in an ANN are known as neurons, nodes serve as inputs, outputs or internal processing units. These neurons communicate and pass signals among themselves using what are known as synapses, which neurons are organized and connected define the architecture of the network. The Multi Layer Perceptron (MLP), the information moves in one direction, forward, from inputs nodes, through the hidden nodes to outputs nodes. The gradient descent with momentum (GDM) is equivalent to the backpropagation algorithm, which is used when the connection weights are updated. The scaled conjugate gradient (SCG) is used to more efficiently guide the error down the error surface and provide faster solutions for an extra second derivative of error information and internal adjustments that are made to the learning parameters [38].

4 Experiments

We have performed experiments with 10-fold cross validation to evaluate the hybrid model, that consist portioning the original set of samples of heartbeats into training set to train each classifier in hybrid model, and test set, herein to validate them. In each basic module unit, we've trained the three classifiers, selected different parameters and tested. For fuzzy KNN algorithm, we used k = 3, 4, and 5. For the second and third classifiers, MLP-GDM, we used 140 input neurons, 50, 100 and 150 hidden neurons, 10,000 epochs, learning rate 0.3 and momentum 0.5, 10 output neurons. For the third classifier, MLP-SCG, 50, 100 and 150 hidden neurons, 10,000 epochs, learning rate 0.001, 10 input neurons. The output neurons represent the classes of arrhythmias and normal beats. We selected the best results for each architecture mentioned to represent the classifiers. The outputs of the three classifiers form the integration matrix, herein the input for the type-1 FIS to combine the output results of the classifiers. We do the same process for both basic module units. Finally, we used other type-1 FIS to combine the output results for the two basic module units in the hybrid model, the output result of this type-1 FIS represents the global classification of the hybrid model.

4.1 MIT-BIH Arrhythmia Database

The MIT-BIH Arrhythmia Database contains 48 half-hour excerpts of two-channel ambulatory ECG recordings, obtained from 47 subjects studied by the Beth Israel Hospital (BIH) Arrhythmia Laboratory between 1975 and 1979. Approximately 60% of these recordings were obtained from inpatients. The subjects were 25 men aged 32 to 89 years, and 22 women aged 23 to 89 years. Two or more cardiologists independently annotated each record; disagreements were resolved to obtain the computer-readable reference annotations for each beat (approximately 110,000 annotations) included with the database.

4.2 Results

4.2.1 Basic Module Unit 1. First Classifier: Fuzzy KNN Algorithm

Firstly, we show the results for the basic module unit 1. In the results for the first classifier in basic module unit 1, we notice best result used $k = 4$ where we obtained 90.3% of classification rate, see confusion matrix in Table 1.

4.2.2 Basic Module Unit 1. Second Classifier: MLP-GDM

The best result of MLP-GDM, second classifier, we used 150 hidden neurons, we obtained 83.9% of classification rate, the results are presented in the confusion matrix of Table 2.

4.2.3 Basic Module Unit 1. Third Classifier: MLP-SCG

For MLP-SCG, third classifier, we obtained best result used 50 hidden neurons with an 92.80% of classification rate, see details in Table 3.

4.2.4 Basic Module Unit 1. Fuzzy System

Finally, we combined the output results of the three classifiers with a type-1 FIS in basic module unit 1, and the global classification rate is of 92.90%. The results for the type-1 FIS is shown in Table 4.

4.2.5 Basic Module Unit 2. First Classifier: Fuzzy KNN Algorithm

In this part we present the results for the basic module unit 2. In the results for the first classifier into basic module unit 2, we notice a best result used $k = 4$ and we obtained an 89.10% of classification rate in 10-fold cross validation, see confusion matrix in Table 5.

4.2.6 Basic Module Unit 2. Second Classifier: MLP-GDM

In the second classifier, MLP-GDM in basic module unit 2, we notice a best result in 10-fold cross validation used 150 hidden neurons with a 76.10% of classification rate. For more details of these results in Table 6.

Table 1 Results for Fuzzy KNN algorithm in 10-fold cross validation, with 4 nearest neighbors, basic module unit 1

Class	N	LBBB	RBBB	PVC	FPN	APB	AAPB	FVNB	VEB	PB
N	**97**	0	0	0	0	1	0	2	0	0
LBBE	0	**92**	1	0	0	1	2	2	2	0
RBBE	0	2	**96**	0	1	0	1	0	0	0
PVC	0	3	0	**83**	3	0	1	2	2	6
FPN	1	0	2	2	**85**	0	1	6	0	3
APB	0	1	0	0	0	**92**	7	0	0	0
AAPB	0	4	0	0	0	10	**80**	1	5	0
FVNB	3	4	1	2	2	1	1	**86**	0	0
VEB	0	0	0	0	0	0	5	0	**95**	0
PB	0	0	0	1	2	0	0	0	0	**97**

Classification rate = 90.3%
Bold indicates "Best results"

Table 2 Results for MLP-GDM in 10-fold cross validation, 150 hidden neurons, basic module unit 1

Class	N	LBBB	RBBB	PVC	FPN	APB	AAPB	FVNB	VEB	PB
N	**97**	0	0	0	0	0	0	3	0	0
LBBB	1	**69**	0	0	0	0	22	4	4	0
RBBB	4	3	**73**	0	18	1	0	0	1	0
PVC	0	3	0	**80**	4	2	1	0	3	7
FPN	0	1	3	2	**90**	1	0	1	0	2
APB	0	0	0	0	0	**99**	1	0	0	0
AAPB	0	5	3	0	1	18	**67**	0	6	0
FVNB	2	20	0	3	0	2	0	**73**	0	0
VEB	0	0	0	0	0	0	5	0	**95**	0
PB	0	0	1	1	2	0	0	0	0	**96**

Classification rate = 83.9%
Bold indicates "Best results"

Table 3 Results for MLP-SCG in 10-fold cross validation, 50 hidden neurons, basic module unit 1

Class	N	LBBB	RBBB	PVC	FPN	APB	AAPB	FVNB	VEB	PB
N	**99**	0	0	0	0	0	0	1	0	0
LBBB	3	**87**	3	0	1	0	1	3	2	0
RBBB	0	1	**97**	0	0	0	0	1	1	0
PVC	0	4	0	**86**	2	0	3	2	1	2
FPN	0	1	1	1	**93**	0	1	0	0	3
APB	0	0	0	0	0	**97**	3	0	0	0
AAPE	1	2	2	0	0	5	**88**	0	2	0
FVNB	1	1	0	2	1	0	1	**94**	0	0
VEB	3	0	0	0	0	0	7	1	**89**	0
PB	0	0	0	0	2	0	0	0	0	**98**

Classification rate = 92.80%
Bold indicates "Best results"

Table 4 Results for type-1 fuzzy inference system in 10-fold cross validation, basic module unit 1

Class	N	LBBB	RBBB	PVC	FPN	APB	AAPB	FVNB	VEB	PB
N	**99**	0	0	0	0	0	0	1	0	0
LBBB	0	**91**	0	0	1	1	2	3	2	0
RBBB	0	2	**96**	0	1	0	1	0	0	0
PVC	0	4	0	**86**	3	0	3	1	0	3
FPN	0	0	2	1	**92**	0	1	2	0	2
APB	0	0	0	0	0	**94**	6	0	0	0
AAPB	0	4	0	0	0	6	**84**	1	5	0
FVNB	1	2	0	1	1	1	0	**94**	0	0
VEB	0	0	0	0	0	0	4	0	**96**	0
PB	0	0	0	0	3	0	0	0	0	**97**

Classification rate = 92.90%
Bold indicates "Best results"

Table 5 Results for fuzzy KNN algorithm in 10-fold cross validation, with 3 nearest neighbor, basic module unit 2

Class	N	LBBB	RBBB	PVC	FPN	APB	AAPB	FVNB	VEB	PB
N	**91**	0	1	0	0	1	6	0	1	0
LBBB	0	**93**	0	2	1	0	0	3	1	0
RBBB	0	0	**99**	0	0	0	0	0	1	0
PVC	2	9	1	**67**	3	8	5	2	3	0
FPN	0	1	0	0	**97**	0	1	0	0	1
APB	6	0	0	2	0	**87**	3	1	1	0
AAPB	6	0	0	2	2	4	**86**	0	0	0
FVNB	2	13	1	3	2	2	0	**77**	0	0
VEB	1	1	1	0	0	0	0	0	**95**	2
PB	0	0	0	0	1	0	0	0	0	**99**

Classification rate = 89.10%
Bold indicates "Best results"

Table 6 Results for MLP-GDM in 10-fold cross validation, 150 hidden neurons, basic module unit 2

Class	N	LBBB	RBBB	PVC	FPN	APB	AAPB	FVNB	VEB	PB
N	**72**	0	0	0	0	23	0	0	5	0
LBBB	0	**90**	3	0	7	0	0	0	0	0
RBBB	2	0	**97**	0	0	0	0	0	1	0
PVC	6	41	1	**18**	2	14	6	5	7	0
FPN	0	0	0	1	**97**	0	0	0	0	2
APB	10	0	0	0	0	**89**	1	0	0	0
AAPB	15	1	10	0	12	14	**46**	0	2	0
FVNB	0	25	1	0	5	11	3	**55**	0	0
VEB	0	0	2	0	0	0	1	0	**97**	0
PB	0	0	0	0	0	0	0	0	0	**100**

Classification rate = 76.10%
Bold indicates "Best results"

4.2.7 Basic Module Unit 2. Third Classifier: MLP-SCG

For the MLP-SCG, third classifier in basic module unit 2, we notice a best result used 50 hidden neurons a 91.20% of classification rate, we present a list with details in Table 7.

4.2.8 Basic Module Unit 2 with Type-1 FIS

Finally, we combining the outputs for the three classifiers used a type-1 FIS, and we obtained a final result in basic module unit 2 with a 92.40% of classification rate, see more details in Table 8.

4.2.9 Hybrid Model with a Type-1 Fuzzy System

Finally, we combined the outputs for the both basic module units 1 and 2 through a type-1 FIS, and we obtained a global result for the complete hybrid model with a 93.80% of classification rate, see more details in Table 9.

5 Conclusions

In the experiments of this work, we have good classification rate with the all classifiers into the two basic module units, and combining their output results with fuzzy logic we have obtained a better achievement. We obtained best results using a type-1 FIS to combine the outputs of the classifiers with a 92.90% of classification rate in basic module unit 1. Respect to the basic module unit 2, we obtained 92.40% of classification rate used type-1 FIS combining the out results of the three classifiers.

We used two basic module units in hybrid model, where each one was trained with an electrode signal of the MIT-BIH arrhythmia database and we combined the output results of the two basic module units used a type-1 fuzzy system, the results shown the increment of the global classification rate in the hybrid model to achieve a 93.80%. We found out that there are samples of heartbeats was classified by one basic module unit as misclassification however when combining their outputs with 2-lead using two basic module units some samples with misclassification was fixed and classified correctly by other basic module unit using fuzzy system, we have obtained

Table 7 Results for MLP-SCG in 10-fold cross validation, 50 hidden neurons, basic module unit 2

Class	N	LBBB	RBBB	PVC	FPN	APB	AAPB	FVNB	VEB	PB
N	**89**	0	1	1	0	3	4	0	2	0
LBBB	0	**99**	0	0	0	0	0	1	0	0
RBBB	0	0	**98**	0	0	0	1	0	1	0
PVC	0	4	2	**70**	0	6	9	4	5	0
FPN	0	1	0	1	**95**	0	0	1	0	2
APB	3	1	0	0	0	**95**	1	0	0	0
AAPB	0	0	0	3	2	5	**90**	0	0	0
FVNB	0	1	1	2	1	2	1	**92**	0	0
VEB	0	1	1	0	0	0	0	0	**97**	1
PB	0	0	0	0	2	0	0	1	10	**87**

Classification rate = 91.20%
Bold indicates "Best results"

Table 8 Results for type-1 FIS in 10-fold cross validation, combining the outputs of the three classifiers in basic module unit 2

Class	N	LBBB	RBBB	PVC	FPN	APB	AAPB	FVNB	VEB	PB
N	**96**	0	0	1	0	1	2	0	0	0
LBBB	0	**94**	1	1	1	0	0	3	0	0
RBBB	0	0	**99**	0	0	0	0	0	1	0
PVC	1	6	1	**73**	2	4	5	3	5	0
FPN	0	1	0	1	**96**	0	0	1	0	1
APB	4	0	0	0	0	**96**	0	0	0	0
AAPB	2	0	0	4	2	3	**88**	0	1	0
FVNB	1	2	1	4	1	1	0	**90**	0	0
VEB	1	1	1	0	0	0	0	0	**95**	2
PB	0	0	0	0	3	0	0	0	0	**97**

Classification rate = 92.40%
Bold indicates "Best results"

Table 9 Results for global type-1 FIS in 10-fold cross validation, combining the outputs of the two basic module units

Class	N	LBBB	RBBB	PVC	FPN	APB	AAPB	FVNB	VEB	PB
N	**97**	0	0	0	0	0	2	1	0	0
LBBB	0	**95**	0	0	0	0	1	2	2	0
RBBB	0	2	**97**	0	0	0	1	0	0	0
PVC	0	5	1	**84**	0	2	4	2	1	1
FPN	0	1	0	0	**97**	0	0	2	0	0
APB	1	0	0	0	0	**92**	7	0	0	0
AAPB	1	3	0	1	0	1	**91**	1	2	0
FVNB	2	3	1	0	0	0	0	**94**	0	0
VEB	1	0	1	0	0	0	2	0	**95**	1
PB	0	0	0	0	4	0	0	0	0	**96**

Classification rate = 93.80%
Bold indicates "Best results"

better performance and classification rate using a hybrid model to resolve cardiac arrhythmia classification problem. We will work to extend our hybrid system to use multi-lead arrhythmia classification using other databases that contains 12-lead to be able to determine a complete medical diagnosis. Also, different applications may be considered like in [39–46].

References

1. J. Amezcua, P. Melin, A modular LVQ neural network with fuzzy response integration for arrhythmia classification, in *IEEE Conference on Norbert Wiener in the 21st Century* (2014)
2. P. Melin, J. Amezcua, F. Valdez, O. Castillo, A new neural network model based on the LVQ algorithm for multi-class classification of arrhythmias. Inf. Sci. **279**, 483–497 (2014)
3. A.F. Khalaf, M.L. Owis, I.A. Yassine, A novel technique for cardiac arrhythmia classification using spectral correlation and support vector machines. Expert Syst. Appl. **42**(21), 8361–8368 (2015). Elsevier
4. M.R. Homaeinezhad, S.A. Atyabi, E. Tavakkoli, H.N. Toosi, A. Ghaffari, R. Ebrahimpour, ECG arrhythmia recognition via a neuro-SVM-KNN hybrid classifier with virtual QRS image-based geometrical features. Expert Syst. Appl. **39**(2), 2047–2058 (2012). Elsevier
5. J.S. Wang, W.C. Chiang, Y.L. Hsu, Y.T. Yang, ECG arrhythmia classification using a probabilistic neural network with a feature reduction method. Neurocomputing **116**, 38–45 (2013). Elsevier
6. R.J. Martis, U.R. Achayra, H. Prasad, C.K. Chua, Application of higher order statistics for atrial arrhythmia classification. Biomed. Signal Process. Control (2013). Elsevier
7. D. Gaetano, S. Panunzi, F. Rinaldi, A. Risi, M. Sciandrone, A patient adaptable ECG beat classifier based on neural networks. Appl. Math. Comput. **213**(1), 243–249 (2009). Elsevier
8. E. Ramirez, O. Castillo, J. Soria, Hybrid system for cardiac arrhythmia classification with Fuzzy K-Nearest Neighbors and neural networks combined by a fuzzy inference system, in *Softcomputing for Recognition Based on Biometrics, Studies in Computational Intelligence*, vol. 312 (Springer, 2010), pp. 37–53, ISBN 978-3-642-15110-1
9. M. Javadi, S.A. Asghar, A. Sajedin, R. Ebrahimpour, Classification of ECG arrhythmia by a modular neural network based on mixture of experts and negatively correlated learning. Biomed. Signal Process. Control **8**, 289–296 (2013). Elsevier
10. M.M. Al Rahhal, Y. Bazi, H. Alhichri, N. Alajlan, F. Melgani, R.R. Yager, Deep learning approach for active classification of electrocardiogram signals. Inf. Sci. **345**, 340–354 (2016). Elsevier
11. P. Melin, E. Ramirez, G. Prado-Arechiga, Cardiac arrhythmia classification using computational intelligence: neural networks and fuzzy logic techniques. European Heart J. **38**, P6388 (2017). Oxford academic
12. P. Melin, G. Prado-Arechiga, I. Miramontes, M. Medina, A hybrid intelligent model based on modular neural network and fuzzy logic for hypertension risk diagnosis. J. Hypertens. **34** (2016)
13. S.M. Jadhav, S.L. Nalbalwar, A.A. Ghatol, ECG arrhythmia classification using modular neural network model, in *IECBES* (2012). ISBN 978-1-4244-7600-8
14. Y. Ozbay, G. Tezel, A new method for classification of ECG arrhythmias using neural network with adaptive activation function. Digital Signal Process. **20**, 1040–1049 (2010). Elsevier
15. C. Zopounidis, M. Doumpos, Multicriteria classification and sorting methods: a literature review. Eur. J. Oper. Res. **138**, 229–246 (2002). Elsevier
16. J.M. Keller, M.R. Gray, J.A. Givens, A fuzzy K-Nearest neighbor algorithm. IEEE Trans. Fuzzy Syst. Man Cybern. **15**, 580–585 (1985)

17. P. Melin, O. Castillo, A review of the applications of type-2 fuzzy logic in classification and pattern recognition. Expert Syst. Appl. **40**(13), 5413–5423 (2013)
18. P. Melin, O. Castillo, A review on the applications of type-2 fuzzy logic in classification and pattern recognition. Expert Syst. Appl. **40**, 5413–5423 (2013). Elsevier
19. R. Ceylan, Y. Ozbay, B. Karlik, A novel approach for classification of ECG arrhythmias: type-2 fuzzy clustering neural network. Expert Syst. Appl. **36**(3), 6721–6727 (2009). ACM
20. T.W. Chua, W.W. Tan, Interval type-2 fuzzy system for ECG arrhythmia classification, in *Fuzzy Systems in Bioinformatics and Computational Biology*, vol. 242 (2009, Springer), pp. 297–314. ISBN 978-3-540-89968-6
21. W.W. Tan, C.L. Foo, T. Chua, Type-2 fuzzy system for ECG arrhythmic classification, in *FYZZ-IEEE* (2007). ISBN 1-4244-1209-9
22. Y. Ozbay, R. Ceylan, B. Karlik, A fuzzy clustering neural network architecture for classification of ECG arrhythmias. Comput. Biol. Med. **36**, 376–388 (2005). Elsevier
23. E.J.S. Luz, W.R. Schwartz, G. Camara-Chavez, D. Menotti, ECG-based heartbeat classification for arrhythmia detection: a survey. Comput. Methods Programs Biomed. **127**, 144–164 (2015). Elsevier
24. M. Wozniak, M. Grana, E. Corchado, A survey of multiple classifier systems as hybrid systems. Inf. Fusion **16**, 3–17 (2014). Elsevier
25. M.S.A. Megat, A.H. Jahidin, A.N. Norall, Hybrid multilayered perceptron network classification of bundle branch blocks, in *IEEE 2012 International Conference on Biomedical Engineering Icobe*. ISBN 978-1-4577-1991-2
26. O. Castillo, P. Melin, E. Ramirez, J. Soria, Hybrid intelligent system for cardiac arrhythmia classification with fuzzy K-Nearest Neighbors and neural networks combined with a fuzzy system. Expert Syst. Appl. **39**, 2947–2955 (2012)
27. S. Osowki, T. Markiewicz, L.T. Hoal, Recognition and classification systems of arrhythmia using ensemble of neural networks. Measurement **41**(6), 610–617 (2018). Elsevier
28. S. Osowksi, K. Siwek, R. Siroic, Neural system for heartbeats recognition using genetically integrated ensemble of classifiers. Comput. Biol. Med. **41**(3), 173–180 (2011). Elsevier
29. T.M. Nazmy, H. EL-Messiry, B. AL-Bokhity, Classification of cardiac arrhythmia based on hybrid system. Int. J. Comput. Appl. **2** (2010)
30. Y.E. Shao, C.D. Hou, C.C. Chiu, Hybrid intelligent modeling schemes for heart disease classification. Appl. Soft Comput. **14**, 47–52 (2014). Elsevier
31. F.A. Elhaj, N. Salim, A.R. Harris, T.T. Swee, T. Ahmed, Arrhytmia recognition and classification using combined linear and nonlinear features of ECG signals. Comput. Methods Programs Biomed. **127**, 52–63 (2016). Elsevier
32. A. Gacek, W. Pedrycz, in *ECG Signal Processing, Classification and Interpretation, a Comprehensive Framework of Computational Intelligence* (Springer, 2012). ISBN 978-0-85729-867-6
33. A. Jovic, N. Bogunovic, Evaluating and comparing performance of feature combinations of heart rate variability measures for cardiac rhythm classification. Biomed. Signal Process. Control **7**, 245–255 (2012). Elsevier
34. R.J. Martis, U.R. Achayra, L.C. Min, ECG beat classification using PCA, LDA, ICA, and discrete wavelet transform. Biomed. Signal Process. Control **8**, 437–448 (2013). Elsevier
35. R. Jayasinghe, in *ECG Workbook* (Churchill Livingstone, Elsevier, 2012). ISBN 978-0-7295-4109-1
36. J.L. Martindale, D.F.M. Brown, in *A Visual Guide to ECG Interpretation*, 2nd edn. (Wolters Kluwer, 2017), ISBN 978-1-4963-2153-4
37. J.R. Hampton, D. Adlam, in *The ECG in Practice*, 6th edn. (Churchill Livingstone, Elsevier, 2013), ISBN 978-0-7020-4643-8
38. C.M. Bishop, in *Neural Network for Pattern Recognition* (Oxford, UK, Clarendon Press)
39. C. Leal Ramírez, O. Castillo, P. Melin, A. Rodríguez Díaz, Simulation of the bird age-structured population growth based on an interval type-2 fuzzy cellular structure. Inf. Sci. **181**(3), 519–535 (2011)
40. N.R. Cázarez-Castro, L.T. Aguilar, O. Castillo, Designing Type-1 and Type-2 fuzzy logic controllers via Fuzzy Lyapunov Synthesis for nonsmooth mechanical systems. Eng. Appl. AI **25**(5), 971–979 (2012)

41. O. Castillo, P. Melin, Intelligent systems with interval type-2 fuzzy logic. Int. J. Innov. Comput. Inf. Control **4**(4), 771–783 (2008)
42. G.M. Mendez, O. Castillo, Interval type-2 TSK fuzzy logic systems using hybrid learning algorithm, in *The 14th IEEE International Conference on Fuzzy Systems, FUZZ'05* (2005), pp. 230–235
43. P. Melin, O. Castillo, Intelligent control of complex electrochemical systems with a neuro-fuzzy-genetic approach. IEEE Trans. Ind. Electron. **48**(5), 951–955 (2001)
44. E. Rubio, O. Castillo, F. Valdez, P. Melin, C. I. González, G. Martinez, An extension of the fuzzy possibilistic clustering algorithm using type-2 fuzzy logic techniques. Adv. Fuzzy Syst. **2017**, 7094046:1–7094046:23 (2017)
45. L. Aguilar, P. Melin, O. Castillo, Intelligent control of a stepping motor drive using a hybrid neuro-fuzzy ANFIS approach. Appl. Soft Comput. **3**(3), 209–219 (2003)
46. Patricia Melin, Oscar Castillo, Adaptive intelligent control of aircraft systems with a hybrid approach combining neural networks, fuzzy logic and fractal theory. Appl. Soft Comput. **3**(4), 353–362 (2003)

Comparative Study of Bio-inspired Algorithms Applied in the Optimization of Fuzzy Systems

Ivette Miramontes, Patricia Melin and German Prado-Arechiga

Abstract In the medical area, it is very important to have accurate results in diagnosis of diseases that people may suffer. This is why, there is a need to perform the optimization of the fuzzy classifier which provides the nocturnal blood pressure profile of patients, and which is important, due that with this diagnosis we may know if the patient is prone to have a cardiovascular event. This fuzzy system is designed using different membership functions, which are trapezoidal and Gaussian membership functions, in order to select the fuzzy system that provides better results when making the classification. Two bioinspired algorithms are used separately to test their performance, which are the Crow Search Algorithm and Chicken Swarm Optimization. Thirty experiments were performed varying the parameters in the algorithms and from which it can be concluded that the CSO provides better results when optimizing fuzzy systems with both types of membership functions.

Keywords Fuzzy systems · Optimization · Bio-inspired algorithm · Nocturnal blood pressure profile

1 Introduction

Fuzzy systems have been widely used in the medical area to provide diagnosis of different types of diseases [1–4]. For this work a fuzzy classifier is designed, which provides the nocturnal blood pressure profile of a patient, this diagnosis is important, since, of this as a result, the cardiologist can know if the patient has a risk of suffering a cardiovascular event, and thus be able to avoid a future event.

Different bioinspired algorithms have been used in many areas for the optimization of different types of problems [5–7]. For the optimization of membership functions of fuzzy systems, different algorithms have been used, such as Genetic Algorithms, Particle Swarm Optimization, among others [8–11]. For this case, the Crow Search Algorithm (CSA) and the Chicken Swarm Optimization (CSO) are used, to perform

I. Miramontes · P. Melin (✉) · G. Prado-Arechiga
Tijuana Institute of Technology, Tijuana, BC, Mexico
e-mail: pmelin@tectijuana.mx

© Springer Nature Switzerland AG 2020
O. Castillo and P. Melin (eds.), *Hybrid Intelligent Systems in Control,
Pattern Recognition and Medicine*, Studies in Computational Intelligence 827,
https://doi.org/10.1007/978-3-030-34135-0_15

the necessary adjustment in a fuzzy system for classification of nocturnal blood pressure profile.

This paper has been organized as follows: in Sect. 2 the literature review is presented, in Sect. 3 a methodology description is presented, in Sect. 4 the results and discussions are presented, in Sect. 5 the statistical test is presented and in Sect. 6 the conclusions obtained after carrying out the tests are presented.

2   Literature Review

This section presents the basic concepts necessary to understand the proposed method.

2.1   Chicken Swarm Optimization

The CSO was proposed by Meng [12] in 2014, mimicking the hierarchical order and the behaviors of the chicken swarm, including roosters, hens and chicks.

The rules in this algorithm are the following:

1. In the chicken swarm, there are different groups, which are comprised of a dominant rooster, a couple of hens, and chicks.
2. The fitness value of the chickens is what determines how to divide the chicken swarm and the identity of the chickens, this means, the chicken with the best fitness are the rooster, the chicken with the worst fitness are the chicks and the rest are hens.
3. The hierarchical order, dominance relationship and mother-child relationship in a group will remain unchanged. These status only update every several time steps.
4. Chickens follow their groupmate rooster to search for food, while they can prevent them from eating their own food. We assume chickens randomly steal good food that others already found.

2.2   Crow Search Algorithm

CSA was proposed by Askarazadeh in 2016 [13]. This algorithm is inspired by the intelligence of the crows, referring to how they observe other birds with the goal of knowing where they hide their food and steal their food. They also use their own experience as thieves to carry their hiding places from being stolen.

The main principles of CSA are the follows:

The crows live in a flock.

The crows memorize the position of their hiding places.

Crows chase each other to steal.

Crows protect their hiding places from being robbed with a certain probability value.

2.3 Blood Pressure and Nocturnal Blood Pressure Profile

Blood pressure is the measure of the force applied against the walls of the blood vessel, this is important, because with this force the nutrients and oxygen are carried throughout our body. Blood pressure has two components, the first is the systolic pressure, which is when the heart contracts and the diastolic pressure, when heart relaxes between beats, and is measured in millimeters of mercury (mmHg) [14–16].

Blood pressure normally decreases during the night. The fall occurring in this period is called Dipper, and is defined as the difference between daytime mean systolic pressure and nighttime mean systolic pressure expressed as a percentage of the day value [17]. Dipper profile has a decrease between 10 and 20% this percenter is considered normal. When the decrease is less than 10% the medical doctors consider them as absent and as a predictor of an adverse cardiovascular disease, this profile is called as non-dipper [18–20].

There exist another two nocturnal blood pressure profiles, when the blood pressure falls more than 20% is called extreme dipper and when the resting blood pressure is higher than the daytime period is then called riser profile [21].

3 Methodology

A neuro-fuzzy hybrid model is designed for the medical diagnosis based on the patient's blood pressure presented in [22–24] from which the fuzzy classifier that provides the patient's nocturnal blood pressure profile [25] is taken, in order to be optimized. A Mamdani type fuzzy system is designed, based on [26] and on the experience of the expert. For this fuzzy system two inputs are considered, the first is the systolic blood pressure and the second is the diastolic blood pressure, which range is between 0.4 and 1.3, this is due to the quotients of the blood pressures respectively [26].

Both inputs are granulated into 4 membership functions, with the linguistic variables "GreaterFall", "Fall", "Increase" and "GreaterIncrease".

For the output, the linguistic values used are "Extreme Dipper", "Dipper", Non Dipper" Riser". In order to perform comparative tests, the fuzzy system is designed with trapezoidal and Gaussian membership functions, and in Fig. 1 the fuzzy system with Gaussian membership functions is presented.

The rule set of the fuzzy system contains four rules, which are listed below:

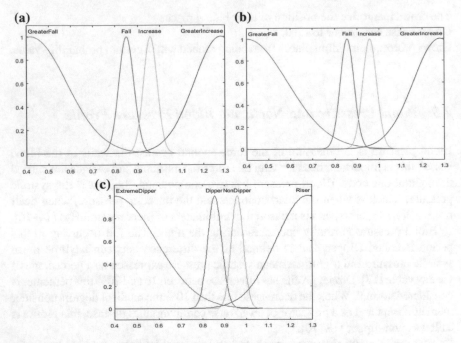

Fig. 1 a Input for systolic. **b** Input for diastolic. **c** Output for profile

1. If systolic is *GreaterFall* and diastolic is *GreaterFall* then Profile is *ExtremeDipper*.
2. If systolic is *Fall* and diastolic is *Fall* then Profile is *Dipper*.
3. If systolic is *Increase* and diastolic is *Increase* then Profile is *NonDipper*.
4. If systolic is *GreaterIncrease* and diastolic is *GreaterIncrease* then Profile is *GreaterIncrease*.

In order to improve the classification rate of the fuzzy systems, the optimization is carried out using the CSO and CSA algorithms respectively.

These algorithms are used to make the necessary parameter adjustment in the membership functions.

The mean square error was used as the fitness function, with which it is aimed at minimizing the classification error for obtaining the best solution.

$$MSE = \frac{1}{2} \sum_{i=1}^{n} \left(\widehat{Y}_i - Y_i \right)^2 \tag{1}$$

4 Results and Discussion

Thirty experiments were carried out, in the first instance with the CSA algorithm, as mentioned in the previous section; they were performed by varying the membership functions, for which Trapezoidal and Gaussian are used.

The parameters of the algorithm were changed in each experiment; these values are presented in Table 1, where:

- N represents the number of individuals.
- AP represents the awareness probability, which was varied between 0 and 1.
- $Iter$ represents the iterations.
- DT represents the dimensions in the fuzzy system with trapezoidal membership functions.
- DG represents the dimensions in the fuzzy system with Gaussian membership functions.
- Fl is the flight length, when random numbers between 0 and 2 were used.

In the same way, 30 experiments were carried out by varying the parameters of the algorithm in each experiment. The parameters are presented in Table 2, where:

- Pop represents the population
- M represents the iterations
- DT represents the dimension in the fuzzy system with trapezoidal membership functions
- DG represents the dimension in the fuzzy system with trapezoidal membership functions
- G represents how often the chickens swarm can be update
- $r\%$ represents the population size of roosters
- $h\%$ represents the population size of hens
- $m\%$ represents the population size of mother hens.

The different fuzzy systems obtained by the CSO and CSA algorithms are tested, for this, 80 patients are used in order to test the correct classification percentage to perform the comparison between methods. In Table 3, the classification obtained is shown by the fuzzy systems with trapezoidal membership functions. Likewise, the classification obtained using the Gaussian membership functions are presented in Table 4.

Table 5 summarizes the averages of the 30 experiments and with each method, in which it is observed that the CSO algorithm provides better classification results in both cases, and that for the FIS with Gaussian membership it obtains an average of 91.46% of correct classification, while for the fuzzy system with Gaussian membership functions it only obtains on average the 87.59% of correct classification.

Table 1 CSA parameters for each experiment and percentage of success

No.	Parameters of CSA					
	N	AP	Iter	DT	DG	fl
1	5	0.5	4000	48	24	2
2	10	0.3	2000	48	24	0.5
3	15	0.7	1333	48	24	0.8
4	18	0.9	1111	48	24	1
5	20	0.2	1000	48	24	1.5
6	23	0.8	869	48	24	2
7	26	0.1	769	48	24	0.4
8	30	0.4	666	48	24	1.3
9	33	0.6	606	48	24	1.8
10	35	0.7	571	48	24	0.7
11	38	0.8	526	48	24	0.3
12	40	0.5	500	48	24	1
13	43	0.3	465	48	24	1.2
14	47	0.2	425	48	24	0.1
15	50	0.4	400	48	24	0.9
16	52	0.6	384	48	24	1.7
17	55	0.2	363	48	24	2
18	60	0.1	333	48	24	0.3
19	64	0.9	312	48	24	0.6
20	67	0.7	298	48	24	1.6
21	70	0.4	285	48	24	1.4
22	72	0.5	277	48	24	0.2
23	75	0.3	266	48	24	0.7
24	80	0.4	250	48	24	1.1
25	85	0.2	235	48	24	2
26	87	0.6	229	48	24	0.5
27	90	0.1	222	48	24	0.8
28	92	0.8	217	48	24	1
29	96	0.9	208	48	24	0.4
30	100	0.5	200	48	24	0.7

5 Statistical Test

Statistical tests for the comparison between methods are performed, for this a Z-test is used, and the first case presented is when the trapezoidal membership functions are used.

Table 2 CSO parameters for each experiment and percent of success

No.	Parameters of CSO							
	pop	M	DT	DG	G	r%	h%	m%
1	5	4000	48	24	14	0.15	0.6	0.3
2	10	2000	48	24	9	0.15	0.8	0.5
3	15	1333	48	24	3	0.15	0.7	0.3
4	18	1111	48	24	15	0.15	0.6	0.4
5	20	1000	48	24	16	0.15	0.5	0.2
6	23	869	48	24	20	0.15	0.7	0.1
7	26	769	48	24	15	0.15	0.8	0.5
8	30	666	48	24	16	0.15	0.6	0.2
9	33	606	48	24	4	0.15	0.5	0.5
10	35	571	48	24	14	0.15	0.8	0.1
11	38	526	48	24	9	0.15	0.6	0.4
12	40	500	48	24	9	0.15	0.7	0.3
13	43	465	48	24	16	0.15	0.8	0.2
14	47	425	48	24	12	0.15	0.5	0.4
15	50	400	48	24	9	0.15	0.6	0.4
16	52	384	48	24	16	0.15	0.6	0.5
17	55	363	48	24	3	0.15	0.7	0.2
18	60	333	48	24	8	0.15	0.8	0.4
19	64	312	48	24	11	0.15	0.5	0.1
20	67	298	48	24	2	0.15	0.7	0.1
21	70	285	48	24	5	0.15	0.6	0.4
22	72	277	48	24	17	0.15	0.8	0.2
23	75	266	48	24	7	0.15	0.7	0.1
24	80	250	48	24	6	0.15	0.6	0.4
25	85	235	48	24	19	0.15	0.8	0.3
26	87	229	48	24	4	0.15	0.6	0.3
27	90	222	48	24	9	0.15	0.7	0.1
28	92	217	48	24	19	0.15	0.8	0.5
29	96	208	48	24	13	0.15	0.6	0.1
30	100	200	48	24	10	0.15	0.5	0.3

The 30 samples of the performed experiments with the CSO and CSA algorithms are taken when the trapezoidal membership functions are used, using the classification percentage obtained from each of the fuzzy systems designed by the two aforementioned methods.

As mentioned above, a Z-test is used, with the following expression:

Table 3 Percent of success
using trapezoidal MFs

% success trapezoidal MFs	
CSO	CSA
92.5	86.25
90	81.25
88.75	87.5
91.25	87.5
87.5	81.25
90	88.75
92.5	88.75
91.25	86.25
91.25	86.25
91.25	86.25
90	87.5
90	88.75
91.25	88.75
87.5	87.5
90	86.25
93.75	88.75
91.25	87.5
93.75	87.5
91.25	87.5
95	87.5
93.75	88.75
93.75	88.75
91.25	86.25
93.75	87.5
90	88.75
92.5	88.75
91.25	87.5
92.5	86.25
90	88.75
95	88.75

$$Z = \frac{(\bar{x}_1 - \bar{x}_2) - (\mu_1 - \mu_2)}{\sigma_{\bar{x}_1 - \bar{x}_2}} \qquad (2)$$

where:

$\bar{x}_1 - \bar{x}_2$: It is the observed difference

$\mu_1 - \mu_2$: It is the expected difference.

$\sigma_{\bar{x}_1 - \bar{x}_2}$: Standard error of the differences.

Table 4 Percent of success using Gaussian MFs

% success Gaussian MFs	
CSO	CSA
85	72.25
83.75	72.5
83.75	95
85	78.75
85	88.75
90	86.25
87.5	75
90	83.75
91.25	91.25
86.25	76.25
92.5	78.75
87.5	90
87.5	80
88.75	80
88.75	92.5
81.25	76.25
87.5	87.5
87.5	88.75
86.25	90
86.25	83.75
88.75	83.75
88.75	78.75
88.75	75
90	88.75
87.5	95
90	91.25
88.75	92.5
87.5	90
87.5	85
86.25	93.75

Table 5 Average of 30 experiments

CSO		CSA	
Trapezoidal	Gaussian	Trapezoidal	Gaussian
91.46%	87.59%	87.25%	84.7%

Table 6 Parameters of hypothesis test

Parameters	
Confidence interval	95%
Alpha	0.05
Ho	$\mu_1 \leq \mu_2$
Ha	$\mu_1 > \mu_2$
Critical value	$Z = 1.645$

Table 7 Descriptive statistics of Z-test

Variable	Observations	Mean	Std. deviation
CSO	30	91.458	1.944
CSA	30	87.250	1.897

Table 8 Results of the Z-test

Difference	4.208
z (observed value)	8.485
z (critical value)	1.645
p-value	0
Alpha	0.05

The null hypothesis establishes that the mean of the classification in CSO are less than or equal to the mean of the classification in CSA, the alternative hypothesis being that the mean of classification in CSO are greater than the average of classification in CSA, the parameters of the hypothesis test are presented in Table 6.

In Table 7, the descriptive statistics for this test are presented.

In Table 8, the results obtained by applying Eq. 2 are presented.

Since the result of the p-value is less than the level of significance alpha $= 0.05$, we reject the null hypothesis and accept the alternative hypothesis. It can then be concluded that there is enough evidence at the 5% level of significance to support the claim that the averages of the classification in CSO are greater than the classification with CSA.

The statistical test is also performed using the 30 experiments made with the CSO and CSA algorithms when the Gaussian membership functions, using the classification percentage obtained from each of the fuzzy systems designed by the two aforementioned methods.

The null hypothesis establishes that the mean of the classification rates in CSO is less than or equal to the mean of the classification in CSA, the alternative hypothesis being that the mean of classification in CSO is greater than the average of classification in CSA, the parameters of the hypothesis test are presented in Table 9.

In Table 10, the descriptive statistics for this test are presented.

In Table 11, the results obtained by applying Eq. 2 are presented.

Table 9 Parameters of the hypothesis test

Parameters	
Confidence interval	95%
Alpha	0.05
Ho	$\mu_1 \leq \mu_2$
Ha	$\mu_1 > \mu_2$
Critical value	$Z = 1.645$

Table 10 Descriptive statistics of Z-test

Variable	Observations	Mean	Std. deviation
CSO	30	87.500	2.390
CSA	30	84.700	7.030

Table 11 Results of the Z-test

Difference	2.800
z (Observed value)	2.065
z (Critical value)	1.645
p-value	0.0165
Alpha	0.05

Since the result of the p-value is less than the level of significance alpha $= 0.05$, we reject the null hypothesis and accept the alternative hypothesis. It can conclude that there is enough evidence with a 5% level of significance to support the claim that the averages of the classification in CSO are greater than the classification with CSA.

6 Conclusions

In this work, the optimization of the nocturnal blood pressure profile fuzzy system using two bioinspired algorithms is presented, and a comparison to observe which performs a better optimization is also done. The membership functions are optimized with the aim of making the necessary adjustments in it and thus improve its performance. Two fuzzy system are designed, the first with trapezoidal membership functions and the second, with Gaussian membership functions.

30 different experiments are carried out with each algorithm, of which the parameters of the same were varied; likewise these experiments were carried out with the two designed fuzzy systems.

Once the experimentation is finished and the corresponding statistical tests are carried out, it can be concluded that the CSO provided better results by optimizing the fuzzy systems with both Gaussian and trapezoidal membership functions. As future work, the optimization of this same fuzzy classifier will be performed with interval type-2 fuzzy logic [30–32], and in the same way the necessary comparisons will be made, as well as its statistical test. Additionally, a comparison will be made with the type-1 FIS to observe which provides the most accurate diagnosis. Other applications for the method can also be considered [27–29, 33, 34].

Acknowledgements The authors would like to express thank to the Consejo Nacional de Ciencia y Tecnologia and Tecnologico Nacional de Mexico/Tijuana Institute of Technology for the facilities and resources granted for the development of this research.

References

1. S. Kumar, G. Kaur, Detection of heart diseases using fuzzy logic. Int. J. Eng. Trends Technol. (IJETT) **4**(6), 2694–2699 (2013)
2. X.Y. Djam, Y.H. Kimbi, Fuzzy expert system for the management of hypertension. Pac. J. Sci. Technol. **12**(1), 390–402 (2011)
3. Q. Duodu, J.K. Panford, J. Ben Hafron-acquah, Designing algorithm for malaria diagnosis using fuzzy logic for treatment (AMDFLT) in Ghana. Int. J. Comput. Appl. **91**(17) (2014)
4. J.C. Guzman, P. Melin, G. Prado-Arechiga, Design of an optimized fuzzy classifier for the diagnosis of blood pressure with a new computational method for expert rule optimization. Algorithms **10**(3), 79 (2017)
5. X.S. Yang, M. Karamanoglu, X. He, Flower pollination algorithm: a novel approach for multiobjective optimization. Eng. Optim. **46**(9), 1222–1237 (2014)
6. S. Mirjalili, S.M. Mirjalili, A. Lewis, Grey wolf optimizer. Adv. Eng. Softw. **69**, 46–61 (2014)
7. X.-S. Yang, Firefly Algorithm, Lévy flights and global optimization, in *Research and Development in Intelligent Systems XXVI* (2010), pp. 209–218
8. M.L. Lagunes, O. Castillo, J. Soria, Methodology for the optimization of a fuzzy controller using a bio-inspired algorithm, in *Fuzzy Logic in Intelligent System Design* (2018), pp. 131–137
9. J. Perez, P. Melin, O. Castillo, F. Valdez, C. Gonzalez, G. Martinez, Trajectory optimization for an autonomous mobile robot using the bat algorithm, in *Fuzzy Logic in Intelligent System Design* (2018), pp. 232–241
10. C. Peraza, F. Valdez, P. Melin, Optimization of intelligent controllers using a Type-1 and interval Type-2 fuzzy harmony search algorithm. Algorithms **10**(3), 1–17 (2017)
11. O.R. Carvajal, O. Castillo, J. Soria, Optimization of membership function parameters for fuzzy controllers of an autonomous mobile robot using the flower pollination algorithm. J. Autom. Mob. Robot. Intell. Syst. **12**(1), 1–23 (2018)
12. X. Meng, Y. Liu, X. Gao, H. Zhang, A new bio-inspired algorithm: chicken swarm optimization, in *Advances in Swarm Intelligence* (2014), pp. 86–94
13. A. Askarzadeh, A novel metaheuristic method for solving constrained engineering optimization problems: crow search algorithm. Comput. Struct. **169**(Supplement C), 1–12 (2016)
14. J.M. Wilson, Essential cardiology: principles and practice. Tex. Heart Inst. J. **32**(4), 616 (2005)
15. B. Wizner, B. Gryglewska, J. Gasowski, J. Kocemba, T. Grodzicki, Normal blood pressure values as perceived by normotensive and hypertensive subjects. J. Hum. Hypertens. **17**(2), 87–91 (2003)
16. O.A. Carretero, S. Oparil, Essential Hypertension. Circulation **101**(3), 329–335 (2000)
17. D. Bloomfield, Night time blood pressure dip. World J. Cardiol. **7**(7), 373 (2015)
18. M. Brian, A. Dalpiaz, E. Matthews, S. Lennon-Edwards, D. Edwards, W. Farquhar, Dietary sodium and nocturnal blood pressure dipping in normotensive men and women. J. Hum. Hypertens. Hypertens. **31**, 145–150 (2016)

19. L.E. Okamoto et al., Nocturnal blood pressure dipping in the hypertension of autonomic failure. Hypertension **53**(2), 363–369 (2009)
20. E. Grossman, Ambulatory blood pressure monitoring in the diagnosis and management of hypertension. Diab. Care **36**(Supplement 2), S307–S311 (2013)
21. O. Friedman, A.G. Logan, Nocturnal blood pressure profiles among normotensive, controlled hypertensive and refractory hypertensive subjects. Can. J. Cardiol. **25**(9), e312–e316 (2009)
22. I. Miramontes, G. Martínez, P. Melin, G. Prado-Arechiga, A hybrid intelligent system model for hypertension diagnosis, in *Nature-inspired design of hybrid intelligent systems*, ed. by P. Melin, O. Castillo, J. Kacprzyk (Springer International Publishing, Cham, 2017), pp. 541–550
23. I. Miramontes, G. Martínez, P. Melin, G. Prado-Arechiga, A hybrid intelligent system model for hypertension risk diagnosis, in *Fuzzy Logic in Intelligent System Design* (2018), pp. 202–213
24. P. Melin, I. Miramontes, G. Prado-Arechiga, A hybrid model based on modular neural networks and fuzzy systems for classification of blood pressure and hypertension risk diagnosis. Expert Syst. Appl. **107**, 146–164 (2018)
25. P. Melin, G. Prado-Arechiga, I. Miramontes, J.C. Guzman, Classification of nocturnal blood pressure profile using fuzzy systems. J. Hypertens. **36**, e111–e112 (2018)
26. M.D. Feria-carot, J. Sobrino, Nocturnal hypertension. Hipertens. y riesgo Cardiovasc. **28**(4), 143–148 (2011)
27. P. Melin, A. Mancilla, M. Lopez, O. Mendoza, A hybrid modular neural network architecture with fuzzy Sugeno integration for time series forecasting. Appl. Soft Comput. **7**(4), 1217–1226 (2007)
28. P. Melin, O Castillo, *Modelling, Simulation and Control of Non-linear Dynamical Systems: An Intelligent Approach Using Soft Computing and Fractal Theory* (CRC Press, 2001)
29. P. Melin, D. Sánchez, O. Castillo, Genetic optimization of modular neural networks with fuzzy response integration for human recognition. Inf. Sci. **197**, 1–19 (2012)
30. C.I. González, P. Melin, J.R. Castro, O. Mendoza, O. Castillo, An improved sobel edge detection method based on generalized type-2 fuzzy logic. Soft. Comput. **20**(2), 773–784 (2016)
31. C.I. González, P. Melin, J.R. Castro, O. Castillo, O. Mendoza, Optimization of interval type-2 fuzzy systems for image edge detection. Appl. Soft Comput. **47**, 631–643 (2016)
32. E. Ontiveros, P. Melin, O. Castillo, High order α-planes integration: a new approach to computational cost reduction of General Type-2 Fuzzy Systems. Eng. Appl. AI **74**, 186–197 (2018)
33. P. Melin, D. Sánchez, Multi-objective optimization for modular granular neural networks applied to pattern recognition. Inf. Sci. **460–461**, 594–610 (2018)
34. D. Sánchez, P. Melin, O. Castillo, Optimization of modular granular neural networks using a firefly algorithm for human recognition. Eng. Appl. AI **64**, 172–186 (2017)

Design of Interval Type-2 Fuzzy Systems for Classification of Blood Pressure Load

Juan Carlos Guzmán, Patricia Melin and German Prado-Arechiga

Abstract In this work we will design the interval type-2 fuzzy system (FS), which will be based on the definitions and classifications of the blood pressure loads, which are elaborated by cardiology experts and based on table of ranges, the use of an interval type-2 fuzzy system and the reliability of the FOU (footprint of uncertainty) can allows achieving a 100% effective classification, it should be noted that in previous works experiments with type-1 fuzzy systems have been carried out, the goal is to have a good classification of the blood pressure load which is very important nowadays for a cardiologist, since based on this determine a cardiovascular event, the blood pressure load which indicates that the daytime blood pressure load (% of diurnal readings $\geq$ 135/85 mmHg) and the nocturnal blood pressure load (% of nocturnal readings $\geq$ 120/70 mmHg).

Keywords Fuzzy system · Hypertension · Diagnosis · Load blood pressure

1 Introduction

Currently there are many diseases, but one of the most frequent is hypertension, which is a silent disease, which indicates that there are patients who do not know that they suffer from this disease, however this is attacking in silence and one day could cause a cardiovascular event, it is for this reason the importance that should be given to the use of constant check-ups, at home, in the office and most importantly to carry out a 24-h monitoring at least every six months, this with the objective of obtaining information of the functioning of our blood pressure(BP) and see if there is a load of blood pressure in our body and thus dealing with a specialist in these disorders in BP [1–5].

J. C. Guzmán · P. Melin (✉)
Tijuana Institute of Technology, Tijuana, BC, Mexico
e-mail: pmelin@tectijuana.mx

G. Prado-Arechiga
Excel Medical Center, Tijuana, BC, Mexico

© Springer Nature Switzerland AG 2020
O. Castillo and P. Melin (eds.), *Hybrid Intelligent Systems in Control,
Pattern Recognition and Medicine*, Studies in Computational Intelligence 827,
https://doi.org/10.1007/978-3-030-34135-0_16

233

In daily check-ups, people check their blood pressure, in which the doctor tells them if they have high blood pressure or not, but they omit to check all the other hours during the day and night, which is why the 24-h check-up It allows to obtain valuable information both day and night, this helps to know if there are blood pressure loads in these periods of time and to be able to treat them and to prevent possible cardiovascular events in the periods of time with more loads of pressure [6, 7]. This work focuses on the use of fuzzy logic intelligent computing technique, which allows fuzzy systems based on experts, in this case, we have basic information which, given by experts in the area of cardiology [8–12].

The paper is organized as follows: in Sect. 2 Design of interval type-2 fuzzy system is described, in Sect. 3 Results of the experiments of interval type-2 fuzzy system is described, and in Sect. 4 the Conclusion obtained with interval type-2 fuzzy classifier.

2 Design of Interval Type-2 Fuzzy System

2.1 Blood Pressure Load

The measurement of the blood pressure load is a measure that results in the 24-h monitoring device, in which the number of readings is analyzed above the ranges established by Table 1. These categories are shown in Table 1.

In Table 2 we have an example of an analysis of the pressure load.

Table 1 Definitions and classification of the blood pressure load

Blood pressure load categories	Blood pressure load ranges
Normal Load	<20
Intermediate Load	20–40
High Load	>40

Table 2 Example of a blood pressure load record

Day and night period	Time	Interval	Readings	Readings with BP_load	Awake	Asleep
Day period	07–22	20 min	35	3	8.60%	
Night period	22–07	30 min	18	15		83.33%
Day BP load	(% of day readings ≥ 135/85 mmHg)					
Night BP load	(% of night readings ≥ 120/70 mmHg)					

2.2 Interval Type-2 Fuzzy System for the Blood Pressure Load

The interval type-2 fuzzy system has 2 inputs which are named Diurnal_Load and Nocturnal_Load respectively and an output that is called Load_level. The membership functions for the Diurnal_load input are: Normal_Diurnal_Load, Intermediate_Diurnal_Load and High_Diurnal_Load, for the second input called Nocturnal_Load, the membership functions are: Normal_Nocturnal_Load, Intermediate_Nocturnal_Load and High_Nocturnal_Load and finally the membership functions for the output called Load_Level are: Normal, Intermediate and High. The interval type-2 fuzzy system is mamdani type and ten fuzzy rules [13–16].

The structure, membership functions and linguistic values of the interval type-2 fuzzy system are shown in the following figures: Fig. 1 shows the structure of interval type-2 fuzzy system for the classification of blood pressure load, Fig. 2 shows linguistic variable and membership functions of the input "Diurnal_Load", Fig. 3 shows the linguistic variable and membership functions of the input "Nocturnal_Load", Fig. 4 shows the linguistic variable and membership functions of the output "Load_level" [17, 18].

Then, in Fig. 5 shows interval type-2 fuzzy system rules for the classification of blood pressure load.

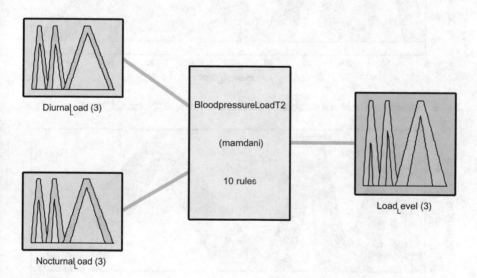

Fig. 1 Structure of the interval type-2 fuzzy system for the classification of blood pressure load

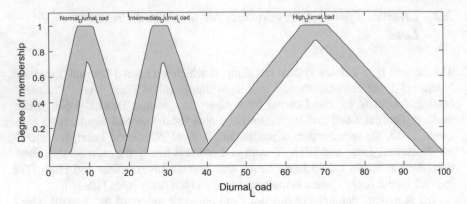

Fig. 2 Linguistic variable and membership functions of the input "Diurnal_Load"

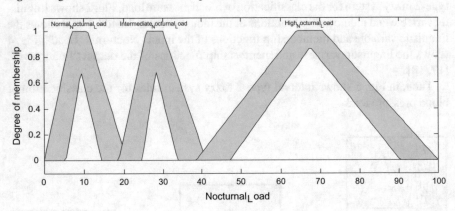

Fig. 3 Linguistic variable and membership functions of the input "Nocturnal_Load"

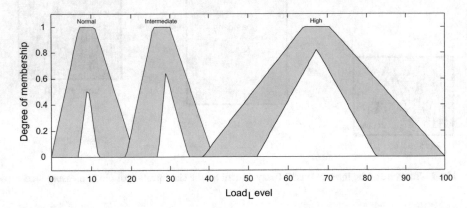

Fig. 4 Linguistic variable and membership functions of the output "Load_level"

1. If (Diurnal_Load is Normal_Diurnal_Load) and (Nocturnal_Load is Normal_Nocturnal_Load) then (Load_Level is Normal) (1)
2. If (Diurnal_Load is Intermediate_Diurnal_Load) and (Nocturnal_Load is Intermediate_Nocturnal_Load) then (Load_Level is Intermediate) (1)
3. If (Diurnal_Load is High_Diurnal_Load) and (Nocturnal_Load is High_Nocturnal_Load) then (Load_Level is High) (1)
4. If (Diurnal_Load is Normal_Diurnal_Load) or (Nocturnal_Load is Intermediate_Nocturnal_Load) then (Load_Level is Intermediate) (1)
5. If (Diurnal_Load is Normal_Diurnal_Load) or (Nocturnal_Load is High_Nocturnal_Load) then (Load_Level is High) (1)
6. If (Diurnal_Load is Intermediate_Diurnal_Load) or (Nocturnal_Load is Normal_Nocturnal_Load) then (Load_Level is Intermediate) (1)
7. If (Diurnal_Load is High_Diurnal_Load) or (Nocturnal_Load is Normal_Nocturnal_Load) then (Load_Level is High) (1)
8. If (Diurnal_Load is Intermediate_Diurnal_Load) or (Nocturnal_Load is High_Nocturnal_Load) then (Load_Level is High) (1)
9. If (Diurnal_Load is High_Diurnal_Load) or (Nocturnal_Load is Intermediate_Nocturnal_Load) then (Load_Level is High) (1)
10. If (Diurnal_Load is High_Diurnal_Load) or (Nocturnal_Load is High_Nocturnal_Load) then (Load_Level is High) (1)

Fig. 5 Fuzzy interval type-2 rules for the classification of blood pressure load

3 Results of the Experiments of Interval Type-2 Fuzzy System

It is important to mention that the diurnal load is independent of the nocturnal load, and this is because the 24 h of the day are divided into two periods of time, which are day and night and based on these periods of time, the pressure load of each of these periods, at the end we have two percentages which are analyzed with the interval type-2 fuzzy classifier [19–21].

The patients that were analyzed are 30, which had to carry the device for a period of 24 consecutive hours, in which 45 samples were obtained, the blood pressure load was analyzed and displayed by period diurnal and nocturnal.

In Table 3, For this work, the tests were performed with 30 patients, from which information was obtained with the 24-h monitoring, which gives the pressure load percentage for the day time period and for the night time period and as results we have to have classified 100% of the patients correctly based on the ranges established by the experts in cardiology, as shown in Table 1.

4 Conclusions

Use of artificial intelligence in the area of medicine has achieved great contributions in decision-making, that is why the use of fuzzy logic and the implementation of interval type-2 fuzzy system is making timely classifications based on the analysis of information given by a device of 24 h, this gives two values, which are represented as day load and night load and based on this, the fuzzy classifier with the help of fuzzy rules and structured membership functions based on knowledge from an expert we can conclude that we are achieving the effective, rapid and timely classifications and this would help to strengthen a diagnosis of hypertension, by blood pressure load, we can see the malfunction that the patient may have in their blood pressure and give a notion of which period of time is the most vulnerable or suffers the greatest cardiovascular problem and through this information can treat the lack of control and avoid a cardiovascular event. To finish we can mention that so far the type-2 classifier has achieved the expected results.

Table 3 Results of the classification of blood pressure load

Patient	Diurnal_Load (%)	Nocturnal_Load (%)	Load_Level
1	2	55.3	High
2	8.3	47.8	High
3	55.2	41.8	High
4	8.0	45.0	Intermediate
5	77.0	82.9	High
6	0.0	2.3	Normal
7	76.0	33.3	High
8	0.0	5.0	Normal
9	53.0	56.1	High
10	32.0	23.0	Intermediate
11	46.9	93.9	High
12	54.2	90.0	High
13	62.4	23.0	High
14	3.3	33.8	Intermediate
15	28.6	57.3	High
16	36.8	21.2	Intermediate
17	7.9	5.3	Normal
18	5.3	14.7	Normal
19	46.8	18.8	High
20	17.7	31.8	Intermediate
21	9.9	27.6	Intermediate
22	36.3	16.7	Intermediate
23	9.3	0.0	Normal
24	7.3	31.3	Intermediate
25	55.5	0.0	High
26	71.6	47.1	High
27	23.6	33.3	Intermediate
28	5.8	15.3	Normal
29	61.0	63.0	High
30	0.0	12.0	Normal

Acknowledgements We would like to express our gratitude to the CONACYT and Tijuana Institute of Technology for the facilities and resources granted for the development of this research.

References

1. L. Kenney, R. Humphrey, D. Mahler, and C. Brayant, *ACSM's Guidelines for Exercise Testing and Prescription* (Williams & Wilkins, 1995)
2. Texas Heart Institute, *High Blood Pressure (Hypertension)* (2017)
3. G. Mancia, G. Grassi, S.E. Kjeldsen, *Manual of Hypertension of the European Society of Hypertension* (Informa Healtcare, UK, 2008)
4. B. Wizner, B. Gryglewska, J. Gasowski, J. Kocemba, T. Grodzicki, Normal blood pressure values as perceived by normotensive and hypertensive subjects. J. Hum. Hypertens. **17**(2), 87–91 (2003)
5. C. Rosendorff, *Essential Cardiology*, 3rd edn. (Springer, Bronx, NY, USA, 2013)
6. E.J. Battegay, G.Y.H. Lip, G.L. Bakris, *Hypertension: Principles and Practices* (Taylor & Francis, Boca Raton, FL, 2005)
7. O.A. Carretero, S. Oparil, Essential hypertension. Circulation **101**(3), 329–335 (2000)
8. L.A. Zadeh, Fuzzy sets. Inf. Control **8**(3), 338–353 (1965)
9. P. Melin, O. Castillo, *Hybrid Intelligent Systems for Pattern Recognition Using Soft Computing* (Springer, Berlin, 2005)
10. Q. Duodu, J.K. Panford, J. Ben Hafron-Acquah, Designing algorithm for malaria diagnosis using fuzzy logic for treatment (AMDFLT) in Ghana. Int. J. Comput. Appl. **91**(17), 22–27 (2014)
11. I. Morsi, Y.Z. Abd El Gawad, Fuzzy logic in heart rate and blood pressure measuring system, in *IEEE Sensors Applications Symposium Proceedings* (2013), pp. 113–117
12. R. Nohria, P.S. Mann, Diagnosis of hypertension using adaptive neuro-fuzzy inference system. Int. J. Comput. Sci. Technol. **8491**, 36–40 (2015)
13. S. Sikchi, S. Sikchi, M. Ali, Design of fuzzy expert system for diagnosis of cardiac diseases. Int. J. Med. Sci. Public Health **2**(1), 56 (2013)
14. O. Oparaku, E. Udo, Fuzzy logic system for fetal heart rate determination. Int. J. Eng. Res. **5013**(4), 60–63 (2015)
15. A.A. Sadat Asl, M.H.F. Zarandi, A Type-2 fuzzy expert system for diagnosis of leukemia, in *Fuzzy Logic in Intelligent System Design* (2018), pp. 52–60
16. S. Sotudian, M.H.F. Zarandi, I.B. Turksen, From Type-I to Type-II fuzzy system modeling for diagnosis of hepatitis. World Acad. Sci. Eng. Technol. Int. J. Comput. Electr. Autom. Control Inf. Eng. **10**(7), 1280–1288 (2016)
17. V. Pabbi, Fuzzy expert system for medical diagnosis. Int. J. Sci. Res. Publ. **5**(1), 1–7 (2015)
18. K.A. Mohamed, E.M. Hussein, Malaria parasite diagnosis using fuzzy logic. Int. J. Sci. Res. **5**(6), 2015–2017 (2016)
19. I. Miramontes, G. Martínez, P. Melin, G. Prado-Arechiga, A hybrid intelligent system model for hypertension risk diagnosis, in *Fuzzy Logic in Intelligent System Design* (2018), pp. 202–213
20. P. Melin, I. Miramontes, G. Prado-Arechiga, A hybrid model based on modular neural networks and fuzzy systems for classification of blood pressure and hypertension risk diagnosis. Expert Syst. Appl. **107**, 146–164 (2018)
21. I. Miramontes, G. Martínez, P. Melin, G. Prado-Arechiga, A hybrid intelligent system model for hypertension diagnosis, in *Nature-Inspired Design of Hybrid Intelligent Systems*, ed. by P. Melin, O. Castillo, J. Kacprzyk (Springer International Publishing, Cham, 2017), pp. 541–550

Tracking of Non-rigid Motion in 3D Medical Imaging with Ellipsoidal Mapping and Germinal Center Optimization

Carlos Villaseñor, Nancy Arana-Daniel, Alma Y. Alanis,
Carlos Lopez-Franco and Roberto Valencia-Murillo

Abstract Visualizing physical phenomena is a central tool for nowadays research. In particular, volumetric representations are a critical factor in the diagnosis of diseases and surgery planning. In the last years, rendering techniques have been essential for medical practice, but these approaches are suitable for representing non-rigid motion in tissue and internal organs. In the present chapter, we introduce a mapping algorithm capable of track non-rigid deformations on free-form objects. The proposed method uses k-means for partition algorithm and covariance ellipsoid, afterward the Germinal Center Optimization is used to adapt the ellipsoid parameters. We offer experimental results over the Stanford Repository and tumors.

Keywords Germinal Center Optimization · 3D reconstruction · Internal organs mapping

1 Introduction

In multiple fields of study, it is essential to visualize physical phenomena, for example, in the medical field visualization of internal organs and other tissues is a critical factor for the diagnosis of diseases and surgery planning [1].

C. Villaseñor · N. Arana-Daniel · A. Y. Alanis (✉) · C. Lopez-Franco · R. Valencia-Murillo
Universidad de Guadalajara, Blvd Marcelino García Barragán 1421,
44430 Guadalajara, Mexico
e-mail: almayalanis@gmail.com

C. Villaseñor
e-mail: cavp@outlook.com

N. Arana-Daniel
e-mail: nancy.arana@cucei.udg.mx

C. Lopez-Franco
e-mail: clzfranco@gmail.com

R. Valencia-Murillo
e-mail: vamyur@gmail.com

© Springer Nature Switzerland AG 2020   241
O. Castillo and P. Melin (eds.), *Hybrid Intelligent Systems in Control,*
Pattern Recognition and Medicine, Studies in Computational Intelligence 827,
https://doi.org/10.1007/978-3-030-34135-0_17

In last decade medical image techniques [2], have played a key role in clinical practice, but in last years, volumetric representations are attracting considerable interest due to the huge advantage to understand functions and anomalies which could lead to better diagnosis and treatments [1, 3].

As is set in [4], volumetric medical image is the method of extracting meaningful data for three-dimensional data. Some visualization methods are Multiplanar Reformation (MPR) [5], Surface Rendering (SR) [6] and Volume Rendering (VR) [7].

MPR technique offers a comprehensive model of internal organs but with a memory-expensive solution, SR and VR are rendering techniques based in computer graphics technologies where radiance of object is modeled and rendered with the emission-absorption optical model that lead to the Volume-Rendering Integral in (1) [7].

$$I(D) = I_0 e^{-\int_{s_0}^{D} k(t)dt} + \int_{s_0}^{D} q(s) e^{\int_{s_0}^{D} k(t)dt} ds \tag{1}$$

where, I is the radiance, $k(t)$ is the absorption coefficient in the time t, q is the source of the radiance emission, s_0 is the entry point into the volume, and D is the exit point to the modeled camera.

The principal disadvantage of SR and VR approaches is that they model a camera, then the 3D information is projected into a plane, and the model is not manipulable. Another disadvantage is that solving (1), require high-performance hardware.

Besides 3D models of internal organs, it is also essential to understand them through time to better understand the biological phenomena. The Internal organs and tissues are subject to non-rigid deformations then it is desired to have a mapping algorithm capable of tracking non-rigid motion on free-form objects [8].

In [9], we presented a novel algorithm for robotic mapping using Geometric Algebra G6,3 and based in the Hyperellipsoidal Neuron [10]; this algorithm is capable of adjusting multiple ellipsoids to a point cloud for data compression in robotic tasks. Afterward, we present an algorithm for tracking non-rigid motion in internal organs using 4D Magnetic Resonance Imaging datasets [11], where we used Differential Evolution and Covariance Ellipsoid to adapt the movement.

In [11], the ellipsoid tracks the non-rigid movement of a partition of the object but preserving orientation. In this chapter, we extend [11], by introducing a change of rotation in the ellipsoids and using Germinal Center Optimization (GCO) to find the non-rigid transform.

GCO [12] is a multivariate continuous optimization algorithm inspired in the Germinal Center Reaction [13]. GCO mimics various competitive process in the Germinal Center that we discuss in Sect. 3 and it offers adaptive leadership for particle selection.

To test the proposed approach, we include experimental results with an artificial non-rigid deformation (non-isotropic scale) over the Standford 3D Scanning Repository [14]. Finally, we include experimental results of 3 tumors scans [15].

The chapter is organized as follows: in Sect. 2, we introduce the Covariance ellipsoid and the Mahalanobis distance. In Sect. 3, we describe the GCO algorithm and in Sect. 4, we explain the proposed method and the cost function. In Sect. 5, we show the experimental results and finally in Sect. 6, we include the conclusions of this work.

2 Covariance Ellipsoid

In this section, we briefly introduce to the Covariance Ellipsoid (CE) and the Mahalanobis distance [16]. This tool is used to map a point cloud into an ellipsoid.

To measure a joint probability distribution of two random variables we can use the covariance defined in (2). In the particular case of three variables, we can use the covariance matrix defined in (3).

$$\sigma_{x,y} = \sum_{i=1}^{n} \frac{(x_i - \mu_x)(y_i - \mu_y)}{n} \tag{2}$$

$$S = \begin{bmatrix} \sigma_{x,x} & \sigma_{x,y} & \sigma_{x,z} \\ \sigma_{x,y} & \sigma_{y,y} & \sigma_{y,z} \\ \sigma_{x,z} & \sigma_{y,z} & \sigma_{z,z} \end{bmatrix} \tag{3}$$

We can use spectral decomposition of the covariance matrix shown in (4), where $V \in \mathcal{R}^{3 \times 3}$ defined in (5), is an orthonormal matrix formed by eigenvectors, and $D \in \mathcal{R}^{3 \times 3}$ defined in (6) is a diagonal matrix containing the eigenvalues of the matrix S.

$$S = VDV^T \tag{4}$$

$$V = \begin{bmatrix} V_1 & V_2 & V_3 \end{bmatrix} \tag{5}$$

$$D = \begin{bmatrix} \lambda_1 & 0 & 0 \\ 0 & \lambda_2 & 0 \\ 0 & 0 & \lambda_3 \end{bmatrix} \tag{6}$$

We can obtain CE with center in $\{\mu_x, \mu_y, \mu_z\}$ and orientation given by V and the semi-axes are given by the eigenvalues in $\{\lambda_1, \lambda_2, \lambda_3\}$. The precision matrix is the inverse of the covariance matrix, defined in (7). With this equation, we can get the Mahalanobis Distance (MD) described in (8).

$$S^{-1} = VD^{-1}V^T \tag{7}$$

$$d_M(p_1, p_2) = \sqrt{(p_1 - p_2)^T S^{-1}(p_1 - p_2)} \tag{8}$$

With $\{\lambda_1, \lambda_2, \lambda_3\}$ we can control the size of the ellipsoid and with Eq. (9), where $\{\phi, \theta \psi\}$ are the Tait-Bryan angles and R is the rotation matrix.

$$V' = R(\phi, \theta, \psi)V \tag{9}$$

3 Germinal Center Optimization

In this section, we introduce the Germinal Center Optimization (GCO) algorithm [12], which is a meta-heuristic for solving multivariate optimization problems. This algorithm has been successfully applied to neural control [17, 18]. In this chapter, we use GCO to adapt the ellipsoids parameters to tracking the non-rigid motion.

GCO is based in the germinal center reaction which is a biological phenomenon where lymphocytes B (also known as B-cells) compete to get the best affinity to an antigen in a closed environment. In Fig. 1, we showed a simplified diagram of the germinal center reaction.

Germinal Centers (GCs) are clusters of B-cells in the secondary lymphoid nodes which form in the presence of antigen, and inactive B-cells bound them. Inside the GCs, we can find two zones histologically differentiable, the Dark Zone (DZ) and the Light Zone (LZ). In one hand, in the DZ the B-cells proliferate in a process called *Clonal expansion* and mutate its Antibodies (Abs) through *Somatic hypermutation*. In the other hand, in the LZ the B-cells compete for Antigen (Ag) internalization, in this process, the B-cell catches the Ag with a receptor and decompose it in its

Fig. 1 Germinal Center Reaction is the process through which the B-cell compete to get the best affinity to the Antigen, when a B-cell has a good affinity performance the T-cell give it a life signal to proliferate

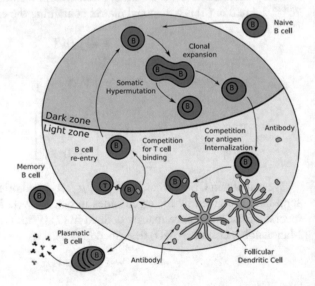

peptides. After this, the B-cells compete for binding with a lymphocyte T CD4$^+$ (T-cell), which gives a life signal to the B-cell. Then the B-cells with better receptors can catch more antigen and have more chances to get the life signal, in consequences, the B-cell with life signal reentry to the DZ and proliferate for more generations.

Algorithm 1: GCO algorithm

Initialize B-cells (B_i)
foreach $k \in \{1, \cdots, Iterations\}$ **do**
  /* *Dark-zone process* */
  foreach $i \in \{1, \cdots, N\}$ **do**
    if $r_1 \in \{1, \cdots, N\}$ **then**
      Add one to B_i cells counter
    else
      if *Cells in* $B_i > 1$ **then**
        Rest one to B_i cells counter
      end
    end
    Calculate distribution $\mathcal{C}$
    Calculate $r_1, r_2, r_3 \sim \mathcal{C}$
    foreach $j \in \{1, \cdots, d\}$ **do**
      if $r \sim U[0,1] < CR$ **then**
        $M(j) \leftarrow B_{r_1}(j) + F * (B_{r_2}(j) - B_{r_3}(j))$
      else
        $M(j) \leftarrow B_i(j)$
      end
    end
    Calculate $f(M)$
    if $f(M) < f(B_i)$ **then**
      $B_i \leftarrow M$
      Add 10 to $\mathcal{L}$ of B_i
      if $f(M) < best$ **then**
        $best \leftarrow f(M)$
        best_index$\leftarrow$i
      end
    end
  end
  /* *Light-zone process* */
  foreach $i \in \{1, \cdots, N\}$ **do**
    Rest 10 units to $\mathcal{L}$ of B_i
    $$\text{fit}_i = \frac{f(B_i) - \max f(B_k)}{\min f(B_k) - \max f(B_k)} \in [0,1]$$
    Add $10 * \text{fit}_i$ to $\mathcal{L}$ of B_i
  end
end

Based in the process explained above, the GC reaction is an iterative refinement of the B-cells that have more chances to catch a particular Ag (the so-called *affinity*). Then, the B-cell with high affinity can go out of the GC and differentiate to a plasmatic cell that secret Ab to overcome the infection. The B-cell could also differentiate into a Memory B-cell which have a long lifespan, in consequence, it carries the information for fighting a particular Ag.

The GCO algorithm mimics the behavior of the GC reaction by modeling the DZ and LZ. In Algorithm 1, we show the GCO pseudocode that we explain below. Each B-cells represent a candidate solution of a continuous multivariate problem, the affinity of a B-cell represent the fitness of the candidate solution, and every Bcell has a life signal modeled as a real value $L \in [0, 100]$, by default the life signal begins in 70. We also count the B-cell multiplicity because each B-cell can have multiple clones with the same information.

In the DZ, the L value we represent the probability of the B-cell to clone or otherwise to die. This affects the B-cell multiplicity counter. The B-cell mutate with the same scheme that Differential Evolution algorithms but with a different parents selection. This selection is implemented with Roulette Selection but not based on the fitness of the candidate solution but the multiplicity distribution C.

Using the multiplicity distribution is an essential feature of the GCO algorithm because the distribution is modeled by the competitive process to get more life signal. This selection rule includes new features that other evolutionary and bioinspired algorithms do not have, which are adaptive leadership and delayed leadership, as is discussed in [12]. The mutated B-cell is better than his parent B-cell we substitute the parent B-cell y the son B-cell and set the multiplicity counter to one and reset the life signal.

In the LZ, the B-cell compete for a better life signal in the T-cell bounding. In the GCO analogy, this process means to evaluate the candidate solution in the cost function, and each B-cell is rewarded proportionally to their fitness, where the best particle get plus ten points of life signal and the worst B-cell get zero points of life signal.

4   Tracking Non-rigid Motion with CE and GCO

In this section, we develop the application of tracking non-rigid motion with Covariance ellipsoids and GCO algorithm.

First, we need to map one time the point cloud using multiple ellipsoids. In order to achieve this goal, we use the k-means algorithm which solves the problem of partitioning a point cloud in the Voronoi sense by minimizing the problem in (10).

$$\min_{C_1,\ldots,C_l} \sum_{i=1}^{l} \sum_{p \in C_i} ||p - \mu_i||_2 \tag{10}$$

After we calculate a CE for each cluster C_i, and recover the parameters $\{\mu_x, \mu_y, \mu_z, \lambda_1, \lambda_2, \lambda_3, \phi, \theta, \psi\}$ using the covariance ellipsoid. Then, to track the non-rigid motion, we need to recalculate these parameters the change of the parameter for each ellipsoid. For the first three parameters, we calculate the centroid of the cluster using

$$
\mu_i = \begin{bmatrix} \mu_x \\ \mu_y \\ \mu_z \end{bmatrix} = \frac{1}{|C_i|} \sum_{p \in C_i} \begin{bmatrix} p_x \\ p_y \\ p_z \end{bmatrix}
\tag{11}
$$

For the rest of the parameters, we use the GCO algorithm to find them. We want to find the parameters $\chi_k = \{\lambda_{1k}, \lambda_{2k}, \lambda_{3k}, \phi_k, \theta_k, \psi_k\}$ that minimize the change of density in the cluster, where k is the iteration. The first step to achieve this goal is to find the precision matrix that represents the rotated ellipsoid with the new semiaxes. This is implemented by rotate (7) with (9). We explicitly show the parameters χ_k for calculating the new precision matrix S_{ik} in (12).

$$
S_{ik}^{-1} = R(\phi_k, \theta_k, \psi_k) V_{k-1} \begin{bmatrix} \frac{1}{\lambda_{1k}} & & \\ & \frac{1}{\lambda_{2k}} & \\ & & \frac{1}{\lambda_{3k}} \end{bmatrix} V_{k-1}^T R(\phi_k, \theta_k, \psi_k)^T
\tag{12}
$$

After, we calculate a penalization function P described in (13). A distance $d_M(p, \mu_i) > 1$ implies that the point p is outside the ellipsoid, then we penalize this state to get the best fit for the semi-axes.

$$
P(p, i) = \begin{cases} \frac{1}{\lambda_{1k}\lambda_{2k}\lambda_{3k}} & \text{if } \sqrt{(p - \mu_i)^T S_{ik}^{-1}(p - \mu_i)} > 1 \\ 0 & \text{otherwise} \end{cases}
\tag{13}
$$

Finally, in (14) we show the optimization problem, where the first factor is the density of the ellipsoid (ratio between the volume of the ellipsoid and the number of points) and the second is the penalization described before over each point of the cluster i. Then, we calculate the tracking of the non-rigid motion by applying Eqs. (10)–(14) for every ellipsoid in the map.

$$
\min_{\chi_k} \frac{4\pi}{|C_i|} \lambda_{1k}\lambda_{2k}\lambda_{3k} + \frac{1}{|C_i|} \sum_{p \in C_i} P(p, i)
\tag{14}
$$

5 Experimental Results

In this section, we present the experimental results. For testing the tracking of nonrigid motion, we use the non-isotropic deformation described in (15) applied over every point of the point cloud, where the scale factors are given by (16)–(18).

$$
\begin{bmatrix} p'_x \\ p'_y \\ p'_z \end{bmatrix} = \begin{bmatrix} s_x & 0 & 0 \\ 0 & s_y & 0 \\ 0 & 0 & s_z \end{bmatrix} \left(\begin{bmatrix} p_x \\ p_y \\ p_z \end{bmatrix} - \begin{bmatrix} \mu_x \\ \mu_y \\ \mu_z \end{bmatrix} \right) + \begin{bmatrix} \mu_x \\ \mu_y \\ \mu_z \end{bmatrix}
\tag{15}
$$

$$
s_x = 1.5 \sin\left(\frac{2k\pi}{m}\right) + 3
\tag{16}
$$

$$
s_y = \sin\left(\frac{4k\pi}{m}\right) + 2
\tag{17}
$$

$$
s_z = 1.25 \sin\left(\frac{6k\pi}{m}\right) + 2.5
\tag{18}
$$

We applied this non-rigid transformation to two different cloud points data sets, the first one contain four-point clouds from the Stanford 3D repository [14], and the second one contains three point clouds of real tumors, courtesy of Andres Serna and Flavio Prieto from their paper in [15].

5.1 First Data Set

The first experiment in this data set is the well known Stanford bunny. In Fig. 2, we see the ellipsoidal map achieve with the k-means algorithm and the covariance

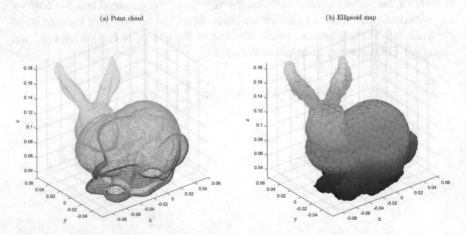

Fig. 2 Ellipsoidal mapping of the Stanford bunny

ellipsoid. In Fig. 3, we observe the non-rigid motion tracking with the GCO. Each subfigure contains two maps in two different moments of the non-rigid motion so that we can compare them (considers that there are limitations in showing dynamic results on paper).

In the second experiment, we show the result for the Stanford armadillo. In Fig. 4 we show the ellipsoidal map and in Fig. 5, we show the non-rigid motion tracking.

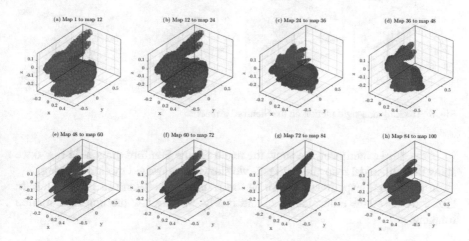

Fig. 3 Tracking non-rigid motion on the Stanford bunny

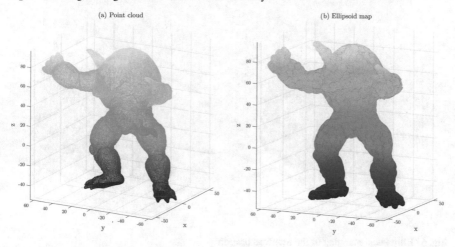

Fig. 4 Ellipsoidal mapping of the Stanford armadillo

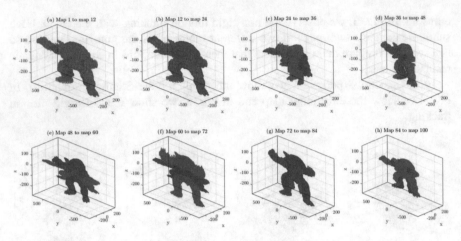

Fig. 5 Tracking non-rigid motion on the Stanford armadillo

In the third experiment, we show the result for the Stanford dragon. In Fig. 6 we show the ellipsoidal map and in Fig. 7, we show the non-rigid motion tracking.

In the second experiment, we show the result for the Stanford Lucy statue. In Fig. 8 we show the ellipsoidal map and in Fig. 9, we show the non-rigid motion tracking.

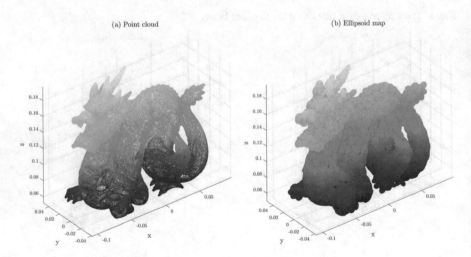

Fig. 6 Ellipsoidal mapping of the Stanford dragon

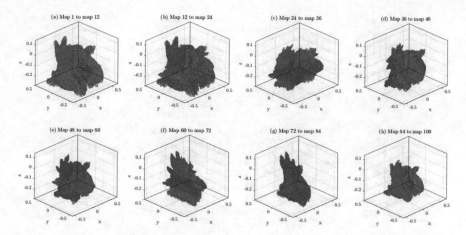

Fig. 7 Tracking non-rigid motion on the Stanford dragon

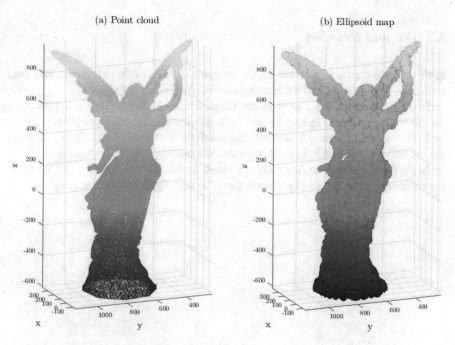

Fig. 8 Ellipsoidal mapping of the Stanford Lucy statue

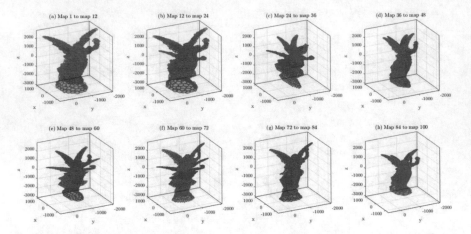

Fig. 9 Tracking non-rigid motion on the Stanford Lucy statue

5.2 Second Data Set

In this second data set, we show the experimental results of three tumors previously studied in [15]. We applied the non-isotropic deformation in (15).

In the first experiment, we show the result for the Tumor 1. In Fig. 10 we show the ellipsoidal map and in Fig. 11, we show the non-rigid motion tracking.

In the second experiment, we show the result for the Tumor 2. In Fig. 12 we show the ellipsoidal map and in Fig. 13, we show the non-rigid motion tracking.

In the third experiment, we show the result for the Tumor 3. In Fig. 14 we show the ellipsoidal map and in Fig. 15, we show the non-rigid motion tracking.

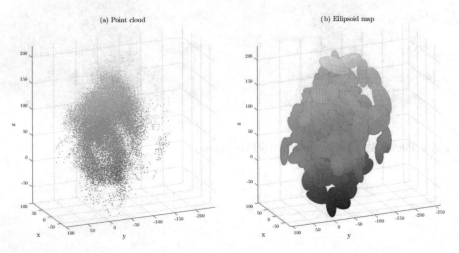

Fig. 10 Ellipsoidal mapping of the Tumor 1

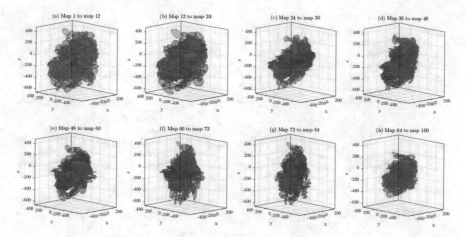

Fig. 11 Tracking non-rigid motion on the Tumor 1

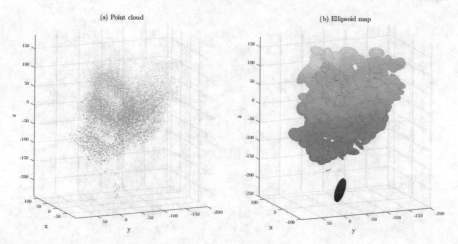

Fig. 12 Ellipsoidal mapping of the Tumor 2

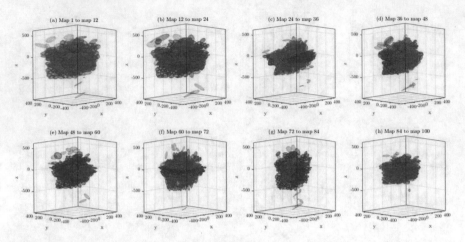

Fig. 13 Tracking non-rigid motion on the Tumor 2

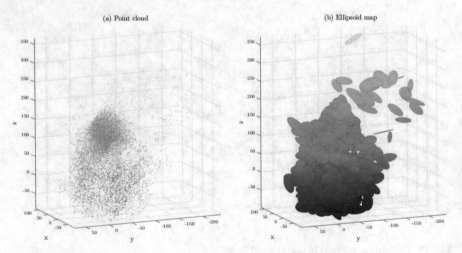

Fig. 14 Ellipsoidal mapping of the Tumor 3

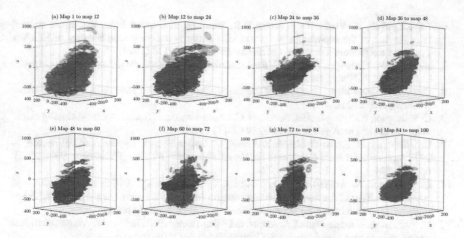

Fig. 15 Tracking non-rigid motion on the Tumor 3

6 Conclusion

In this chapter we have presented a novel algorithm for tracking non-rigid motion in free-form objects, the proposal is applied to 4D point clouds. The proposed algorithm uses k-means algorithm to segment the point cloud and Covariance ellipsoid for mapping the object, this approach offers compact maps (low memory cost) with high representation (free-form objects). Afterward, we use the Germinal Center Optimization for adapting the ellipsoids parameters and track the non-rigid transformation. We offer seven experiments that prove the algorithm capabilities. As future work, is essential to notice that this algorithm is highly parallelizable, we can get low run-times.

References

1. B. Preim, D. Bartz, *Visualization in Medicine. Theory, Algorithms, and Applications* (Morgan Kaufmann, Amserdam, 2007)
2. A. Webb, G.C. Kagadis, Introduction to biomedical imaging. Med. Phys. **30**(8), 2267–2267 (2003)
3. R.A. Robb, 3-dimensional visualization in medicine and biology. Handbook of medical imaging: processing and analysis, pp. 685–712 (2000)
4. Q. Zhang, R. Eagleson, T.M. Peters, Volume visualization: a technical overview with a focus on medical applications. J. Digit. Imaging **24**(4), 640–664 (2011)
5. S.Y. Baek, D.H. Sheafor, M.T. Keogan, D.M. DeLong, R.C. Nelson, Two-dimensional multiplanar and three-dimensional volume-rendered vascular ct in pancreatic carcinoma: interobserver agreement and comparison with standard helical techniques. Am. J. Roentgenol **176**(6), 1467–1473 (2001)

6. J.J. Kulynych, K. Vladar, D.W. Jones, D.R. Weinberger, Gender differences in the normal lateralization of the supratemporal cortex: MRI surface-rendering morphometry of Heschl's gyrus and the planum temporale. Cereb. Cortex **4**(2), 107–118 (1994)
7. K. Engel, M. Hadwiger, J. Kniss, C. Rezk-Salama, D. Weiskopf, *Real-Time Volume Graphics* (AK Peters/CRC Press, Natick, 2006)
8. G. Farin, *Curves and Surfaces for Computer-Aided Geometric Design: A Practical Guide* (Elsevier, Amsterdam, 2014)
9. C. Villasenor, N. Arana-Daniel, A. Alanis, C. Lopez-Franco, J. Gomez-Avila, Multiellip-soidal mapping algorithm. Appl. Sci. **8**(8), 1239 (2018)
10. C. Villasenor, N. Arana-Daniel, A.Y. Alanis, C. Lopez-Franco, Hyperellipsoidal neuron, in *2017 International Joint Conference on Neural Networks (IJCNN)* (IEEE, New York, 2017), pp. 788–794
11. C. Villasenor, N. Arana-Daniel, A.Y. Alanis, C. Lopez-Franco, Differential evolution and covariance ellipsoid for non-rigid transformation tracking of internal organs, in *2018 International Joint Conference on Neural Networks (IJCNN)*. IEEE (in press)
12. C. Villasenor, N. Arana-Daniel, A. Alanis, C. Lopez-Franco, E.A. Hernandez-Vargas, Germinal center optimization algorithm. Int. J. Comput. Intell. Syst. (in press)
13. D. Tarlinton, G. Victora, Editorial overview: Germinal centers and memory b-cells: from here to eternity (2017)
14. The Stanford 3D Scanning Repository. http://graphics.stanford.edu/data/3Dscanrep/. Accessed 27 Sept 2018
15. A. Serna, F. Prieto, Towards a 3d modeling of brain tumors by using endoneurosonography and neural networks. Revista Ingenierías Universidad de Medellín **16**(30), 129–148 (2017)
16. R. De Maesschalck, D. Jouan-Rimbaud, D.L. Massart, The Mahalanobis distance. Chemometr. Intell. Lab. Syst. **50**(1), 1–18 (2000)
17. J.D. Rios, C. Villasenor, A.Y. Alanis, N. Arana-Daniel, C. Lopez-Franco, Germinal center optimization applied to recurrent high order neural network observer. IFAC-PapersOnLine **51**(13), 332–337 (2018)
18. C. Villasenor, J.D. Rios, N. Arana-Daniel, A.Y. Alanis, C. Lopez-Franco, E.A. Hernandez-Vargas, Germinal center optimization applied to neural inverse optimal control for an all-terrain tracked robot. Appl. Sci. **8**(1), 31 (2017)

Hybrid Neural-Fuzzy Modeling and Classification System for Blood Pressure Level Affectation

Martin Vázquez, Patricia Melin and German Prado-Arechiga

Abstract In the recent years, the technological advancements tend to leave people with the necessity to stay in a sedentary position for several hours in order to achieve their work objectives. One of the parameters involving health issues tends to do with blood pressure. These are measurements that involve the circulation of the blood throughout the human body. We are developing a modeling and classification of hybrid system to evaluate and see the tendency of how much does a sedentary state affect while working in front of a computer for more than 8 h. The main idea is that the system learns the behavior or the trend of the different blood pressure that are extracted from individuals that fit this experiment. A modular network will be used for the part of modeling, while using fuzzy module to help us determine the best practical classification process in order to obtain a more precise affection of blood pressure.

Keywords Accelerometer · Blood pressure · Classification · Diastolic pressure · Fuzzy system · Modeling · Modular neural network · Neural system · Sedentary behavior · Sitting time · Systolic pressure

1 Introduction

Nowadays, with technological advances improving on a day to day basis, workers in the computers industry find themselves occupied with less need to get up from their work places in order to fulfill their tasks.

M. Vázquez · P. Melin (✉)
Tijuana Institute of Technology, Tijuana, B.C., Mexico
e-mail: pmelin@tectijuana.mx

G. Prado-Arechiga
Excel Medical Center, Tijuana, Mexico

© Springer Nature Switzerland AG 2020

O. Castillo and P. Melin (eds.), *Hybrid Intelligent Systems in Control,*
Pattern Recognition and Medicine, Studies in Computational Intelligence 827,
https://doi.org/10.1007/978-3-030-34135-0_18

In large metropolis it is very common for most of work activities to be handled inside offices, for which millions of people spend a large amount of hours in a sitting position. This activity tends to be a sedentary position that can be an issue in later stages of that person's health.

Energy is the fuel the body requires in order to make an activity functional. Blood is the responsible to distribute oxygen throughout the human body. Depending on the blood pressure, a subjects reacts a certain way to the same physical demand.

Even though different health awareness programs are known and exposed for the safety of people, they are often not taken seriously of the simple case that most individuals think they are the exception of the rule. Analyzing and putting information to good use is half the battle for a healthier life style. We intend to give that spark of interest to surpass that case and recommend a way to have a better well-being lifestyle.

The process of classification [1, 2] and recognition [3, 4] are two of the characteristics of the human kind that have played a very important role in its evolution. The objective of classification focuses into dividing a data set of items into groups or categories. Data set have to be arranged in classes or categories according to shared qualities or similar characteristics. While recognition focuses on automatically identifying and responding correctly to a related data.

The main contribution is that we want to create awareness for people to consider having a healthier life style that includes regular and constant basic exercise. We try to achieve a good evaluation so that a medical professional can use it as a performance tool in order to make the best decision for a possible treatment.

2 Basic Concepts

Our general objective is to develop a neural-fuzzy hybrid system for the model and classification of the levels of blood pressure in working stations, specifically in front of a computer [5].

Blood pressure levels are measurements obtained by the current state of an individual. In those levels [6], heart rhythms are registered as data that we can process.

The specific objectives are to analyze if there is some variation in blood pressure and base on that determine where affection can occur or not. Also implement a neural system that allows modeling the levels of pressure for the analysis of the tendency. A fuzzy system will be used for the classification of the affections in blood pressure [1, 7–9].

Part of the human behavior in this investigation is based on the person's posture that tends to be on sedentary state which is due to physical inactivity.

The networks of artificial neurons are computer systems composed of multiple basic units called neurons. The neurons are interconnected with each other in order to produce an output stimulus based on a series of inputs that are processed automatically. This is achieved by means of a learning algorithm used to train the network [5].

Certain algorithms and methods of computational learning allow to obtain good prediction results once trained the model with a known number of data, applying some technique of presentation of the same, waiting for the model to produce a response, but identical, if close to the correct.

Due to its excellent adjustment capabilities, RNAs are successfully applied in different scientific, social and technological fields:

- Manufacturing
- Biology
- Finance
- Weather forecast
- Trend analysis
- Patterns

It is intended to implement a modular RNA neural network [10–13]. They are inspired by the way in which biological systems process different information groups in subsystems whose output is then reprocessed.

The main idea is that, instead of the input data being processed by all the input nodes, they are grouped and processed by a group of input nodes and only by them [14, 15].

The output of these is then processed by nodes of higher levels in the same way as it is done with forward-fed networks.

Fuzzy logic is a methodology that provides a simple and elegant way to get a conclusion from vague, ambiguous, inaccurate, noise or incomplete input information. In general, fuzzy logic imitates how a person makes decisions based on information with the mentioned characteristics [16, 17]. Fuzzy logic is a computational intelligence technique that allows work with information with high degree of impression [7, 8], in this it differs from the conventional logic that works with well-defined and precise information. It is a multi-valued logic that allows intermediate values to be able to define evaluations between yes/no, true/false, black/white, hot/cold, among others.

3   Literature Review

The background and related work for this investigation is briefly described in this Section.

In the paper by Guzman et al. [18], a method to optimize the fuzzy rules for classifying the blood pressure level was proposed. The fuzzy classifier is only part of the complete neuro-fuzzy hybrid model, which uses techniques such as: neural networks, fuzzy logic and evolutionary computation. In this latter technique, genetic algorithms are used, which use individuals as possible solutions and thus obtain the best solution, in this case find the appropriate number of fuzzy rules for fuzzy system. This study aims to model blood pressure for 24 h and obtain the trend for each patient, once this trend is obtained, this information enters a fuzzy system based on rules given by an expert, who will be classified into some of the blood pressure levels based on level European guide [19] and finally provide a diagnosis for the patient.

Another important work by Melin and Arechiga [1], describes and explains high blood pressure (HBP) or hypertension, which is a kind of disease that often leads to fatal outcomes, such as heart attack, stroke and renal failure. The HBP seriously threats the health of people worldwide. One of the most dangerous aspects is that people may not know that they have it. In fact, nearly one-third of people who have high blood pressure don't know it. Therefore, a Fuzzy System for the diagnosis of the HBP was developed. Firstly, the input parameters include Systolic Blood Pressure and Diastolic Blood Pressure. Secondly, we have as an output parameter: Blood Pressure Levels (BPL). The input linguistic values include Low, Normal Low, Normal, Normal High, High, Very High, Too High and Isolated Systolic Hypertension. Finally, 14 fuzzy rules were developed to make a diagnosis.

Also as another important relevant related work we have to mention the research of Alhajj and Wiil [5]. This work describes in detail a methodology for Network Modeling Analysis in Health Informatics and Bioinformatics. In addition, illustrative examples are provided for explaining how to develop models for computer science applied in Medicine.

4 Problem Statement and Proposed Method

One goal of this work is to analyze how the excessive use of computer is affecting the heart rate. The aim is to implement a computational model for the affectations in blood levels as shown in Fig. 1, as the proposed method.

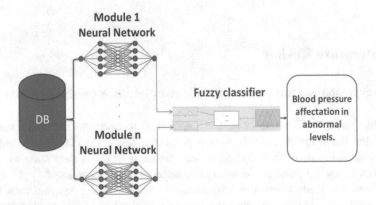

Fig. 1 General proposed method

The general model uses the person's blood pressure as information so that the samples can be used on the modules. The goal was to create a neural-fuzzy system with modules [17, 20, 21] that learn the behavior or tendency of the different levels of pressure. We have decided to use the internal database of postgraduate students, since it complies with the characteristics and variables that will serve as a basis to work with the model. The database will supply the data set which will go into the modules. Systolic, diastolic, and time modules are considered to enter the Modular Neural Network [10, 11, 22]. This will lead us to a trend of affectation after training is completed, the data set will be the input fuzzy classifier which will help us determine the affectations for each case.

5 Methodology

In Table 1 we can appreciate the values of categories for the blood pressure based on the UK charts for adults.

In Fig. 2, the way we interpret this table is to locate the current systolic and/or diastolic number to consider the following: 90 over 60 (90/60) or less indicate a low pressure. More than 90 over 60 (90/60) and less than 120 over 80 (120/80) states that the current blood pressure is in an ideal and healthy state. Higher than 120 over 80 and less than 140 over 90 (120/80–140/90) implies that the data has a normal blood pressure reading but it is a little higher than it should be, and the patient should try to lower it. However, if 140 over 90 (140/90) or higher means you may have high blood pressure, known as hypertension [7, 8, 23–26].

The risk of being hypertensive includes some of these factors [1]:

- Age
- Sex

Table 1 Blood pressure UK chart for adults

Blood pressure category	Systolic pressure (mmHg)		Diastolic pressure (mmHg)
Low blood pressure	Less than 90	or	Less than 60
Normal blood pressure	90–119	and	60–79
Prehypertension	120–139	or	80–89
Hypertension stage 1	140–159	or	90–99
Hypertension stage 2	160 or higher	or	100 or higher
Hypertensive crisis	Higher than 190	or	Higher than 110

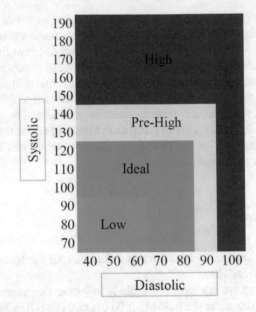

Fig. 2 Blood pressure chart

- Genetic ascendance
- Obesity
- Absence of exercise
- Where the subject smokes or not
- Stress level
- The use of salt
- The utilization of alcohol

In Table 2 we can find an index for the database we are currently using. So far we have 100 subjects, of which 70% will be used for training. The remaining 30% of the data will be used to make tests. The values we are considering for this research are as follows [1, 26, 27]:

- Age.
- Sex.
- Levels in Systolic pressure.
- Levels in Diastolic pressure.

The range of data will be using with is the working time. It should be mentioned that the database does not have marked ranges of which people have been working in front of the computer, which will be used to create the fuzzy system.

Table 2 Database index

Variable	Definition
edad	Age
sexo	Gender
IMC	Body mass index
Sis	Systolic pressure
Dia	Diastolic pressure
Fuma	Smoke (Yes or No)
Padre	Hypertensive parents (0, 1, 2)
EdSys	Age multiply systolic pressure

6 Results and Discussion

We show in the following figures the detail classification structure used with neural network training [25, 26, 28–30]. Figure 3 shows the data scattered graphic as a raw data set. After training, in Fig. 4 we can note how the classification connections are established by SOM weight positions. The optimal weights for the classification and grouping of data with similarity are determined.

In Fig. 5 we can find the overall structure for the neural network training tool. The results of this training can be appreciated in a confusion matrix shown in Fig. 6. Neural Network Training Confusion with 21epochs is displayed in Fig. 7 [25, 27].

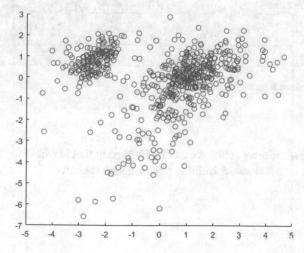

Fig. 3 Data scattered graphic

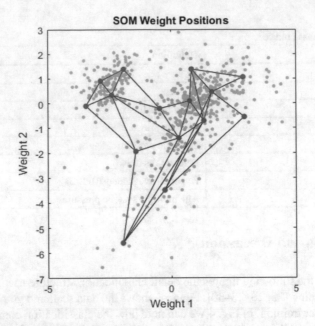

Fig. 4 SOM weight positions for classification

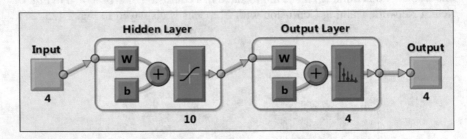

Fig. 5 Neural network training tool

In order to view the error status we need a histogram that is what Fig. 8 establishes. On Fig. 9, the neural network training performance is shown.

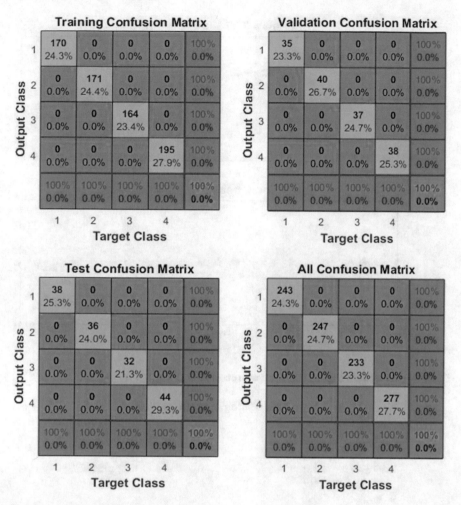

Fig. 6 Neural network training confusion matrices

7 Conclusions and Future Work

This study showed that a population of a data set can be used in order to have an impact on the health awareness programs. We are defined by nature's rules, which relay on the current health status of any living being. We have an upper hand with all the knowledge and tools that have been created throughout the history. Putting this information to good use is an actual lifesaving study. The state of art is only the tip of the iceberg research-wise in order to comprehend the solution for the stated problem. Many of the analysis have been dedicated to this section, which indicates the

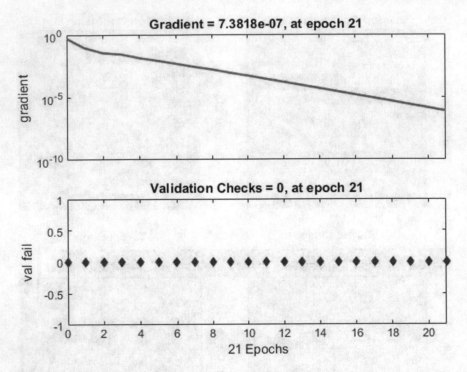

Fig. 7 Neural network training confusion with epoch 21

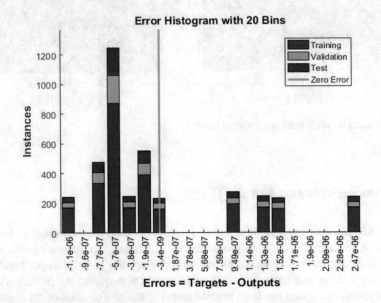

Fig. 8 Error histogram with 20 bins

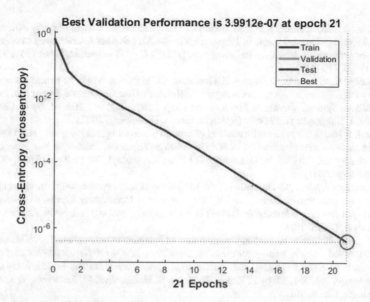

Fig. 9 Neural network training performance

importance of being well informed. For our research work to develop it is necessary to obtain more data in order to make the database grow. Implementing the modular neural and analyzing process of the fuzzy system for the application is a module we will incorporate in order to obtain more precise results. In addition, we could use type-2 fuzzy logic [31] and other applications could be considered [32].

References

1. P. Melin, G. Prado-Arechiga, Fuzzy logic for arterial hypertension classification, in *New Hybrid Intelligent Systems for Diagnosis and Risk Evaluation of Arterial Hypertension* (Springer International Publishing, Berlin, 2018), pp. 5–13
2. O. Otto (ed.) *Information and Classification.* ISBN 978-3-642-50974-2, XI (1993), p. 517
3. A. Shigeo, Pattern classification. 2001. Massimo Volpe. High Blood Pressure & Cardiovascular Prevention. ISSN: 1120-9879 (print version). Journal no. 40292
4. P. Melin, O. Castillo, *Hybrid Intelligent Systems for Pattern Recognition Using Soft Computing* (Springer, Berlin, 2005)
5. R. Alhajj, U.K. Wiil, *Network Modeling Analysis in Health Informatics and Bioinformatics.* ISSN: 2192-6670 (electronic version) Journal no. 13721
6. F. Thomas, J. Blacher, A. Benetos, M.E. Safar, B. Pannier, J. Hypertens. **26**(6), 1072–1077 (2008)
7. S. Das, P.K. Ghosh, Hypertension diagnosis: a comparative study using fuzzy expert system and neuro fuzzy system, in *Proceedings of the IEEE International Conference on Fuzzy Systems (FUZZ-IEEE)*, Hyderabad, India (7–10 July 2013), pp. 1–7
8. M. Neshat, M. Yaghobi, M.B. Naghibi, A. Esmaelzadeh, Fuzzy expert system design for diagnosis of liver disorders, in *Proceedings of the 2008 KAM'08 International Symposium on Knowledge Acquisition and Modeling*, Wuhan, China (21–22 December 2008), pp. 252–256

9. A.A. Abdullah, Z. Zakaria, N.F. Mohammad, Design and development of fuzzy expert system for diagnosis of hypertension, in *Proceedings of the 2011 Second International Conference on Intelligent Systems, Modeling and Simulation (ISMS)*, Kuala Lumpur, Malaysia (25–27 January 2011), pp. 113–117

10. P. Melin, C. Gonzalez, D. Bravo, F. Gonzalez, G. Martinez, Modular neural networks and fuzzy Sugeno integral for pattern recognition: the case of human face and fingerprint, in *Hybrid Intelligent Systems. Studies in Fuzziness and Soft Computing*, vol. 208, ed. by O. Castillo, P. Melin, J. Kacprzyk, W. Pedrycz (Springer, Berlin, Heidelberg, 2007)

11. I. Leal, P. Melin, Time series forecasting of tomato prices and processing in parallel in Mexico using modular neural networks, in *Hybrid Intelligent Systems. Studies in Fuzziness and Soft Computing*, vol. 208, ed. by O. Castillo, P. Melin, J. Kacprzyk, W. Pedrycz (Springer, Berlin, Heidelberg, 2007)

12. J. Amezcua, P. Melin, Optimization of the LVQ network architectures with a modular approach for arrhythmia classification, in *Novel Developments in Uncertainty Representation and Processing. Advances in Intelligent Systems and Computing*, vol. 401, ed. by K. Atanassov, et al. (Springer, Cham, 2016)

13. A. Uriarte, P. Melin, F. Valdez, Optimization of modular neural network architectures with an improved particle swarm optimization algorithm, in *Recent Developments and the New Direction in Soft-Computing Foundations and Applications. Studies in Fuzziness and Soft Computing*, vol. 361, ed. by L. Zadeh, R. Yager, S. Shahbazova, M. Reformat, V. Kreinovich (Springer, Cham, 2018)

14. F. Gaxiola, P. Melin, F. Valdez, J.R. Castro, Person recognition with modular deep neural network using the iris biometric measure, in *Fuzzy Logic Augmentation of Neural and Optimization Algorithms: Theoretical Aspects and Real Applications. Studies in Computational Intelligence*, vol. 749, ed. by O. Castillo, P. Melin, J. Kacprzyk (Springer, Cham, 2018)

15. D. Sánchez, P. Melin, J. Carpio, H. Puga, Comparison of optimization techniques for modular neural networks applied to human recognition, in *Nature-Inspired Design of Hybrid Intelligent Systems. Studies in Computational Intelligence*, vol. 667, ed. by P. Melin, O. Castillo, J. Kacprzyk (Springer, Cham, 2017)

16. S. Sotirov, E. Sotirova, P. Melin, O. Castilo, K. Atanassov, Modular neural network preprocessing procedure with intuitionistic fuzzy intercriteria analysis method, in *Flexible Query Answering Systems 2015. Advances in Intelligent Systems and Computing*, vol. 400, ed. by T. Andreasen, et al. (Springer, Cham, 2016)

17. M. Pulido, P. Melin, G. Prado-Arechiga, A new model based on a fuzzy system for arterial hypertension classification, in *Fuzzy Logic Augmentation of Neural and Optimization Algorithms: Theoretical Aspects and Real Applications. Studies in Computational Intelligence*, vol. 749, ed. by O. Castillo, P. Melin, J. Kacprzyk (Springer, Cham, 2018)

18. J.C. Guzmán, P. Melin, G. Prado-Arechiga, Fuzzy optimized classifier for the diagnosis of blood pressure using genetic algorithm, in *Fuzzy Logic Augmentation of Neural and Optimization Algorithms: Theoretical Aspects and Real Applications. Studies in Computational Intelligence*, vol. 749, ed. by O. Castillo, P. Melin, J. Kacprzyk (Springer, Cham, 2018)

19. F. Thomas, J. Blacher, A. Benetos, M.E. Safar, B. Pannier, Cardiovascular risk as defined in the 2003 European blood pressure classification: the assessment of an additional predictive value of pulse pressure on mortality. J. Hypertens **26**(6), 1072–1077 (2008)

20. P. Melin, Image processing and pattern recognition with Mamdani interval type-2 fuzzy inference systems, in *Combining Experimentation and Theory. Studies in Fuzziness and Soft Computing*, vol. 271, ed. by E. Trillas, P. Bonissone, L. Magdalena, J. Kacprzyk (Springer, Berlin, Heidelberg, 2012)

21. B. González, P. Melin, F. Valdez, G. Prado-Arechiga, Ensemble neural network optimization using a gravitational search algorithm with interval type-1 and type-2 fuzzy parameter adaptation in pattern recognition applications, in *Fuzzy Logic Augmentation of Neural and Optimization Algorithms: Theoretical Aspects and Real Applications. Studies in Computational Intelligence*, vol. 749, ed. by O. Castillo, P. Melin, J. Kacprzyk (Springer, Cham, 2018)

22. P. Melin, V. Ochoa, L. Valenzuela, G. Torres, D. Clemente, Modular neural networks with fuzzy Sugeno integration applied to time series prediction, in *Hybrid Intelligent Systems. Studies in Fuzziness and Soft Computing*, vol. 208, ed. by O. Castillo, P. Melin, J. Kacprzyk, W. Pedrycz (Springer, Berlin, Heidelberg, 2007)
23. R. Poli, S. Cagnoni, R. Livi, G. Coppini, G. Valli, A neural network expert system for diagnosing and treating hypertension. Computer **24**, 64–71 (1991)
24. J.A. Staessen, J. Wang, G. Bianchi, W.H. Birkenhäger, Essential hypertension. Lancet **361**, 1629–1641 (2003)
25. J.C. Guzmán, P. Melin, G. Prado-Arechiga, Neuro-fuzzy hybrid model for the diagnosis of blood pressure, in *Nature-Inspired Design of Hybrid Intelligent Systems. Studies in Computational Intelligence*, vol. 667, ed. by P. Melin, O. Castillo, J. Kacprzyk (Springer, Cham, 2017)
26. J.C. Guzmán, P. Melin, G. Prado-Arechiga, Design of a fuzzy system for diagnosis of hypertension, in *Design of Intelligent Systems Based on Fuzzy Logic, Neural Networks and Nature-Inspired Optimization. Studies in Computational Intelligence*, vol. 601, ed. by P. Melin, O. Castillo, J. Kacprzyk (Springer, Cham, 2015)
27. J.C. Guzmán, P. Melin, G. Prado-Arechiga, A proposal of a fuzzy system for hypertension diagnosis, in *Novel Developments in Uncertainty Representation and Processing. Advances in Intelligent Systems and Computing*, vol. 401, ed. by K. Atanassov, et al. (Springer, Cham, 2016)
28. P. Srivastava, A. Srivastava, A. Burande, A. Khandelwal, A Note on Hypertension Classification Scheme and Soft Computing Decision Making System. ISRN Biomath, **2013** (2013) https://doi.org/10.1155/2013/342970
29. B.B. Sumathi, Pre-diagnosis of hypertension using artificial neural network. Glob. J. Comput. Sci. Technol. **11**, 2 (2011)
30. Z. Abrishami, I. Azad, Design of a fuzzy expert system and a multi-layer neural network system for diagnosis of hypertension. Bull. Environ. Pharmacol. Life Sci. **4**, 138–145 (2015)
31. P. Melin, C.I. González, J.R. Castro, O. Mendoza, O. Castillo, Edge-detection method for image processing based on generalized type-2 fuzzy logic. IEEE Trans. Fuzzy Syst. **22**(6), 1515–1525 (2014)
32. P. Melin, I. Miramontes, G. Prado-Arechiga, A hybrid model based on modular neural networks and fuzzy systems for classification of blood pressure and hypertension risk diagnosis. Expert Syst. Appl. **107**, 146–164 (2018)

Robotic Applications

Environment Recognition for Path Generation in Autonomous Mobile Robots

Ulises Orozco-Rosas, Kenia Picos, Oscar Montiel and Oscar Castillo

Abstract An efficient algorithm for path generation in autonomous mobile robots using a visual recognition approach is presented. The proposal includes image filtering techniques by employing an inspecting camera to sense a cluttered environment. Template matching filters are used to detect several environment elements, such as obstacles, feasible terrain, the target location, and the mobile robot. The proposed algorithm includes the parallel evolutionary artificial potential field to perform the path planning for autonomous navigation of the mobile robot. Our problem to be solved for autonomous navigation is to safely take a mobile robot from the starting point to the target point employing the path with the shortest distance and which also contains the safest route. To find the path that satisfies this condition, the proposed algorithm chooses the best candidate solution from a vast number of different paths calculated concurrently. For achieving efficient autonomous navigation, the proposal employs a parallel computation approach for the evolutionary artificial potential field algorithm for path generation and optimization. Experimental results yield accuracy in environment recognition in terms of quantitative metrics. The proposed algorithm demonstrates efficiency in path generation and optimization.

U. Orozco-Rosas · K. Picos
CETYS Universidad, Centro de Innovación y Diseño (CEID), Av. CETYS Universidad
No. 4, El Lago, 22210 Tijuana, Baja California, Mexico
e-mail: ulises.orozco@cetys.mx

K. Picos
e-mail: kenia.picos@cetys.mx

O. Montiel (✉)
Instituto Politécnico Nacional, CITEDI-IPN, Av. Instituto Politécnico Nacional
No. 1310, Nueva Tijuana, 22435 Tijuana, Baja California, Mexico
e-mail: oross@ipn.mx

O. Castillo
Tecnológico Nacional de México, Calzada Del Tecnológico S/N, Tomas Aquino,
22414 Tijuana, Baja California, Mexico
e-mail: ocastillo@tectijuana.mx

© Springer Nature Switzerland AG 2020
O. Castillo and P. Melin (eds.), *Hybrid Intelligent Systems in Control,
Pattern Recognition and Medicine*, Studies in Computational Intelligence 827,
https://doi.org/10.1007/978-3-030-34135-0_19

Keywords Parallel evolutionary artificial potential field · Path planning · Mobile robots · Template matching · Object recognition

1 Introduction

Nowadays, there is a necessity for autonomous mobile robots in several fields of application, such as cleaning, material transport, taking care of children at school, surveillance, monitoring, driver assistance, military and agricultural applications. These mobile robots must interact with their environment to realize their tasks. Computer vision methods are broadly employed to sense this environment, which is commonly changing. The autonomous mobile robot must interact with the environment and avoid collisions with the obstacles, and it must follow the path to reach its established target. These tasks must be performed without the support of a human operator. Three tasks must be performed by the mobile robot to allow the execution of autonomous navigation. These actions are environment perception that consists in obstacle recognition and modeling; path planning for finding an ordered sequence of objective points and adapt this sequence into a path; the path tracking to control the mobile robot to follow the path to reach the target point. In this work, we present an extension of the work presented in [1], which deals with the problem of autonomous mobile robot navigation, considering the navigation as the displacement from a start point to a target point over a defined path. The extension consists of the implementation of the parallel evolutionary artificial potential field [2] for taking advantages of parallel processors architectures.

The mobile robot environment is modeled through the implementation of an accurate object recognition system. A template matching method is employed to detect obstacles and accurately estimate their location coordinates consistently. To reach autonomous mobile robot navigation, the combination of a recognition system based on a template-matching procedure and a path planning system based on the parallel evolutionary artificial potential field algorithm is proposed. This work presents and discusses the computer simulation results in terms of the accuracy of the recognition system, computational efficiency, and the effectiveness of path planning.

This work is organized as follows, in Sect. 2 is presented the obstacle recognition with template matched filtering, and path planning employing the parallel evolutionary artificial potential field algorithm. In Sect. 3 is described the methodology for obstacle recognition for path planning in autonomous mobile robots. In Sect. 4 is presented the experimental results found with the methodology employing real and simulated environments for robot navigation. Finally, in Sect. 5 the conclusions are summarized.

2 Visual Environment Recognition and Path Planning

For autonomous mobile robot navigation, we study the environment in which several challenges can be presented for feasible path planning. In this section, we describe the fundamentals of the core components of our proposal. First, an automatic visual environment recognition is used for obstacle detection by template matching techniques using a correlation filtering approach; additionally, a path planning method is explained with the approach of the evolutionary artificial potential field for mobile robot navigation.

2.1 Obstacle Recognition by Correlation Filtering

The mobile robot environment is known by the image capture using a monocular camera in a superior position. Let be $f(x, y)$ the input scene which is composed by a feasible workspace for navigation $b(x, y)$. As shown in Fig. 1, the robot can move from the start point to a specific target point in an autonomous fashion. The input image $f(x, y)$ presented in the optical setup also contains several obstacles $t^j(x, y)$, as a challenge for the navigation, placed at an unknown location (x_0, y_0). For scenes of real-world image conditions, additive noise $n(x, y)$ is considered for the representation of the signal model, given as follows [3]

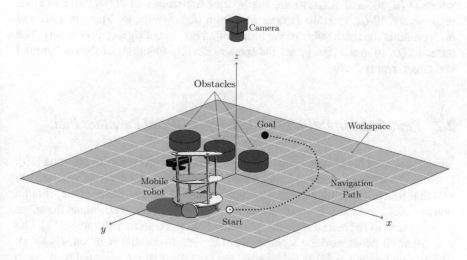

Fig. 1 Representation of a common environment in a mobile robot navigation

$$f(x, y) = \sum_{j=1}^{J} t^j \left(x - x_0^j, y - y_0^j \right) + b(x, y) \sum_{j=1}^{J} w^j \left(x - x_0^j, y - y_0^j \right) + n(x, y),$$

$$(1)$$

where the superposition of the J obstacles $\{t^j(x, y) | j = 1, \ldots, J\}$ forms the scene. The binary function $w^j(x, y)$ represents the inverse support area of the j-th target $t^j(x, y)$.

Template matching is successfully used for target detection using a correlation process [4]. Correlation filters are linear systems, in which the impulse response is computed to produce a high matching value between the input image and a filter $H(\mu, v)$. The designed filter is related to a prior target reference information. These filters can be synthesized by the optimization of specific performance measures [4, 5] in order to produce high matching values considering different image conditions. A well-used measure as the signal-to-noise ratio (SNR) is used for optimal filter design. The generalized matched filter (GMF) [6] considers the SNR to improve target detection. The GMF also optimize the variance of location errors [7, 8]. From the image representation given by Eq. 1, the frequency response of a GMF is depicted as follows [9]

$$H^*(\mu, v) = \frac{T(\mu, v) + m_b W(\mu, v) + m_t W_t(\mu, v)}{|W_b(\mu, v)|^2 * N_b(\mu, v) + N_t(\mu, v)},$$

$$(2)$$

where, $T(\mu, v)$, and $W(\mu, v)$ are the Fourier transforms of $t(x, y)$ and $w(x, y)$, respectively. $W_t(\mu, v)$ is the Fourier transform of $1 - w(x, y)$. The terms m_b and m_t represents the mean value of the background and target signal, respectively. The terms $N_b(\mu, v)$ and $N_t(\mu, v)$ are the spectral density functions of the background and target, respectively.

2.2 Path Planning with Evolutionary Artificial Potential Field

Khatib proposed the artificial potential field method in 1985 for local planning [10]. Initially, the artificial potential field method was employed for obstacle avoidance in real-time and path planning in robot manipulators [11]. The core idea of the artificial potential field method is to create an attractive potential field force around the target point, as well as to create a repulsive potential field force around obstacles [12]. The two potential fields working together form the total potential function, $U_{total}(q)$. The artificial potential field method searches the falling of the potential function to generate a collision-free path, which is generated from the start point to the target point [2]. The total potential function can be found by the summation of the attractive potential and repulsive potential,

$$U_{total}(q) = \frac{1}{2}\left[k_a\left(q - q_f\right)^2 + k_r\left(\frac{1}{\rho} - \frac{1}{\rho_0}\right)^2\right], \tag{3}$$

where q denotes the mobile robot position vector in a two-dimensional workspace, $q = [x, y]^T$. The vector q_f denotes the target point and k_a is a positive scalar-constant that denotes the proportional gain of the function. The expression $q - q_f$ is associated with the linear distance between the mobile robot and the target point. The repulsive potential function was given by Khatib [10], and it has a limited range of influence; this prevents the movement of the mobile robot from being affected by a remote obstacle, where ρ_0 denotes the limit distance of influence of the potential field and ρ is the shortest distance to the obstacle; the positive scalar-constant k_r denotes the repulsive proportional gain.

The total force $F_{total}(q)$ which is employed to drive the mobile robot, it is found by the negative gradient of the total potential function [13], this force is stated as follows

$$F_{total}(q) = -\nabla U_{total}(q). \tag{4}$$

In Eq. 3, all the parameters are known except for the proportional gains of attraction k_a and repulsion k_r. Several methods can be employed to find the suitable value of this proportional gains, the most common methods are mathematical analysis and approximate methods, e.g., evolutionary algorithms. In the evolutionary artificial potential field method, the artificial potential field is combined with evolutionary algorithms [14], in specific with a genetic algorithm to obtain the optimal values for the proportional gains. A genetic algorithm is an adaptive heuristic search algorithm premised on the evolutionary ideas of natural selection and genetic [15]. The elementary concept of a genetic algorithm is intended to simulate the process in natural systems required for the evolution [16], wherein the most basic form, a genetic algorithm can be algorithmically modeled for computer simulation employing the difference equation stated as follows

$$P(t + 1) = s(v(P(t))), \tag{5}$$

where t denotes the time, the next population $P(t + 1)$ is found from the existing population $P(t)$ after it was worked by random variation v, and selection s [17].

3 Path Planning Using Parallel Evolutionary Artificial Potential Field

The reference block diagram shown in Fig. 2 describes the methodology for autonomous mobile robot navigation. The employed method is specified into three

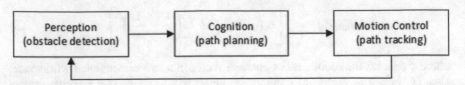

Fig. 2 Reference block diagram for autonomous mobile robot navigation considering the three basic subsystems

basic subsystems: obstacle detection, path planning, and path tracking described as follows.

3.1 Obstacle Detection Algorithm

To avoid collisions between the mobile robot and the obstacles, a correlation filtering approach is applied. In Fig. 3, an obstacle detection process is presented. Firstly, an input image $f(x, y)$ enters the system given by Eq. 1, in which, we assume that the scene contains a feasible workspace $b(x, y)$ for mobile robot navigation with some obstacles $t^j(x, y)$. After that, a filter is build using a GMF from Eq. 2 trained with a template reference of $t(x, y)$ and the statistical parameters of $f(x, y)$. Lastly, a cross-correlation between the input scene and the designed filter is computed. This is to obtain a correlation plane

$$c(x, y) = \text{IFT}\{F(\mu, \nu)H^*(\mu, \nu)\}. \tag{6}$$

The output correlation plane yields quantity levels of the best match. In this case, several peaks will be produced where the area of each obstacle coincides. To

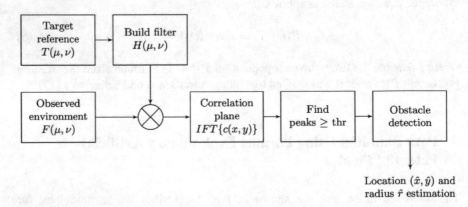

Fig. 3 Block diagram for obstacle detection using template matched filters in frequency domain

conclude, the estimated location coordinates $\left(x_0^j, y_0^j\right)$ of each detected obstacle are computed as

$$(\hat{x}_0, \hat{y}_0) = \underset{(x,y)}{\operatorname{argmax}}\left\{|c(x, y)|^2\right\}.\qquad(7)$$

For radii estimation, area A of each obstacle is computed employing image binarization. The estimated radius is computed as $\hat{r} = \sqrt{A\pi}$.

3.2 Path Planning Algorithm

Figure 4 shows the flowchart of the parallel evolutionary artificial potential field for path planning in mobile robots. A genetic algorithm composes the parallel evolutionary artificial potential field algorithm, that is defined in the flowchart by the blocks without color background, and by the artificial potential field method defined by the colored background blocks. The backbone of the parallel evolutionary artificial potential field algorithm is the evolutionary computation for finding dynamically the optimal proportional gains k_a and k_r values required in Eq. 3, and parallel computation for taking advantages of novel parallel processors architectures.

Evolutionary computation permits the mobile robot to navigate without being stuck in local minima, making the parallel evolutionary artificial potential field algorithm appropriate to work in complex environments. The artificial potential field method acts as a fitness function in the parallel evolutionary artificial potential field, and as a path planner when the genetic algorithm has found the optimal proportional

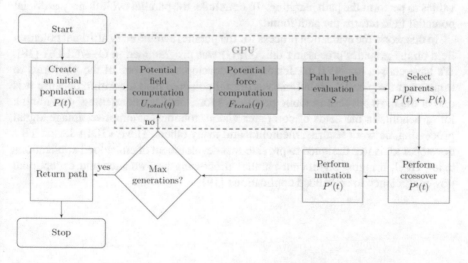

Fig. 4 Flowchart of the parallel evolutionary artificial potential field for path planning

gains k_a and k_r values. In Fig. 4, the parallel evolutionary artificial potential field starts with the construction of an initial random population $p(t)$ of individuals, where the individuals are codified with a pair of the proportional gains, one k_a, and k_r. The resultant path given by the parallel evolutionary artificial potential field will be composed by the optimal pair of proportional gains found. After the initial population is created, the parallel evaluation process starts. The fitness function evaluation contains three steps. Firstly, the potential field $U_{total}(q)$ is calculated using Eq. 3. After that, the potential force $F_{total}(q)$ is calculated using Eq. 4. Lastly, the distance to travel by the mobile robot is measured by the path length function

$$S = \sum_{i=0}^{n} \|q_{i+1} + q_i\| \tag{8}$$

where, n is the number of mobile robot configurations to reach the target point.

When the evaluation has been completed, the procedure continues with the selection process, $P'(t) \leftarrow P(t)$. In the selection process, the best individuals are chosen according to their fitness value. The selection process drives the parallel evolutionary artificial potential field to improve the population fitness over the successive generations [15]. Next, the crossover process is performed over the $P'(t)$. The crossover process roughly mimics biological recombination between two single-chromosomes organisms [16]. Then, the mutation process is performed over the $P'(t)$, where random mutations alter a certain percentage of the bits in the list of individuals. The mutation process tends to distract the evolution from converging on a popular solution [15]. The parallel evolutionary artificial potential field iterates until the maximum number of generations is achieved. Hence, the parallel evolutionary artificial potential field evolves the proportional gains k_a and k_r to obtain the corresponding optimal values to perform the path planning. To conclude, the parallel evolutionary artificial potential field returns the path found.

In this work, the evaluation process for the parallel evolutionary artificial potential field on the graphics processing unit (GPU) was programmed in C++/CUDA [18]. We have chosen the NVIDIA Jetson TK1 developer kit as one of the platforms to implement the parallel evolutionary artificial potential field algorithm because it is one of the most advanced mobile processors for embedded computing, providing a lot of benefits in the fields of computer vision, robotics, automotive, image signal processing, network security, medicine, and many others. The NVIDIA Jetson TK1 developer kit is was designed to provide fast development of embedded applications using GPU, bringing significant parallel processing performance and exceptional power efficiency to embedded applications [19].

3.3 Path Tracking

The path found by the parallel evolutionary artificial potential field algorithm is a sequence of ordered objective points from the start to the target point. This sequence of ordered objective points is converted to rotate and advance motion commands to let the mobile robot moves from the start to the target point.

4 Experimental Results

In this section, the experimental results of obstacle recognition and path planning for autonomous mobile robot navigation are presented. The algorithms for obstacle detection Fig. 3, and for path planning Fig. 4 where implemented on two different computers: (1) a quad-core Intel i7-4710HQ running Ubuntu 16.04 distribution of Linux (with CUDA 8.0), and a GPU GeForce GTX 860M with 640 CUDA cores, (2) a quad-core ARM Cortex-A15 running Ubuntu 14.04 distribution of Linux (with CUDA 6.5), and a GPU NVIDIA Kepler with 192 CUDA cores (Jetson TK1). To compare the path planning performance on two different parallel architectures, we used a Logitec C920HD webcam for obstacle detection.

The experimental results are evaluated in terms of (1) accuracy of obstacle recognition, (2) path planning effectiveness and (3) path planning performance in parallel architectures. Figure 5 shows the test environments employed for each experiment (map0, map1, … , map5) to evaluate the obstacle detection algorithm and the path planning algorithm. The results in Fig. 5 show the effectiveness and efficiency of the algorithms graphically to perform the obstacle detection and path planning for a feasible and safe mobile robot navigation in the different test environments.

4.1 Obstacle Recognition Results

To quantify the accuracy of obstacle recognition, we evaluate the correlation filter with an objective metric known as discrimination capability (DC). The DC measures the ability of a filter to recognize a target among false artifacts. A DC representation in a percentage of the quality of detection is given by

$$DC(\%) = \left(1 - \frac{|c^b|^2}{|c^t|^2}\right) \times 100\%, \tag{9}$$

where c^b and c^t are the maximum value produced in the background and the target, respectively. In Fig. 6 is shown the performance of the algorithm in terms of DC.

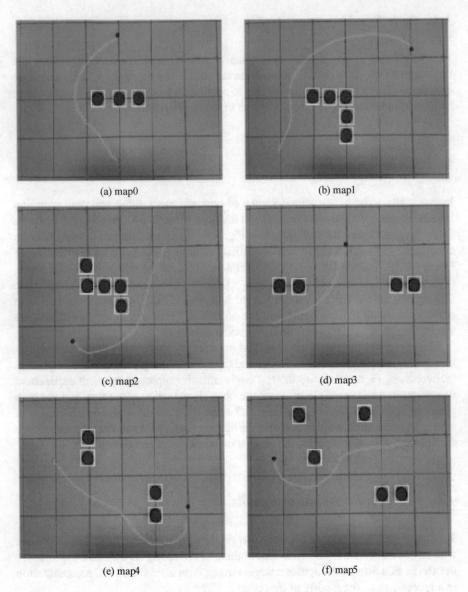

(a) map0

(b) map1

(c) map2

(d) map3

(e) map4

(f) map5

Fig. 5 Test environments employed for the experiments, obstacle detection and path planning

Observe that the algorithm exhibits good accuracy for obstacle detection obtaining an average of $DC = 92.68 \pm 6.98\%$ for the entire test environment set, from map0 to map5.

The location estimation of the obstacle presented in each test environment was computed by the location error (LE), which is given by

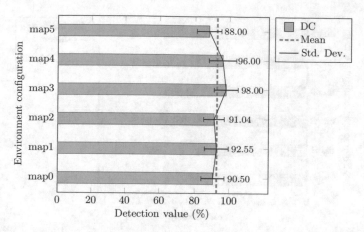

Fig. 6 Detection performance in terms of DC metric for different test environment configuration

$$LE = \sum_{j=1}^{J} \sqrt{\left(x_0^j - \hat{x}_0^j\right)^2 + \left(y_0^j - \hat{y}_0^j\right)^2}, \tag{10}$$

where $\left(x_0^j, y_0^j\right)$ and $\left(\hat{x}_0^j, \hat{y}_0^j\right)$ are the real and estimated coordinates of the obstacles presented in the current map. As shown in Fig. 7, the location error is calculated for the jth map. Observe that the algorithm yields high accuracy obtaining an average of $LE = 1.53 \pm 0.24$ pixels. The radius error (RE) is also calculated as

$$RE = \sum_{j=1}^{J} \left\| r^j - \hat{r}^j \right\|, \tag{11}$$

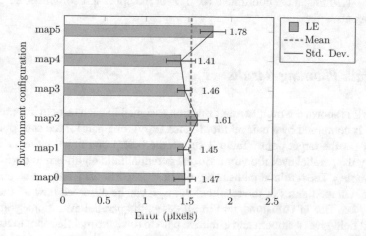

Fig. 7 Performance in terms of errors of location estimation for the different test environments

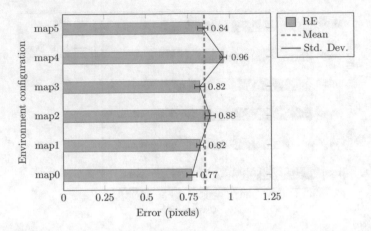

Fig. 8 Performance in terms of errors of radius estimation for the different test environments

where r^j and $\hat{r}^j$ are the real and estimated radii. The algorithm performs an average of $RE = 0.8483 \pm 0.027$ pixels, as shown in Fig. 8. The results of obstacle detection for the entire test environment set can be seen in Fig. 5.

In Table 1 is shown the number of obstacles, the position, and radius of the obstacles detected in each test environment. The first column indicates the environment tested. The second column indicates the number of obstacles detected by the obstacle detection algorithm, described in Fig. 3. The third column contains the information of each obstacle detected in pixels, the information is ordered in the coordinates (x, y), and its radius r. We are considering a workspace of 300×200 pixels (images), where the coordinate $(0, 0)$ is at the upper left corner. The fourth column contains information about the obstacles in meters (real world). It is the same information that the third column but converted to meters. Therefore, we are considering a workspace of 1.80×1.20 m and the coordinate $(0, 0)$ is at the upper left corner for all the test environments.

4.2 Path Planning Results

In Table 2 is shown the mobile robot mission assigned in each test environment, the mission is composed by a pair of coordinates (x, y), one pair for the start point and one pair for the target point. Table 2 shows the best proportional gains k_a and k_r found by the parallel evolutionary artificial potential field algorithm to perform the path planning. The resultant paths using the best proportional gains can be observed in Fig. 5. On the figures of the resultant paths, it can be observed how all the paths are safe, i.e., free of collisions, and in all cases, the parallel evolutionary artificial potential field gives a smooth and effective path to drive the mobile robot to its target point. Table 2 also shows the statistical results for thirty independent tests in each

Table 1 Number of obstacles, position and radius of the obstacles detected in each test environment

Test environment	Number of obstacles detected	Obstacles $j(x, y, r)$ in pixels	Obstacles $j(x, y, r)$ in meters
map0	3	(120, 101, 7.96), (152, 101, 7.76), (181, 101, 7.60)	(0.72, 0.61, 0.05), (0.91, 0.61, 0.05), (1.09, 0.61, 0.05)
map1	5	(102, 100, 7.98), (128, 101, 7.82), (152, 101, 7.72), (152, 150, 7.92), (153, 126, 7.67)	(0.61, 0.60, 0.05), (0.77, 0.61, 0.05), (0.91, 0.61, 0.05), (0.91, 0.90, 0.05), (0.92, 0.76, 0.05)
map2	5	(101, 75, 8.10), (102, 100, 8.02), (128, 101, 7.82), (152, 101, 7.74), (153, 126, 7.72)	(0.61, 0.45, 0.05), (0.61, 0.60, 0.05), (0.77, 0.61, 0.05), (0.91, 0.61, 0.05), (0.92, 0.76, 0.05)
map3	4	(50, 102, 8.04), (79, 103, 7.86), (225, 101, 7.65), (252, 101, 7.72)	(0.30, 0.61, 0.05), (0.47, 0.62, 0.05), (1.35, 0.61, 0.05), (1.51, 0.61, 0.05)
map4	4	(101, 78, 8.08), (102, 52, 8.00), (202, 122, 8.00), (202, 151, 7.76)	(0.61, 0.47, 0.05), (0.61, 0.31, 0.05), (1.21, 0.73, 0.05), (1.21, 0.91, 0.05)
map5	5	(78, 25, 8.14), (101, 79, 7.86), (176, 23, 7.74), (201, 126, 7.74), (231, 125, 7.76)	(0.47, 0.15, 0.05), (0.60, 0.47, 0.05), (1.06, 0.14, 0.05), (1.21, 0.76, 0.05), (1.39, 0.75, 0.05)

Table 2 Results for each robot mission, best resultant proportional gains, and path planning results for each test environment

	map0	map1	map2	map3	map4	map5
Start	(0.90, 1.08)	(0.30, 1.02)	(1.32, 0.30)	(0.24, 0.90)	(0.30, 0.48)	(1.50, 0.36)
Target	(0.90, 0.12)	(1.50, 0.24)	(0.48, 1.02)	(0.90, 0.30)	(0.30, 0.48)	(0.24, 0.48)
k_a	7.21	4.98	15.08	20.63	2.03	3.31
k_r	66.28	69.26	78.70	73.76	16.61	43.70
Best	1.25	1.75	1.30	0.93	1.59	1.46
Mean	1.32	1.82	1.32	0.94	1.74	1.48
Worst	1.74	2.18	1.44	0.95	2.77	1.51
S. Dev.	0.12	0.10	0.03	0.00	0.31	0.01

All data are expressed in meters except for the proportional gains

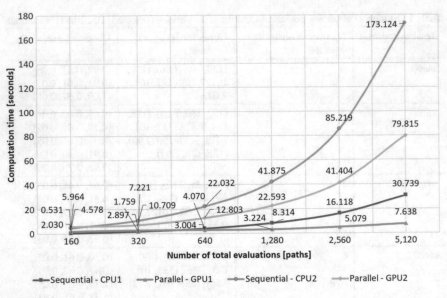

Fig. 9 Performance

environment. The shortest path length found in meters, the average path length, the worst path length found for each test environment, and the standard deviation of the tests.

4.3 Path Planning Performance Results

Figure 9 shows the path planning performance results. The evaluation was performed on two different computers, and thirty independent tests were carried out for each number of total evaluations (paths). The performance results show that the parallel path evaluation on GPU1 accelerates the evaluation process by a factor of 4.0× for the bigger population tested in comparison with sequential path evaluation on CPU1. Also, the performance results show that the parallel path evaluation on GPU2 accelerates the evaluation process by a factor of 2.2× for the bigger population tested in comparison with sequential path evaluation on CPU2. We can see the advantage of using the parallel path evaluation on GPU, as well as we can see that this advantage applies for onboard computers like the Jetson TK1.

5 Conclusions

An accurate obstacle recognition for path planning in autonomous mobile robot navigation is presented. Given an input scene, a mobile robot navigates in a feasible workspace avoiding several obstacles. We employ a correlation filtering approach to solve obstacle recognition. The proposed algorithm yields a good accuracy of detection performance in terms of DC. The obtained results also perform efficiently in obstacle location and radius estimation.

We have seen through the results how the proposed parallel evolutionary artificial potential field algorithm solves the path planning problem for the different test environments exposed. The parallel evolutionary artificial potential field algorithm takes the information of the environment given by the obstacle detection algorithm; then the parallel evolutionary artificial potential field can perform the path planning to give us a result, a path to perform the offline navigation by the mobile robot.

The overall results demonstrate the effectiveness and the efficiency of the proposal to perform the obstacle recognition for path planning in autonomous mobile navigation. The low variability demonstrates the stability of the proposed parallel evolutionary artificial potential field algorithm in the results; the standard deviation is in the order of the tenths even in some cases in the hundredths, that is good.

Related to the path planning performance results, we have presented the parallel path evaluation on GPU using the parallel evolutionary artificial potential field programmed in C++/CUDA. The performance results show the advantage of using the parallel path evaluation on GPU, as well as we can see that this advantage applies for onboard computers like the Jetson TK1.

Acknowledgements This work was supported in part by the Coordinación de Investigación of CETYS Universidad, in part by the Consejo Nacional de Ciencia y Tecnología (CONACYT, Mexico).

References

1. U. Orozco-Rosas, K. Picos, O. Montiel, R. Sepúlveda, V. Díaz-Ramírez, Obstacle recognition for path planning in autonomous mobile robots, in *Optics and Photonics for Information Processing* (2016)
2. O. Montiel, R. Sepúlveda, U. Orozco-Rosas, Optimal path planning generation for mobile robots using parallel evolutionary artificial potential field. J. Intell. Robot. Syst. **79**(2) (2014)
3. V.H. Diaz-Ramirez, V. Contreras, V. Kober, K. Picos, Real-time tracking of multiple objects using adaptive correlation filters with complex constraints. Opt. Commun. **309**, 265–278 (2013)
4. B.V.K. Vijaya Kumar, L. Hassebrook, Performance measures for correlation filters. Appl. Opt. **29**(20), 2997–3006 (1990)
5. R. Kerekes, B. Vijaya-Kumar, Correlation filters with controlled scale response. IEEE Trans. Image Process. **15**(7), 1794–1802 (2006)
6. B. Javidi, J. Wang, Design of filters to detect a noisy target in nonoverlapping background noise. J. Opt. Soc. Am. A **11**, 2604–2612 (1994)

7. V.H. Diaz-Ramirez, K. Picos, V. Kober, Target tracking in nonuniform illumination conditions using locally adaptive correlation filters. Opt. Commun. **323**, 32–43 (2014)
8. V. Kober, J. Campos, Accuracy of location measurement of a noisy target in a nonoverlapping background. J. Opt. Soc. Am. A **13**(8), 1653–1666 (1996)
9. K. Picos, V.H. Diaz-Ramirez, V. Kober, A.S. Montemayor, J.J. Pantrigo, Accurate three-dimensional pose recognition from monocular images using template matched filtering. Opt. Eng. **55**(6), 1–11 (2016)
10. O. Khatib, Real-time obstacle avoidance for manipulators and mobile robots, in *IEEE International Conference on Robotics and Automation* (1985)
11. Q. Zhang, D. Chen, T. Chen, An obstacle avoidance method of soccer robot based on evolutionary artificial potential field. Energy Proc. **16-C**, 1792–1798 (2012)
12. U. Orozco-Rosas, O. Montiel, R. Sepúlveda, Pseudo-bacterial potential field based path planner for autonomous mobile robot navigation. Int. J. Adv. Rob. Syst. **12**(81) (2015)
13. O. Montiel, U. Orozco-Rosas, R. Sepúlveda, Path planning for mobile robots using bacterial potential field for avoiding static and dynamic obstacles. Expert Syst. Appl. **42**(12), 5177–5191 (2015)
14. P. Vadakkepat, C.T. Kay, M.-L. Wang, Evolutionary artificial potential fields and their application in real time robot path planning, in *Congress on Evolutionary Computation* (2000)
15. S.N. Sivanandam, S.N. Deepa, *Introduction to Genetic Algorithms* (Springer, Heidelberg, 2008)
16. M. Mitchell, *An introduction to Genetic Algorithms* (Bradford, Cambridge, MA, 2001)
17. D.B. Fogel, An introduction to evolutionary computation, in *Evolutionary Computation: The Fossil Record* (Wiley-IEEE Press, 1998), pp. 1–28
18. U. Orozco-Rosas, O. Montiel, R. Sepúlveda, Parallel bacterial potential field algorithm for path planning in mobile robots: a GPU implementation, in *Fuzzy Logic Augmentation of Neural and Optimization Algorithms: Theoretical Aspects and Real Applications* (Springer, Berlin, 2018), pp. 207–222
19. U. Orozco-Rosas, O. Montiel, R. Sepúlveda, Embedded implementation of a parallel path planning algorithm based on EAPF for mobile robotics, in *Avances recientes en Ciencias Computacionales - CiComp 2016* (Ensenada, 2016)

Optimization of Fuzzy Controllers for Autonomous Mobile Robots Using the Grey Wolf Optimizer

Eufronio Hernández, Oscar Castillo and José Soria

Abstract Through the advance of technology, every day new methods or computational techniques emerge that allow us to solve problems in different areas, such as medicine, engineering, even in any industrial process. Optimization is of vital importance in this industry, the main objective being to find the best possible solution to the problem. In this work we propose to use the Grey Wolf Optimizer (GWO), which is a metaheuristic, which is inspired by the hunting behavior and leadership hierarchy of grey wolves, in addition to analyzing and explaining the proposed methodology for the optimization of fuzzy controllers for mobile autonomous robots.

Keywords Grey wolf optimizer (GWO) · Autonomous mobile robots · Fuzzy controllers · Optimization · Fuzzy system · Bio-inspired algorithm

1 Introduction

Bio-inspired computing is based on using analogies with natural or social systems to solve complex problems. The bio inspired algorithms simulate the behavior of natural systems for the design of non-deterministic heuristic methods of search, learning and behavior [1].

The optimization is characterized by finding the best solution for any problem computationally speaking whether maximization or minimization, for its realization we need bio inspired algorithms that simulate in a natural way some task that seeks to obtain a solution. In addition to having certain stages, such as inputs (variables), development or process (fitness) and an output (cost) [2].

There are several bio inspired algorithms that help us solve these problems, such as: flower pollination algorithm [3], bat algorithm [4], firefly algorithm [5], cuckoo search [6], bacterial foraging [7], artificial bee colony [8], ant colony optimization [9], particle swarm [10], genetic algorithm [11], mention the most used in this area.

E. Hernández · O. Castillo (✉) · J. Soria
Tijuana Institute of Technology, Tijuana, BC, Mexico
e-mail: ocastillo@tectijuana.mx

© Springer Nature Switzerland AG 2020
O. Castillo and P. Melin (eds.), *Hybrid Intelligent Systems in Control,
Pattern Recognition and Medicine*, Studies in Computational Intelligence 827,
https://doi.org/10.1007/978-3-030-34135-0_20

This research is being carried out in order to present a different way of performing fuzzy controller's optimization using the gray wolf algorithm, thus improving its performance.

This paper has been ordered as follows: in Sect. 2 the literature review, in Sect. 3 the proposed method is shown, in Sect. 4 a methodology details is presented, in Sect. 5 fuzzy logic controllers, in Sect. 6 the results and discussions short summary, in Sect. 7 the conclusions obtained after obtaining the results made with fuzzy system.

2 Literature Review

In this section the basic concepts are shown to obtain an overview of the proposed method.

2.1 Grey Wolf Optimizer

Seyedali Mirjalili proposed the GWO in 2014, basically is responsible for simulating the natural behavior of a pack of wolves to hunt their prey, in other words, mimics the leadership hierarchy and hunting mechanism of the gray wolf, you can say that within this herd there are different kinds of wolves like: alpha, beta, delta and omegas. As shown in Fig. 1 [12].

The hunting phases of GWO are listed as follows [13]:

- Tracking, chasing, and approaching the prey.
- Pursuing, encircling, and harassing the prey until it stops moving.
- Attack towards the prey.

In this section some concepts are detailed to obtain a better panorama about the proposed method:

Fig. 1 Pyramid of hierarchy

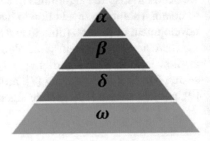

2.2 Encircling Prey

Wolves surround the prey during the hunt. To mathematically represent this behavior, the following equations are presented (1) and (2):

$$D = \left| C\, X_p(t) - X(t) \right| \tag{1}$$

$$X(t+1) = X_p(t) - A D \tag{2}$$

: indicates the current iteration, : are coefficients, X_p: is the position vector of the prey, : indicates the position vector of a grey wolf. The coefficients, are calculated as follows, by using Eqs. (3) and (4):

$$A = 2a\, r_1 - a \tag{3}$$

$$C = 2\, r_2 \tag{4}$$

a: is a number that linearly decrease from 2 to 0 over the course of iterations, r_1 and r_2 are random numbers in [0, 1].

2.3 Hunting Prey

Gray wolves naturally have the ability to recognize the location of the prey and surround them; the alpha wolf is sometimes responsible for guiding the hunt, sometimes involving the beta and delta wolves. To mathematically simulate this, we consider that alpha (best solution) beta and delta have a better knowledge about the possible location of the dam. As may be seen in Eqs. (5), (6) and (7).

So, we keep the first three best solutions obtained so far and force the other search agents (omegas) to update their positions according to the position of the best search agents.

In essence, the following formulas are proposed.

$$D_\alpha = |C_1 X_\alpha - X| \; D_\beta = |C_2 X_\beta - X| \; D_\delta = |C_1 X_\delta - X| \tag{5}$$

$$X_1 = X_\alpha - A_1(D_\alpha) \; X_2 = X_\beta - A_2(D_\beta) \; X_3 = X_\delta - A_3(D_\delta) \tag{6}$$

$$X(t+1) = \frac{X_1 + X_2 + X_3}{3} \tag{7}$$

Fig. 2 Pseudo code GWO

Initialize the gray wolf population $X_i(i = 1, 2, ..., n)$

Initialize a, A and C

Calculate the fitness of each search agent

X_α = The best search agent

X_β = The second best search agent

X_δ = The third best search agent

While (t < Maximum number of iterations)

    For, each search agent

    Update the position of the current search agent using equation (7)

    End For

Update a, A, and C

Calculate the fitness of all search agents

Update X_α, X_β, X_δ

$t = t + 1$

End While

Return X_α

2.4 Attacking Prey

Gray wolves attack the prey when it stops moving. To mathematically model this characteristic, the value of the parameter "" is decreased within the algorithm. When the values of "" are within the range $[-1, 1]$.

(a) If $\| < 1$, then the prey is attacked (Exploitation)
(b) If $\| > 1$, then the prey is searched (Exploration)

Pseudo code of the GWO:
See Fig. 2.

2.5 Fuzzy Logic

Fuzzy logic was proposed by LA Zadeh in 1965, from the University of California, Berkeley, based on a work on the theory of fuzzy sets, allows the representation of human knowledge, to treat inaccurate information by combining rules or linguistic values, based on the theory of fuzzy sets, such as: "very low", "low", "medium", "high", "very high" [14, 15, 16].

Traditional logic only accepts the use of two values, completely true or completely false, for example the proposition "tomorrow will be warm" must be true or false. However, the information that people use contains a certain degree of uncertainty. In fuzzy logic there is the possibility of having a degree of truth within a range of 0 to 1, for example in the proposition "the president of Mexico is old", this can have a degree of accuracy of 0.7, then the fuzzy logic try to measure that degree and help the computer understand that information. It is worth mentioning that fuzzy systems transform these into mathematical models, with the aim of facilitating the work of the designer and the computer, resulting in more real representations.

2.6 Autonomous Mobile Robot

They are systems capable of interacting with a high degree of autonomy within the limits of their environment, being able to move in any environment according to their programming. This leads to its use in areas such as: space exploration, wastewater treatment, and tasks that seem tedious or heavy for humans [17, 18].

Robots in the industry have been stealthily evolving from stationary machines to refined mobile platforms for the performance of numerous automation jobs. That is why robotics has achieved its greatest success in the world of industrial manufacturing. Currently the robot arms, or manipulators, are configured in a specific position on the assembly line, in order to perform jobs that require high speed and precision, such as moving heavy products, accommodating some type of material in order, etc.

Autonomous mobile robots are able to navigate difficult environments to make a path. One of the advantages is that they require little data such as external input to maneuver, which highlights an important capacity. The construction areas are inherently rough. Some innovations are the integration of intelligence systems, such as neural networks, bio inspired algorithms, some vision tools; the latter is equipped with multiple sensors helping them to perceive a dynamic environment in real time. Autonomous mobile robots have great industrial potential so it is very feasible to implement them in various areas.

3 Proposed Method

In particular, the work will focus on obtaining the best values for the fuzzy system. In Fig. 3 the proposed model is presented [19].

4 Methodology

A system is created to optimize fuzzy controllers for mobile autonomous robots using the gray wolf optimizer. Through the combination of 2 methods, the use of a bio-inspired algorithm and a fuzzy system will allow to reach the desired trajectory for the robots.

As a first step we will obtain the parameters of the membership functions to optimize, and then we execute the GWO algorithm with the fuzzy system parameters

Fig. 3 Proposed method

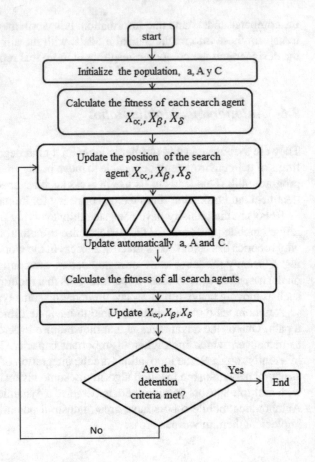

for optimization, so we will obtain the optimized fuzzy controller and finally we will see a simulated fuzzy controller, marking the desired path, as shown in Fig. 4.

Fig. 4 Methodology method

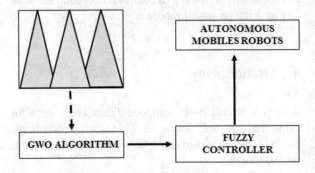

5   Fuzzy Logic Controller

These rules of the fuzzy system describe a relation between the linguistic variables based on the membership functions, taking into account the inputs and outputs [20–23].

A fuzzy Mamdani type system is designed containing two linear error (ev) and angular error (ew) inputs, which comprise a range of $[-1\ 1]$, which contain three membership functions, two trapezoidal at the ends and one triangular in the middle. So it has two outputs Torque 1 and Torque 2, $[-1\ 1]$. They contain three triangular membership functions, as shown in Fig. 5 [17, 24–26].

This fuzzy controller has 9 if-then rules, which are shown Fig. 6 [27–29].

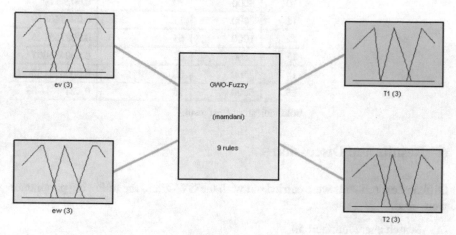

Fig. 5   Autonomous mobile robot fuzzy system

Fig. 6   Rules robot fuzzy system

1. If (ev is N) and (ew is N) then (T1 is N) (T2 is N)
2. If (ev is N) and (ew is Z) then (T1 is N) (T2 is Z)
3. If (ev is N) and (ew is P) then (T1 is N) (T2 is P)
4. If (ev is Z) and (ew is N) then (T1 is Z) (T2 is N)
5. If (ev is Z) and (ew is Z) then (T1 is Z) (T2 is Z)
6. If (ev is Z) and (ew is P) then (T1 is Z) (T2 is P)
7. If (ev is P) and (ew is N) then (T1 is P) (T2 is N)
8. If (ev is P) and (ew is Z) then (T1 is P) (T2 is Z)
9. If (ev is P) and (ew is P) then (T1 is P) (T2 is P)

Table 1 GWO parameters for each experiment and results of error

No.	Iterations	Number agents	Error
1	500	55	0.000741061
2	400	50	7.29585E−05
3	1000	60	0.003882946
4	600	45	0.064997856
5	450	65	0.018010558
6	500	45	0.000626393
7	800	40	0.021905084
8	550	56	0.019565891
9	4000	60	3.19E−04
10	3000	70	0.0021119
11	2000	75	0.001740619
12	1000	65	0.008165245
13	800	55	0.000436007
14	700	60	0.000417135
15	**5000**	**50**	**0.42498E−05**

Bold indicates "Best results"

6 Results and Discussion

Different experiments were carried out with the GWO, and the following parameters were used:

(a) Search agent number: 50
(b) Iterations: 5000

The mean square error was used as the aptitude function, as see Eq. (8), with which it is intended to minimize the error of the desired trajectory.

$$MSE = \frac{1}{2} \sum_{i=1}^{n} \left(\widehat{Y}_i - Y_i \right)^2 \tag{8}$$

Next, Table 1 is presented, where it is observed that 15 experiments were performed, of which the best result that it gave us was number 15, with 5000 iterations, with a total of 50 search agents, giving as well as error 0.000044249828665182.

7 Conclusions and Future Work

In this work, until now the algorithm of the wolves was analyzed, need to deepen and perform more tests. The tests were carried out with the fuzzy system in the mobile autonomous robot plant; the optimization has given good results. To carry out the optimization of the fuzzy systems, the GWO was used, and with it 15 different experiments were carried out, changing the value of the parameters, so that can say, partially, that good results have been obtained. As future work, will try to perform some tests for simulation, so can make comparisons with other meta-heuristics, also perform the statistical test. In addition, we could use type-2 fuzzy logic as in [30–34]. We can also consider applying the proposed method in different areas of application like in [35–39].

Acknowledgements It is widely appreciated to the Consejo Nacional de Ciencia y Tecnologia and Tecnologico Nacional de Mexico/Tijuana Institute of Technology for the time, resource, space and facilities provided for the development of this work.

References

1. X.-S. Yang, M. Karamanoglu, Swarm intelligence and bio-inspired computation: an overview. Swarm Intell. Bio-Inspired Comput. pp. 3–23 (Yang)
2. P. Gupta, R. Cambini, S.S. Appadoo, Recent advances in optimization theory and applications. Ann. Oper. Res. (2018). https://doi.org/10.1007/s10479-018-2984-yDO
3. X.S. Yang, M. Karamanoglu, X. He, Flower pollination algorithm: a novel approach for multiobjective optimization. Eng. Optim. **46**(9), 1222–1237 (2014)
4. X.-S. Yang, Bat algorithm: literature review and applications. Int. J. Bio-Inspired Comput. **5**(3), 141–149 (2013)
5. X.-S. Yang, Firefly algorithm, Lévy flights and global optimization, in *Research and Development in Intelligent Systems XXVI* (2010), pp. 209–218
6. X.S. Yang, S. Deb, Cuckoo search via Lévy flights, in *Proceeding of World Congress on Nature and Biologically Inspired Computing (NaBIC 2009)*, pp. 210–214
7. S. Das, A. Biswas, S. Dasgupta, A. Abraham, Bacterial foraging optimization algorithm: theoretical foundations, analysis, and applications, in *Foundations of Computational Intelligence vol. 3: Global Optimization* (Springer, Berlin, 2009)
8. D. Karaboga, B. Basturk, A powerful and efficient algorithm for numerical function optimization: artificial bee colony (ABC) algorithm. J. Glob. Optim. **39**, 459–471 (2007)
9. M. Dorigo, K. Socha, An introduction to ant colony optimization, in *Handbook of Metaheuristics*, vol 26, no 1 (IRIDIA, 2006, Brussels), ISSN 1781-3794
10. J. Kennedy, R. Eberhart, Particle swarm optimization, in *Proceedings of IEEE International Conference on Neural Networks* (1995), pp. 1942–1948
11. D. Goldberg, *Genetic Algorithms in Search, Optimization and Machine Learning* (Addison-Wesley, Boston, MA, 1987)
12. S. Mirjalili, S.M. Mirjalili, A. Lewis, Grey wolf optimizer. Adv. Eng. Softw. **69**, 46–61 (2014)
13. C. Muro, R. Escobedo, L. Spector, R. Coppinger, Wolf-pack (Canis lupus) hunting strategies emerge from simple rules in computational simulations. Behav. Process. **88**, 192–197 (2011)
14. L. Zadeh, Fuzzy logic. IEEE Comput. Mag. **1**, 83–93 (1988)
15. L. Zadeh, The concept of a linguistic variable and its application to approximate reasoning—I. Inform. Sci. **8**, 199–249 (1975)
16. L. Zadeh, Fuzzy sets. Inf. Control **8**, 338–353 (1965)
17. O. Castillo, P. Melin, O. Montiel, R. Sepulveda, W. Pedrycz, *Theoretical Advances and Applications of Fuzzy Logic and Soft Computing* (Springer, Tijuana, BC, 2007)

18. R. Martinez, O. Castillo, L.T. Aguilar, Optimization of interval type-2 fuzzy logic controllers for a perturbed autonomous wheeled mobile robot using genetic algorithms. Inf. Sci. **179**(13), 2158–2174 (2009)

19. L. Rodriguez, O. Castillo, J. Soria, A study of parameters of the grey wolf optimizer algorithm for dynamic adaptation with fuzzy logic, in *Nature-Inspired Design of Hybrid Intelligent Systems* (Springer, Tijuana, Mexico, 2017), pp. 371–390

20. O. Castillo, R. Martínez-Marroquín, P. Melin, F. Valdez, J. Soria, Comparative study of bio-inspired algorithms applied to the optimization of type-1 and type-2 fuzzy controllers for an autonomous mobile robot. Inf. Sci. **192**, 19–38 (2012)

21. O. Castillo, H. Neyoy, J. Soria, M. García, F. Valdez, *Dynamic fuzzy logic parameter tuning for ACO and its application in the fuzzy logic control of an autonomous mobile robot* (Int. J. Adv. Robot, Syst, 2013)

22. O. Carvajal, O. Castillo, J. Soria, Optimization of membership function parameters for fuzzy controllers of an autonomous mobile robot using the flower pollination algorithm. J. Autom. Mob. Robot. Intell. Syst. **12**(1), 44–49 (2018)

23. M. Sanchez, O. Castillo, J. Castro, Generalized type-2 fuzzy systems for controlling a mobile robot and a performance comparison with interval type-2 and type-1 fuzzy systems. Expert Syst. Appl. **42**, 5904–5914 (2015)

24. C. Soto, F. Valdez, O. Castillo, A review of dynamic parameter adaptation methods for the firefly algorithm, in *Nature-Inspired Design of Hybrid Intelligent Systems* (Springer, Tijuana, BC, 2007), pp. 285–295

25. L. Astudillo, P. Melin, O. Castillo, *Chemical Optimization Algorithm for Fuzzy Controller Design* (Springer, Tijuana, Mexico, 2014)

26. A. Sombra, F. Valdez, P. Melin, O. Castillo, A new gravitational search algorithm using fuzzy logic to parameter adaptation, in *IEEE Congress on Evolutionary Computation*, Cancun, México (2013), pp. 1068–1074

27. F. Valdez, P. Melin, O. Castillo, Evolutionary method combining particle swarm optimization and genetic algorithms using fuzzy logic for decision making, in *IEEE International Conference on Fuzzy Systems* (2009), pp. 2114–2119

28. F. Olivas, F. Valdez, O. Castillo, C. Gonzales, G. Martinez, P. Melin, Ant colony optimization with dynamic parameter adaptation based on interval type-2 fuzzy logic systems. Appl. Soft Comput. **53**, 74–87 (2016)

29. T.Y. Abdalla, A. Abdulkareem, A PSO optimized fuzzy control scheme for mobile robot path tracking. Int. J. Comput. Appl. **76**(2), 11–17 (2013). https://doi.org/10.5120/13217-0608

30. C. Leal Ramírez, O. Castillo, P. Melin, A. Rodríguez Díaz, Simulation of the bird age-structured population growth based on an interval type-2 fuzzy cellular structure. Inf. Sci. **181**(3), 519–535 (2011)

31. N.R. Cázarez-Castro, L.T. Aguilar, O. Castillo, Designing type-1 and type-2 fuzzy logic controllers via fuzzy Lyapunov synthesis for nonsmooth mechanical systems. Eng. Appl. AI **25**(5), 971–979 (2012)

32. O. Castillo, P. Melin, Intelligent systems with interval type-2 fuzzy logic. Int. J. Innov. Comput. Inf. Control **4**(4), 771–783 (2008)

33. C.I. González, P. Melin, J.R. Castro, O. Mendoza, O. Castillo, An improved Sobel edge detection method based on generalized type-2 fuzzy logic. Soft. Comput. **20**(2), 773–784 (2016)

34. C.I. González, P. Melin, J.R. Castro, O. Castillo, Olivia Mendoza: optimization of interval type-2 fuzzy systems for image edge detection. Appl. Soft Comput. **47**, 631–643 (2016)

35. L. Aguilar, P. Melin, O. Castillo, Intelligent control of a stepping motor drive using a hybrid neuro-fuzzy ANFIS approach. Appl. Soft Comput. **3**(3), 209–219 (2003)

36. P. Melin, O. Castillo, Adaptive intelligent control of aircraft systems with a hybrid approach combining neural networks, fuzzy logic and fractal theory. Appl. Soft Comput. **3**(4), 353–362 (2003)

37. P. Melin, J. Amezcua, F. Valdez, O. Castillo, A new neural network model based on the LVQ algorithm for multi-class classification of arrhythmias. Inf. Sci. **279**, 483–497 (2014)
38. P. Melin, O. Castillo, *Modelling, Simulation and Control of Non-linear Dynamical Systems: An Intelligent Approach Using Soft Computing and Fractal Theory* (CRC Press, Boca Raton, 2001)
39. P. Melin, D. Sánchez, O. Castillo, Genetic optimization of modular neural networks with fuzzy response integration for human recognition. Inf. Sci. **197**, 1–19 (2012)

Towards a Control Strategy Based on Type-2 Fuzzy Logic for an Autonomous Mobile Robot

Felizardo Cuevas, Oscar Castillo and Prometeo Cortes

Abstract The main purpose considered in this paper is to maintain a specific location and behavior for a robot that uses type-2 fuzzy logic for controlling its behavior. In this work, we propose a combination of behaviors by following a trajectory without leaving or losing it and avoiding obstacles in an omnidirectional mobile platform. The results of the simulation show the advantages of the proposed approach. We describe the previous knowledge about type-2 fuzzy logic, the virtualization of the mobile robot and its modeling according to real situations. The proposed control system is developed in Matlab-Simulink, the system can model and guide a mobile robot, successfully in simulated and real environments.

Keywords AMR (Autonomous mobile robots) · T2FS (Type-2 fuzzy systems) · T2FLC (Type-2 fuzzy logic controller) · OMR (Omnidirectional movil robot)

1 Introduction

One of the basic problems that a mobile robot faces when navigating from one place to another using sensors in real environments is the uncertainty, which is due to the large number of inaccuracies in the readings obtained in a real environment [1].

The imprecision that results from the information provided by the different sensors in the mobile robot, will result in inefficient behavior.

A mobile robot must be capable of performing correctly in a real environment, which means navigating autonomously and this requires a control strategy that allows it to handle the uncertainty that exists in the environment where it performs, especially in real time, with low computational load [2, 3].

F. Cuevas · O. Castillo (✉) · P. Cortes
Tijuana Institute of Technology, Tijuana, BC, Mexico
e-mail: ocastillo@tectijuana.mx

P. Cortes
e-mail: Prometeo.cortes@gmail.com

© Springer Nature Switzerland AG 2020
O. Castillo and P. Melin (eds.), *Hybrid Intelligent Systems in Control,
Pattern Recognition and Medicine*, Studies in Computational Intelligence 827,
https://doi.org/10.1007/978-3-030-34135-0_21

(a)   **(b)**

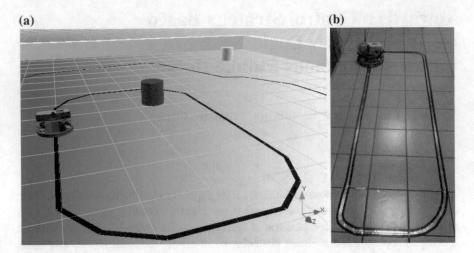

Fig. 1 Navigation scenario where mobile platform moves. **a** Simulation. **b** Real life

Therefore the navigation of a mobile robot must take into consideration:

- The location of the robot in initial position.
- The perception of the information provided by the sensors is almost entirely unreliable. the sensors with their low range that in combination with the characteristics of the environment and other conditions induce noise, errors in the responses of the sensors leading us to inaccurate data.

Currently, one of the main challenges of mobile robotics is the design of models that reliably perform complex tasks in face of the uncertainty of the environment [4]. For this reason, it is interesting to carry out research works in this area. Of course, this makes the realization of works in this field interesting.

Figure 1 shows a navigation scenario where the mobile platform moves and follows the path shown by the line to be pursued.

Today in the design of controllers for mobile robotics, is one of the areas where the use of fuzzy logic is playing a very important role for the design of controllers that are based on behavior and expert knowledge. Allowing robustness due to the vagueness of the information and disturbances that affect the correct performance of the sensors [5].

Behavior is understood as a small part of the control designed to achieve an objective in a set of situations with restrictions. It is here where the need arises in mobile robots when performing a specific task, the development and cooperation of diverse behaviors.

Each behavior is implemented for the control of a task such as, the following of a line, evasion of obstacles and search for an objective.

The main contribution of this work is the implementation of a Type-2 Fuzzy Logic Controller for path tracking and obstacle avoidance present in the development of an autonomous mobile robot (AMR) at low speeds. Starting from the study of the kinematic model and adapting the technique of linear and angular speed control, with the objective of minimizing the orientation and position error within the line that serves as a path to the vehicle. The mathematical models and the control routines were simulated and validated in Matlab/ Simulink, and later implemented in an Omnidirectional Mobil Robot with the same trajectory in the simulation and analyzed the results.

This work is structured as follows: Sect. 2 presents the reference of the 3-wheel omnidirectional robot kinematics. Section 3 presents the design, implementation and simulation of the interval type-2 fuzzy logic controller of the mobile robot. Section 4 presents results and discussions of the achievements. Section 5 concludes with some final clarifications, conclusions and future work.

2 Kinematic Model

The most practical way to get the information of a movement and its characteristics, is by its geometric representation, which has its application in the analysis of movements in mobile robots. Among the applications of kinematics is the possibility of its application as an initial mathematical model for modeling the controller, to raise the equations for calculating its odometry, or its simulation that generates its kinetic behavior of the mobile robot. The following statements show as the limitations in the construction of the kinematic model [6]:

- The displacement of the robot is assumed to be on a flat surface.
- The structure of the robot does not have flexible elements.
- The steering axes of the wheels are almost always present on the wheels, making them perpendicular to the surface where they are applied.
- Friction is ignored.

Table 1 contains the orientation, position and distribution of the omnidirectional wheels that take the point of symmetry that generally coincides with the center of the robotic platform [7].

Table 1 initial configuration parameters of kinematics omnidirectional platform

	Wheel 1	Wheel 2	Wheel 3
α_i	180°	60°	−60°
β_i	0°	0°	0°
γ_i	0°	0°	0°
δ_i	(0, 0, 0)	(0, 0, 0)	(0, 0, 0)
λ_i	$(-L,\ 0, 0)$	$\left(\frac{L}{2}, \frac{L\sqrt{3}}{2}, 0\right)$	$\left(\frac{L}{2}, \frac{-L\sqrt{3}}{2}, 0\right)$

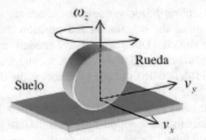

Fig. 2 Representation of wheel contact with the ground

Fig. 3 Omnidirectional wheels types

2.1 Kinematic Behavior

The kinematics of the wheels can be viewed as a principle where the wheels when making contact with the surface of the ground acts as a point of union in a plane with three degrees of freedom [8].

The wheels act as a flat union and this is described below. The wheel is considered to be a rigid element, and it is always in contact with the ground in a unique position that serves as the origin of the reference system explained in Fig. 2. The directions v_x and v_y determine the direction or direction of the wheel, v_x and v_y represent the linear velocities x-y, and w_z is the angular velocity when the robot turns. The conventional wheel in question, has an element v_x is null, although there are other wheels that provide a different behavior, as shown in Fig. 3.

An omnidirectional wheel is constituted as a standard wheel, one that is equipped with a ring with rollers, 120° one from the other, with active actuator, passive direction and the rotation axes are perpendicular to the forward direction [9, 10].

2.2 Kinematic Model

Location and position of the omnidirectional robot is represented by a vector (x, y, Ø)^T. The global speed of OMR is represented with the vector ($\dot{x}$, $\dot{y}$, $\dot{\varnothing}^T$) and the

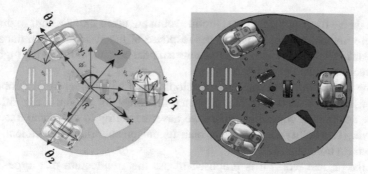

Fig. 4 Kinematic scheme of the robotic platform

angular speed for each wheel, is presented with the vector $(\dot{\theta}1, \dot{\theta}2, \dot{\theta}3)^T$, as shown in Fig. 4 based on the model detailed in [5], and the kinematic model of the robotic platform is presented as follows:

$$P(\phi) = \begin{bmatrix} -\sin(\phi + \varphi_1)\cos(\phi + \varphi_1)R \\ -\sin(\phi + \varphi_1)\cos(\phi + \varphi_1)R \\ -\sin(\phi + \varphi_1)\cos(\phi + \varphi_1)R \end{bmatrix} \quad (1)$$

$$\begin{bmatrix} \dot{\theta}_1 \\ \dot{\theta}_2 \\ \dot{\theta}_3 \end{bmatrix} = \frac{1}{r}\begin{bmatrix} V_1 \\ V_2 \\ V_3 \end{bmatrix} = \frac{1}{r}P = \frac{1}{r}P\begin{bmatrix} \dot{x} \\ \dot{y} \\ \dot{\phi} \end{bmatrix} \quad (2)$$

With:

$\dot{\theta}_n$: Angular speed of the wheel n
V_i: Linear speed of the wheel i
r: Radial distance of the wheel
R: Difference between the wheel and the center of the platform
ϕ: Angular speed
φ_n: Angular delimitation of the wheel n
$P(\phi)$: Change matrix calculation through the angular speed of the wheels and the global speed array $(\dot{x}, \dot{y}, \dot{\phi})^T$.

3 Interval Type-2 Fuzzy Systems

A fundamental requirement of autonomous mobile robots is a good behavior to avoid obstacles. This action helps the mobile robot to move without colliding in an unstructured environment [11]. In this work, the behaviors of avoiding obstacles and path tracking, are basic behaviors that use three infrared sensors and an inductive sensor, in the front and side of the mobile robot. The response signals of the sensors located on the front and side of the mobile robot are taken into account in this work and these imprecise values that are the response of the system, which uncertainty must be taken into account in order to create a model capable of tolerating high levels of imprecision in its environment.

In the interest of considering the development of an efficient mobile robot controller, a T2FLS is used [12]. It is expected that the T2FLS control algorithm will produce an efficient controller, where we have the ability to overcome uncertainty and having this ability the robot can plan its movements running efficiently in an unknown environment [13].

For the present work, it is proposed to use the architecture of a type-2 fuzzy Inference System, as illustrated in Fig. 2.

The upper and lower MFs for interval type-2 fuzzy set can be written in Eqs. (4) and (5). Assuming that we have N rules in a fuzzy rule base of Type 2, they take the following form [15] of Eq. (6):

$$F_A^-(x) = \begin{cases} 0, x < l_1 \\ (x - l_1)/(p_1 - l_1), \ l_1 < x < 0 \\ 1, \ x \geq 0 \\ (r_2 - x)/((r_2 - p_2), \ x \geq 0 \\ 0, \ x \geq r_2 \end{cases} \tag{4}$$

$$F_{\_A}(x) = \begin{cases} 0, \ x < l_2 \\ (x - l_1)/(p_1 - l_2), \ x \leq \frac{r_1(p_2 - l_2)}{(p_2 - l_2)} + \frac{l_2(r_1 - p_1)}{(r_1 - p_1)} \\ \\ (r_2 - x)/((r_2 - p_2), \ x > \frac{r_1(p_2 - l_2)}{(p_2 - l_2)} + \frac{l_2(r_1 - p_1)}{(r_1 - p_1)} \\ 0, \ x > r_2 \end{cases} \tag{5}$$

$$\begin{aligned} R^i : & \ If x_1 \ is \ X_i^l \ and \ \dots and \ \dots, \ x_p \ is \ X_p^l \\ & \ then \big(y \ is \ Y^l \big) \end{aligned} \tag{6}$$

where $X_i^l (i = 1, \dots p)$ and Y^l are type $- 2$ fuzzy sets, and x
$= (x^1, \dots, x^l)$ and y are linguistic variables.

The firing set is defined by interval:

$$F^i(x) = [f_{\_}^i(x), f^{-i}(x) \equiv [f_{\_}^i, f^{-i}] \tag{7}$$

- This IT2FLS handles the uncertainty in the system, while the T1FLS simply does not. Although it is described similarly as a T1FLS, an IT2FLS has some differences when it is described in a block diagram, presented in Fig. 2, the change being shown as it can be seen in the output zone or area, where a Type-Reducer reduces IT2FS to a T1FS, which is then introduced to a Defuzzifier to obtain a crisp result [14] (Fig. 5).

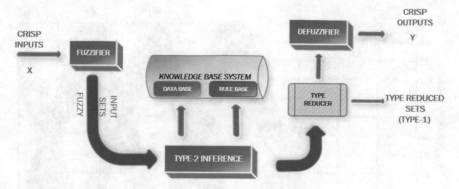

Fig. 5 IT2FLS architecture

There are many ways to implement an IT2FLS type reducer and at the moment [15], one of these techniques is the most used, the center of gravity (cos), where Y_{cos} is an interval, described by the points $\left[Y_l^i, Y_r^i\right]$, which are calculated with Eqs. (8) and (9).

Fuzzification

$$y_l = \sum_{i=1}^{M} f_l^i y_l^i / f_l^i \tag{8}$$

$$y_l = \sum_{i=1}^{M} f_r^i y_r^i / f_r^i \tag{9}$$

The defuzzification technique is performed by Eq. (10), which achieves a crisp output value.

$$y_{(x)} = \frac{y_l + y_r}{2} \tag{10}$$

The variability in each of the actions is modeled in interval type-2 fuzzy sets (T2FS). The T2FS linguistic inputs variables and their ranges are used for path tracking and orientation obstacle avoidance, as shown in Fig. 6a, b, with two outputs which are the linear and angular speed, are shown in Fig. 6c, d.

We know that the T2FS is located in a region built by a main type-1 membership function (T1MF). T2FS is obtained by using fuzzy sets to partition the input domains of the base line T1FS with footprint of uncertainty (FOU) as shown in Fig. 6.

Consequently, the T1MF extends to T2MF by adding the FOU in the antecedent and consequent parts of each rule.

Therefore, membership functions have values distributed with uncertainty. As well as those that belong to the antecedents also in the consequent parts. From the

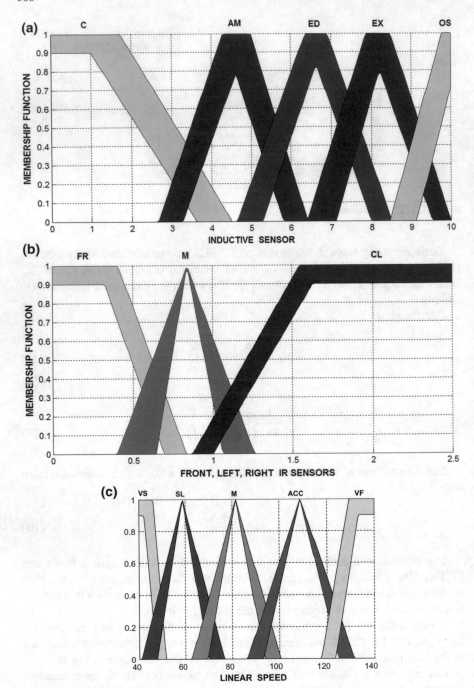

Fig. 6 **a** Inductive sensor as input MFs. **b** Infrared left, front and right sensors input MFs. **c** Linear speed output MFs. **d** Angular speed as output MFs

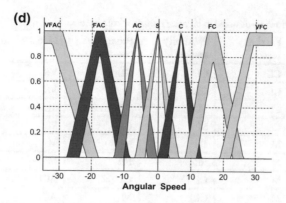

Fig. 6 (continued)

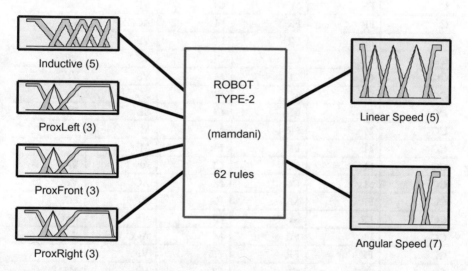

System ROBOT TYPE-2: 4 inputs, 2 outputs, 62 rules

Fig. 7 Interval type-2 fuzzy system with 4 inputs and 2 outputs

design, five MFs are for the Inductive sensor input, 3 MFs for the Infrared, 5 MFs for the Linear speed output and 7 MFs for the Angular Speed output, are used.

Fixed values are assigned in a range of 0–10 V for Inductive sensor, 0–2.5 V for Infrared sensors to measure obstacle distance, 30–120 mm/s for Linear velocity and −30° to 30° for Angular velocity depending on the displacement of the mobile robot.

In this work, seventy-four rules were developed by T2FLS in path tracking behavior and to avoid obstacles. Rules are used in the control of angular and linear speed, shown in Table 2.

Table 2 Path following-obstacle avoidance behavior rule base

Inductive sensor	Proximity sensor	Proximity sensor	Proximity sensor	Linear speed	Angular speed
C	CL	CL	CL	VS	VFAC
C	CL	CL	M	VS	VFC
C	CL	M	FR	SL	C
C	CL	M	CL	SL	VFC
C	M	FR	M	M	S
C	M	FR	FR	M	S
C	M	CL	CL	ACC	VAC
C	M	CL	M	ACC	VFC
C	FR	M	FR	VF	FC
C	FR	M	CL	VF	FAC
C	FR	FR	M	ACC	S
C	FR	FR	FR	VF	S
CC	CL	CL	CL	VS	VFAC
CC	CL	CL	M	VS	VFC
CC	CL	M	FR	SL	C
CC	CL	M	CL	SL	VFC
CC	M	FR	M	M	S
CC	M	FR	FR	M	S
CC	M	CL	CL	ACC	VAC
CC	M	CL	M	ACC	VFC
CC	FR	M	FR	VF	FC
CC	FR	M	CL	VF	FAC
CC	FR	FR	M	ACC	S
CC	FR	FR	FR	VF	S
FL	CL	CL	CL	VS	VFAC
FL	CL	CL	M	VS	VFC
FL	CL	M	FR	SL	C
FL	CL	M	CL	SL	VFC
FL	M	FR	M	M	S
FL	M	FR	FR	M	S
FL	M	CL	CL	ACC	VAC
FL	M	CL	M	ACC	VFC
FL	FR	M	FR	VF	FC
FL	FR	M	CL	VF	FAC
FL	FR	FR	M	ACC	S

(continued)

Table 2 (continued)

Inductive sensor	Proximity sensor	Proximity sensor	Proximity sensor	Linear speed	Angular speed
FL	FR	FR	FR	VF	S
E	CL	CL	CL	VS	VFAC
E	CL	CL	M	VS	VFC
E	CL	M	FR	SL	C
E	CL	M	CL	SL	VFC
E	M	FR	M	M	S
E	M	FR	FR	M	S
E	M	CL	CL	ACC	VAC
E	M	CL	M	ACC	VFC
E	FR	M	FR	VF	FC
E	FR	M	CL	VF	FAC
E	FR	FR	M	ACC	S
E	FR	FR	FR	VF	S
FU	CL	CL	CL	VS	VFAC
FU	CL	CL	M	VS	VFC
FU	CL	M	FR	SL	C
FU	CL	M	CL	SL	VFC
FU	M	FR	M	M	S
FU	M	FR	FR	M	S
FU	M	CL	CL	ACC	VAC
FU	M	CL	M	ACC	VFC
FU	FR	M	FR	VF	FC
FU	FR	M	CL	VF	FAC
FU	FR	FR	M	ACC	S
FU	FR	FR	FR	VF	S

So if an obstacle comes too close to the robot, it must be able to avoid it and change its behavior by slowing down, stopping if necessary and turning.

In this work, the number of rules are Sixty in total. These are obtained from the combination of four inputs with five and three membership functions as shown in Table 2.

4   Results and Discussion

In this work an evaluation is made to analyze the performance of the mobile robot based on T2FLS shown in Fig. 5, when compared to T1FLS with similar number of rules [13]. The responsiveness of the movement of the mobile robot to follow the path and avoid obstacle is done through the use of simulation. The evaluation is carried out in the environment of the simulator of the omnidirectional FESTO platform [15]. The experiments in each environment will have their own parameters whose specification is be explained later.

The data that should be stored in this simulation is the voltage of the input sensors, the linear and angular velocities. All the data is obtained from the real experiment and the simulation with the help of Simulink, each output data is recorded per unit of time, which in this case is 1 mm/s.

The control surface of T2FLS with 60 rules is included, whose inputs are an inductive and infrared sensors and the output is the linear speed. In the development of T2FLS a smooth surface is obtained, which is characterized by the slopes on its surface and each slope has a gradual change represented by the linear velocity, which is shown in Fig. 8 (Fig. 7).

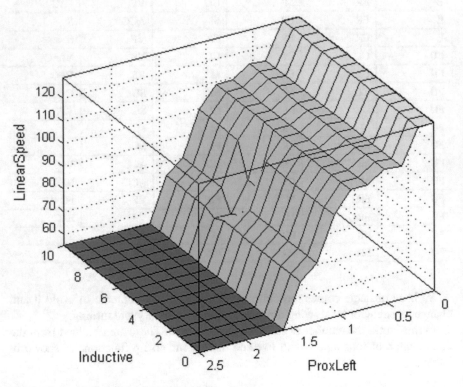

Fig. 8 The control surface of T2FLS with 74 rules

5   Conclusions

The proposed approach implements a type-2 fuzzy controller that produces the linear and angular velocity values in path tracking, with the ability to avoid obstacles that arise in its development using a self-contained omnidirectional mobile robot, so that the robot can travel the road without losing it and with the least possible error. In the initial tests, the effectiveness of the T2FLC is demonstrated when implementing it in an omnidirectional mobile robot, and its use provides an advantage in the efficiency of the route line, the smoothness in its follow-up and the decision on the correct acceleration and braking according to the type of direction change of the line being pursued. In the path that the robot follows with the T2FLC it is possible to observe how the values of linear and angular velocity are changing, as well as conditioned of the inductive sensor signal. In future work we will use Genetic Algorithm and PSO [16, 17, 18] to optimize the membership functions of the fuzzy systems [19–22] and we will work with neural networks and perform more experiments [23–25]. Also, type-2 fuzzy logic could be added as in [26–29].

References

1. S. Oltean, M. Dulau, Position control of Robotino mobile robot using fuzzy logic, in *2010 IEEE International Conference on Automation, Quality and Testing, Robotics* (2010), pp. 1–6
2. T.T. Mac, C. Copot, R.D.E. Keyser, T.D. Tran, T. Vu, MIMO fuzzy control for autonomous mobile robot. J. Autom. Control Eng. **4**, 65–70 (2016)
3. F. Cuevas, O. Castillo, Design and implementation of a fuzzy path optimization system for omnidirectional autonomous mobile robot control in real-time, in *Studies in computational intelligence*, vol. 749 (2018)
4. A.P. Moon, K.K Jajuiwar, Design of adaptive fuzzy tracking controller for autonomous navigation system. Int. J. Recent Trend Eng. Res. **2**, 268–275 (2016) (ijrter.com)
5. X. Li, A. Zell, Motion control of an omnidirectional mobile robot, in *Informatics in Control, Automation and Robotics* (Springer Berlin Heidelberg, Berlin, Heidelberg), pp. 181–193
6. D.R. Parhi, B.B.V.L. Deepak, Kinematic model of three wheeled mobile robot. Mech. Eng. Res. **3**, 307–318 (2011)
7. Y. Zhao, S.L. BeMent, Kinematics, dynamics and control of wheeled mobile robots, in Proceedings of 1992 IEEE International Conference on Robotics and Automation (1992), pp 91–96
8. D. Garcia Sillas, E. Gorrostieta Hurtado, E. Vargas Soto, J. Rodríguez Reséndiz, S. Tovar Arriaga, Kinematics modeling and simulation of an autonomous omni-directional mobile robot. Ing. e Investig. **35**, 74–79 (2015)
9. J.B. Song, K.S. Byun, Design and control of an omnidirectional mobile robot with steerable omnidirectional wheels, in *Mobile Robots, Moving Intelligence* (2006), pp. 224–240
10. K. Kanjanawanishkul, Omnidirectional wheeled mobile robots: wheel types and practical applications. Int. J. Adv. Mechatron. Syst. **6**, 289 (2015)
11. M. Njah, (ICCAT, M.J.-C.A.T., Wheelchair obstacle avoidance based on fuzzy controller and ultrasonic sensors, in *International Conference on Computer Applications Technology (ICCAT)* (2013), pp 1–5
12. O. Castillo, P. Melin, A review on the design and optimization of interval type-2 fuzzy controllers. Appl. Soft Comput. J. **12**, 1267–1278 (2012)

13. O. Castillo, P. Melin, W. Pedrycz, Design of interval type-2 fuzzy models through optimal granularity allocation. Appl. Soft Comput. J. **11**, 5590–5601 (2011)
14. D. Wu, On the fundamental differences between interval Type-2 and Type-1 fuzzy logic controllers. IEEE Trans. Fuzzy Syst. **20**(5), 832–848 (2012)
15. Festo Didactic GmbH & Co. KG, in *Robotino Manual* (Denkendorf, 2007), pp. 7–25
16. P. Ochoa, O. Castillo, J. Soria, *Differential Evolution with Dynamic Adaptation of Parameters for the Optimization of Fuzzy Controllers*. Presented at the (2014)
17. L. Rodríguez, O. Castillo, J. Soria, P. Melin, F. Valdez, C.I. Gonzalez, G.E. Martinez, J. Soto, A fuzzy hierarchical operator in the grey wolf optimizer algorithm. Appl. Soft Comput. J. **57**, 315–328 (2017)
18. R. Martínez-Soto, O. Castillo, L.T. Aguilar, A. Rodriguez, A hybrid optimization method with PSO and GA to automatically design Type-1 and Type-2 fuzzy logic controllers. Int. J. Mach. Learn. Cybern. **6**(2), 175–196 (2015)
19. L. Amador-Angulo, O. Castillo, Optimization of the Type-1 and Type-2 fuzzy controller design for the water tank using the Bee Colony Optimization, in *2014 IEEE Conference on Norbert Wiener in the 21st Century (21CW)* (IEEE, 2014), pp. 1–8
20. D.V. Bhoyar, P.B.J. Chilke, S.S. Kemekar, in *Design and Analysis of fuzzy PID Controllers using Genetic Algorithm* (2016), pp. 135–138
21. C. Caraveo, F. Valdez, O. Castillo, Optimization of fuzzy controller design using a new bee colony algorithm with fuzzy dynamic parameter adaptation. Appl. Soft Comput. **43**, 131–142 (2016)
22. J. Pérez, F. Valdez, O. Castillo, Modification of the bat algorithm using type-2 fuzzy logic for dynamical parameter adaptation, in *Studies in Computational Intelligence* (2017)
23. O. Castillo, R. Martínez-Marroquín, P. Melin, F. Valdez, J. Soria, Comparative study of bio-inspired algorithms applied to the optimization of type-1 and type-2 fuzzy controllers for an autonomous mobile robot. Inf. Sci. (Ny) **192**, 19–38 (2012)
24. L. Amador-Angulo, O. Mendoza, J. Castro, A. Rodríguez-Díaz, P. Melin, O. Castillo, Fuzzy sets in dynamic adaptation of parameters of a bee colony optimization for controlling the trajectory of an autonomous mobile robot. Sensors **16**, 1458 (2016)
25. P. Melin, L. Astudillo, O. Castillo, F. Valdez, M. Garcia, Optimal design of type-2 and type-1 fuzzy tracking controllers for autonomous mobile robots under perturbed torques using a new chemical optimization paradigm. Expert Syst. Appl. **40**, 3185–3195 (2013)
26. Patricia Melin, Claudia I. González, Juan R. Castro, Olivia Mendoza, Oscar Castillo, Edge-Detection method for image processing based on generalized Type-2 fuzzy logic. IEEE Trans. Fuzzy Syst. **22**(6), 1515–1525 (2014)
27. Claudia I. González, Patricia Melin, Juan R. Castro, Oscar Castillo, Olivia Mendoza, Optimization of interval type-2 fuzzy systems for image edge detection. Appl. Soft Comput. **47**, 631–643 (2016)
28. Claudia I. González, Patricia Melin, Juan R. Castro, Olivia Mendoza, Oscar Castillo, An improved sobel edge detection method based on generalized type-2 fuzzy logic. Soft. Comput. **20**(2), 773–784 (2016)
29. Emanuel Ontiveros, Patricia Melin, Oscar Castillo, High order α-planes integration: a new approach to computational cost reduction of General Type-2 fuzzy systems. Eng. Appl. of AI **74**, 186–197 (2018)

Implementation of a Fuzzy Controller for an Autonomous Mobile Robot in the PIC18F4550 Microcontroller

Oscar Carvajal and Oscar Castillo

Abstract Soft Computing has been gaining popularity in real world applications in many fields, an example of the area that has a wide variety of applications of these techniques is the Robotics area. In this work, we introduce the design of a hardware system for an autonomous mobile robot and the development of a Fuzzy Logic Controller for the control of the motion of a robot to follow a trajectory. We consider the error in the distance to the path as the unique input to the fuzzy controller and as the outputs, the linear velocity of each of the two wheels of the robot. We also show the development of the firmware in the PIC18F4550 microcontroller as the implementation of the Fuzzy Logic Controller.

Keywords Fuzzy logic · Microcontroller · Firmware · Robotics

1 Introduction

In recent days, soft computing and robotics have been achieving a good symbiosis [1–5], and this makes sense because the Robotics area is trying to simulate the behavior of humans. For this reason, this kind of works need an intelligent technique such as fuzzy logic, which is based on the human expert knowledge of a person on a particular subject [6–9].

Fuzzy Logic has a significant advantage in the control area, due to the fact that it is more intuitive than classic control methods, which require to know the mathematical model of the components of the system and in many cases there does not exist such mathematical model that represents the system. So Fuzzy logic enhances the productivity of the design in this kind of systems.

There are some platforms to implement a Fuzzy Logic Controller (FLC) it can be a personal computer or a laptop, but in some cases, it is no reliable to invest in a machine for certain types of works that can be used at a significant scale. In instances when the project needs less cost investment and more portability, it is preferred to use other

O. Carvajal · O. Castillo (✉)
Tijuana Institute of Technology, Tijuana, BC, Mexico
e-mail: ocastillo@tectijuana.mx

© Springer Nature Switzerland AG 2020
O. Castillo and P. Melin (eds.), *Hybrid Intelligent Systems in Control,*
Pattern Recognition and Medicine, Studies in Computational Intelligence 827,
https://doi.org/10.1007/978-3-030-34135-0_22

315

devices such as a microcontroller. In this work, we use a PIC18F4550 microcontroller to implement the fuzzy logic controller. In some applications, like in this case, it is preferred to produce a custom HW (Hardware) than using development boards for general purpose. Examples of these general purpose prototypes are boards like *Arduino*, which is a development platform that includes an *Atmega* microcontroller by the *Atmel* company, which was recently purchased by Microchip Technology, the same company that designs the PIC18F4550 microcontroller which is used in this work. Therefore, with the PIC18F4550 microcontroller, we can develop our own Arduino. Some professionals that are not focused on electronic design are attracted to this type of Arduino boards, which in this case would be beneficial for them due to the fact they do not have to deal with the electronic design.

There are other alternatives for implementing fuzzy controllers, such as FPGAs (Field Programmable Gate Arrays), and PLCs (Programmable Logic Controller) that are different worlds when compared to microcontrollers, but can be also a good option for this type of applications. There are some HW that contain an SoC (System on Chip) that have been gaining popularity, for example, *Raspberry Pi* which has a 32-bit ARM architecture which has the characteristic of having faster processing speed. *Raspberry Pi* works with a Linux-based operating system, so it allows more applications when many tasks need to run simultaneously, the advantage of having an operating system is communicating to the hardware indirectly, for example connecting a USB device, and the developer does not have to worry about programming how to read the USB port data, etc.

On the other hand, a PIC18F4550 microcontroller requires the firmware developments of each function for each peripheral, and requires the specific knowledge of them, but using an operating system needs more consumption of resources, and in some cases, it could cause delays in the application. Comparing it with a PIC18f4550 microcontroller that would produce fewer delays and it is better for applications which require more precision on its modules, for example, using the PWM (Pulse Width Modulation).

The structure of the paper is the following, in Section two we describe the model of the autonomous mobile robot, in Section three the hardware design of the robot is introduced, in Section four we describe the control of the FLC with one input and two outputs, in Section five we describe the firmware development process. Finally, Section six offers some conclusions of this work.

2   Mathematical Model

To design controllers sometimes it is useful to find out what is the behavior of the system with the simulation of its mathematical model [2, 8–11].

In Fig. 1 the representation of the robot signals is introduced, the oval part is the robot, and the rectangles are the wheels.

Fig. 1 Robot representation with its dimensions

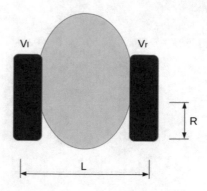

This representation is more common for implementation because it is more natural for a microcontroller controlling the two motors than the model that represents the whole system.

The PWM can be used for dealing with the linear velocity for each of the wheels. The wheels can turn at different rates for moving the robot around by changing v_r and v_l, that are the linear velocities for each servomotor respectively. If the robot wheels have the same speed, the robot is going to run straight-ahead, and when the motor rate is slower than the other, the robot is going to turn towards the region in which the delayed wheel is. The only two parameters needed are R, which is the Radius of the wheels and L which is the separation of them and these parameters can be easily obtained [12].

The following three Equations represent the model of Fig. 1.

$$\dot{x} = \frac{R}{2}(v_r + v_l)\cos(\phi) \tag{1}$$

$$\dot{y} = \frac{R}{2}(v_r + v_l)\sin(\phi) \tag{2}$$

$$\dot{\phi} = \frac{R}{L}(v_r - v_l) \tag{3}$$

The above representation is not usual for simulation because an essential thing about the robot is its position in the x and y plane, and the orientation of the robot, which is represented as an angle ϕ. Figure 2 shows this representation and Eqs. 4–6 are in terms of the variables of the robot in general, where v is the linear velocity and ω angular velocity of the robot [13–17].

$$\dot{x} = v \cdot \cos(\phi) \tag{4}$$

$$\dot{y} = v \cdot sen(\phi) \tag{5}$$

$$\dot{\phi} = \omega \tag{6}$$

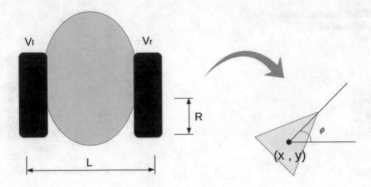

Fig. 2 Representations of the robot on the plane

As we need v_r and v_l for implementation purpose it is useful for obtaining the Equations with the equalization method for solving the Eqs. (1)–(4) and (3)–(6) the following Equations are derived.

$$v = \frac{R}{2}(v_r + v_l) \tag{7}$$

$$\omega = \frac{R}{L}(v_r - v_l) \tag{8}$$

Solving for v_r and v_l we have the following Equations.

$$v_r = \frac{2v + \omega L}{2R} \tag{9}$$

$$v_l = \frac{2v - \omega L}{2R} \tag{10}$$

3 Hardware Design

The hardware is composed of mechanical and electrical-electronics parts. First, as mechanical parts, we have the chassis of the robot and the three wheels of the robot, and as electrical-electronic parts, we have two servomotors and the electronic board that contains the integrated circuits such as the PIC18F4550 microcontroller, the driver LD293, CD40106 Schmitt Triggers, and the CNY70 optical sensors.

The wheels to be controlled are the two which have a servomotor; it means the active wheels; the passive or castor wheel is only for the stability of the robot. Figure 3

Fig. 3 Chassis of the robot

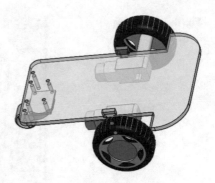

shows the chassis of the autonomous mobile robot. The electronic board design is based on it. This chassis is usually available in the international market at a relatively low cost.

The PCB (Printed Circuit Board) of Fig. 4 was designed with two layers with a dimension of 10 cm × 6 cm using an Electronic Design Automation (EDA) software called EAGLE to bring the robot to life. The PCB is based on the schematics of Fig. 5, which shows the connections with the labels of all the components of the board.

4 Fuzzy Controller

The implemented FLC is introduced in Fig. 6 and is of Mamdani type which has one input and two outputs due to the availability of the memory consumption of the microcontroller, so is preferred to use fewer variables as possible. The input of the FLC is the error of the robot with respect to the reference, so it means how far or near it is compared with the reference, and the outputs are the linear velocity of the right and the left wheels and all the membership functions are of triangular type.

The error is calculated in the microcontroller internally by mapping the logical values obtained by the five optical sensors, and the outputs are the linear velocity applied to the two servomotors using the PWM module of the microcontroller.

Fuzzy rules can be considered as the knowledge of an expert in a specific area, and they are expressed in the IF-THEN sequence to connect a condition through linguistics variable [2, 7, 18–20]. So, we have three rules for the FLC which are listed as follows:

If (ep is Z) then (V_l is High) (V_r is High)
If (ep is N) then (V_l is Medium) (V_r is High)
If (ep is P) then (V_l is High) (V_r is Medium).

Fig. 4 PCB design

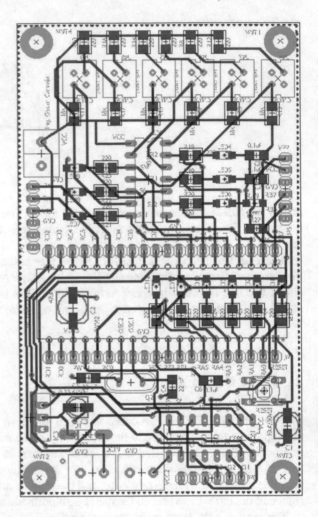

5 Firmware Development

In Fig. 7 the block diagram shows the process of the control of the robot. The four steps are obtaining the position of the robot by using the optical sensors then calculating the error, next evaluating the fuzzy logic controller implemented in the microcontroller, finally, for the evaluation of the FLC we apply the linear velocity to the servomotors and the process continue in a loop.

The process of the fuzzy controller implementation for the robot is illustrated in Fig. 8, and we use the Microchip official IDE for developing the firmware. It was developed on a high-level programming language C, then the XC8 compiler translates the C sources files and C header files to Assembly source file and continue until the creation of the Executable file with.exe extension which is the file that

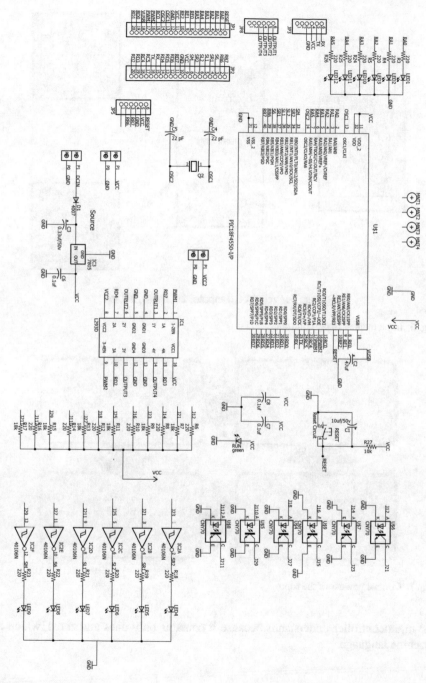

Fig. 5 Schematics of the robot

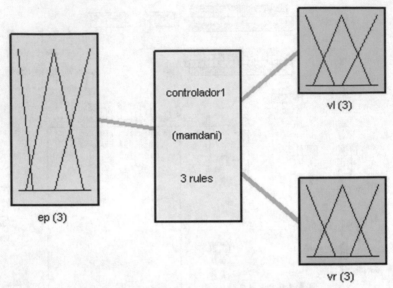

Fig. 6 Implemented fuzzy controller

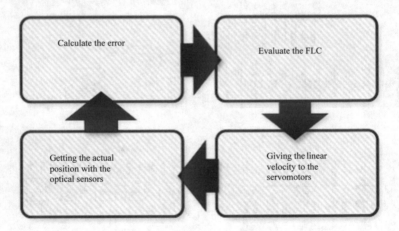

Fig. 7 Control process of the robot

the microcontroller understands because it contains only ones and zeros, which is machine language.

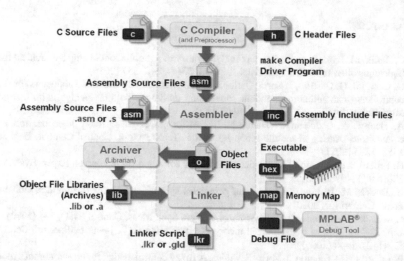

Fig. 8 Firmware development process

6 Conclusions

In this work, we presented the design of an embedded fuzzy controller for an autonomous mobile robot with one input and two outputs for achieving a desired tracking trajectory. We can conclude that the microcontrollers are suitable for implementation purposes of fuzzy logic techniques when the fuzzy controller has not too many variables, they allow productivity on electronic design projects in the control area and others areas. Besides, we presented the firmware development process of the PIC18F4550 which was written in C language. Finally, we presented the Printed Circuit Board of the electronic design, and we presented a way of creating a custom design for a robotic application and the importance of developing controllers using fuzzy logic, due to the fact that they are more intuitive than traditional control systems. As a future work, we are going to add the obstacle avoidance part to the fuzzy controller of the robot, and make the optimization of the parameters of the membership functions, using a bio-inspired algorithm, and implement the fuzzy controller using other technologies such a Raspberry Pi, FPGAs, PLCs or even combine them with the PIC microcontroller to observe the performance of the fuzzy controller on different platforms. In addition, we can perform more experiments with variations in values of some parameters and we could use type-2 fuzzy logic as in [21–25]. In addition, we can deal with other types of applications, like in [26–31].

References

1. C. Rekik, M. Jallouli, N. Derbel, Optimal trajectory of a mobile robot using hierarchical fuzzy logic controller. Int. J. Comput. Appl. Technol. **53**(4), 348–357 (2016)
2. O. Carvajal, O. Castillo, J. Soria, Optimization of membership function parameters for fuzzy controllers of an autonomous mobile robot using the flower pollination algorithm. J. Autom. Mob. Robot. Intell. Syst. **12**(1), 44–49 (2018)
3. A. Hechri, A. Ladgham, F. Hamdaoui, A. Mtibaa, Design of fuzzy logic controller for autonomous parking of mobile robot. Int. J. Sci. Tech. Autom. Control Comput. Eng. **5**(2), 1558–1575 (2011)
4. H. Erdem, Application of neuro-fuzzy controller for sumo robot control. Expert Syst. Appl. **38**(8), 9752–9760 (2011)
5. G. Dudek, M. Jenkin, *Computational principles of mobile robotics* (Cambridge University Press, Cambridge, 2010)
6. L.A. Zadeh, A rationale for fuzzy control. J. Dyn. Syst. Meas. Control **94**(1), 3–4 (1972)
7. L.A. Zadeh, Toward a generalized theory of uncertainty (GTU)—an outline. Inf. Sci. (Ny) **172**(1–2), 1–40 (2005)
8. E. Lizarraga, O. Castillo, J. Soria, F. Valdez, A fuzzy control design for an autonomous mobile robot using ant colony optimization, in *Recent Advances on Hybrid Approaches for Designing Intelligent Systems* (Springer, Berlin), pp. 289–304
9. L. Amador-Angulo, O. Castillo, J.R. Castro, A generalized type-2 fuzzy logic system for the dynamic adaptation the parameters in a Bee Colony Optimization algorithm applied in an autonomous mobile robot control, in *2016 IEEE International Conference on Fuzzy Systems (FUZZ-IEEE)* (2016), pp. 537–544
10. P. Glotfelter, M. Egerstedt, A parametric MPC approach to balancing the cost of abstraction for differential-drive mobile robots. arXiv Prepr. arXiv1802.07199 (2018)
11. M. Egerstedt, X. Hu, H. Rehbinder, A. Stotsky, Path planning and robust tracking for a car-like robot, in *Proceedings of the 5th Symposium on Intelligent Robotic Systems* (1997), pp. 237–243
12. V.M. Peri, D. Simon, Fuzzy logic control for an autonomous robot, in *NAFIPS 2005. Annual Meeting of the North American Fuzzy Information Processing Society* (2005), pp. 337–342
13. P. Melin, L. Astudillo, O. Castillo, F. Valdez, M. Garcia, Optimal design of type-2 and type-1 fuzzy tracking controllers for autonomous mobile robots under perturbed torques using a new chemical optimization paradigm. Expert Syst. Appl. **40**(8), 3185–3195 (2013)
14. C. Caraveo, F. Valdez, O. Castillo, Optimization of fuzzy controller design using a new bee colony algorithm with fuzzy dynamic parameter adaptation. Appl. Soft Comput. **43**, 131–142 (2016)
15. C.T. Kilian, *Modern Control Technology: Components and Systems* (Delmar/Thomson Learning, 2006)
16. J. Perez, P. Melin, O. Castillo, F. Valdez, C. Gonzalez, G. Martinez, Trajectory optimization for an autonomous mobile robot using the bat algorithm, in *North American Fuzzy Information Processing Society Annual Conference* (2017), pp. 232–241
17. O. Castillo, J. Soria, J. Kacprzyk, Optimization of reactive control for mobile robots based on the CRA using type-2 fuzzy logic, in *Nature-Inspired Design of Hybrid Intelligent Systems* (Springer, 2017), pp. 505–515
18. P. Melin, O. Castillo, Fuzzy controllers for autonomous mobile robots, in *Springer Handbook of Computational Intelligence* (Springer, 2015), pp. 1517–1531
19. M.A. Sanchez, O. Castillo, J.R. Castro, Generalized type-2 fuzzy systems for controlling a mobile robot and a performance comparison with interval type-2 and type-1 fuzzy systems. Expert Syst. Appl. **42**(14), 5904–5914 (2015)
20. O. Castillo, H. Neyoy, J. Soria, P. Melin, F. Valdez, A new approach for dynamic fuzzy logic parameter tuning in ant colony optimization and its application in fuzzy control of a mobile robot. Appl. Soft Comput. **28**, 150–159 (2015)

21. C. Leal Ramírez, O. Castillo, P. Melin, A. Rodríguez Díaz, Simulation of the bird age-structured population growth based on an interval type-2 fuzzy cellular structure. Inf. Sci. **181**(3), 519–535 (2011)
22. N.R. Cázarez-Castro, L.T. Aguilar, O. Castillo, Designing Type-1 and Type-2 fuzzy logic controllers via fuzzy lyapunov synthesis for nonsmooth mechanical systems. Eng. Appl. AI **25**(5), 971–979 (2012)
23. O. Castillo, P. Melin, Intelligent systems with interval type-2 fuzzy logic. Int. J. Innov. Comput. Inf. Control **4**(4), 771–783 (2008)
24. G.M. Mendez, O. Castillo, Interval type-2 TSK fuzzy logic systems using hybrid learning algorithm, in *The 14th IEEE International Conference on Fuzzy Systems. FUZZ'05* (2005), pp. 230–235
25. P. Melin, C.I. González, J.R. Castro, O. Mendoza, O. Castillo, Edge-detection method for image processing based on generalized type-2 fuzzy logic. IEEE Trans. Fuzzy Syst. **22**(6), 1515–1525 (2014)
26. L. Cervantes, O. Castillo, D. Hidalgo, R. Martinez-soto, Fuzzy dynamic adaptation of gap generation and mutation in genetic optimization of type 2 fuzzy controllers, in *Advances in Operation Research*, vol. 2018 (Hindai, 2018)
27. P. Melin, O. Castillo, Intelligent control of complex electrochemical systems with a neuro-fuzzy-genetic approach. IEEE Trans. Ind. Electron. **48**(5), 951–955 (2001)
28. E. Rubio, O. Castillo, F. Valdez, P. Melin, C. I. González, G. Martinez. An extension of the fuzzy possibilistic clustering algorithm using type-2 fuzzy logic techniques. Adv. Fuzzy Syst. **2017**, 7094046:1–7094046:23 (2017)
29. P. Melin, A. Mancilla, M. Lopez, O. Mendoza, A hybrid modular neural network architecture with fuzzy Sugeno integration for time series forecasting. Appl. Soft Comput. **7**(4), 1217–1226 (2007)
30. P. Melin, O. Castillo. *Modelling, Simulation and Control of Non-Linear Dynamical Systems: An Intelligent Approach Using Soft Computing and Fractal Theory* (CRC Press, 2001)
31. P. Melin, G. Prado-Arechiga, *New Hybrid Intelligent Systems for Diagnosis and Risk Evaluation of Arterial Hypertension* (Springer, Switzerland, 2018)

Implementation a Fuzzy System for Trajectory Tracking of an Omnidirectional Mobile Autonomous Robot

Jacinto González-Aguilar, Oscar Castillo and Prometeo Cortés-Antonio

Abstract This document presents a problem of tracking lines of an omnidirectional mobile robot, using a fuzzy proportional control system and classical proportional control to compare the result achieved by each control. It is implemented in the Robotino mobile platform, two digital optical sensors are used to control the direction of the angular velocity of the robot and an inductive analog sensor to control the linear velocity in x v_x. For the analysis of the system, Robotino SIM and Simulink is used, which is an additional Matlab tool that allows graphic representation by means of blocks, both linear and non-linear systems. Tests and comparisons are made where the best performance of the fuzzy proportional controller is appreciated.

Keywords Proportional control (p) · Integral control (I) · Derivative control (D) · Fuzzy control · Robotino

1 Introduction

Fuzzy logic is a powerful tool for controlling complex and non-linear systems.

The PID controllers are the most used to control robots because it has a simple structure. The design of this controller only needs three parameters the proportional, integral and derivative, can be tuned using the well-known technique of Ziegler-Nichols [1]. Conventional control systems have required a qualitative and accurate description of the input/output relationship using complex equations [2–5]. The most appropriate fuzzy controller for most applications in control is Mandani [6] since this provides the fuzzification and defuzzification interface that allow to convert real numerical data coming from sensors in fuzzy sets that through the inference engine

J. González-Aguilar · O. Castillo (✉) · P. Cortés-Antonio (✉)
Tijuana Institute of Technology, Tijuana, BC, Mexico
e-mail: ocastillo@tectijuana.mx

P. Cortés-Antonio
e-mail: prometeo.cortes@tectijuana.mx

J. González-Aguilar
e-mail: goaj_7@hotmail.com

© Springer Nature Switzerland AG 2020
O. Castillo and P. Melin (eds.), *Hybrid Intelligent Systems in Control,*
Pattern Recognition and Medicine, Studies in Computational Intelligence 827,
https://doi.org/10.1007/978-3-030-34135-0_23

and the base of rules, allow to generate the appropriate control action to bring the system to the desired conditions giving solution to the control problem. In this paper, we will take the Mandani model to implement the fuzzy controller for line tracking.

The rest of this article is organized as follows. In Sect. 2, the law of kinematic control, description and line monitoring is proposed. Section 3 presents the classification of the conventional control system and the fuzzy proportional control. Section 4 simulations are carried out to compare results of the proposed control systems and Sect. 5 concludes this paper.

2 Literature Review

This section presents the basic concepts to understand work done: Two optical sensors (direct reflection sensors) are used to control the action of turning clockwise or anticlockwise. Also an inductive sensor that detects the metal line on the ground is used.

2.1 Kinematic Model of the Omnidirectional Robot

An omnidirectional robot is one that has three degrees of freedom with a minimum of three motors [7], which allows lateral motions or move and rotate in any direction depending on the velocity of each wheel. The combination of the motion of the three wheels causes motions forward, backward, sideways or rotation of the robot.

In this document, we use an omnidirectional mobile robot [8], called Robotino [9], as vehicular platform to simulate and experience the fuzzy controller for tracking proposed lines [10]. The kinematic model of the robot is presented in Fig. 1, where θ is the rotation angle of the robot in the counterclockwise direction with respect to a global coordinate axis in the floor. Assuming that the coordinate axis (x, y) is in the center of the robot (local coordinates), as shown in Fig. 1.

L is the distance from the center of the robot to the center of the roller omnidirectional, r is the radius of the Omnidirectional rollers (mm), w_1, w_2, w_3 is the velocity of engines 1, 2 and 3 (rpm), θ is the angular velocity of the robot.

Based on Fig. 1 we can determine the kinematic model of the robot represented by Eq. (1) [11].

$$\begin{pmatrix} v_1 \\ v_2 \\ v_3 \end{pmatrix} = \begin{pmatrix} -\sin\theta & -\cos\theta & L \\ -\sin(\pi/3 - \theta) & -\cos(\pi/3 - \theta) & L \\ \sin(\pi/3 + \theta) & -\cos(\pi/3 + \theta) & L \end{pmatrix} \begin{pmatrix} v_x \\ v_y \\ \dot{\theta} \end{pmatrix} \quad (1)$$

Fig. 1 Kinematic model of
the robot

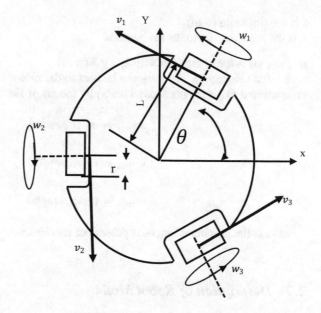

Observing in Fig. 1. We can say that: $\theta = 60$ and by replacing this value in Eq. (1) it is reduced to:

$$\begin{pmatrix} v_1 \\ v_2 \\ v_3 \end{pmatrix} = \begin{pmatrix} -\sqrt{3/2} & 1/2 & L \\ 0 & -1 & L \\ \sqrt{3/2} & 1/2 & L \end{pmatrix} \begin{pmatrix} v_x \\ v_y \\ \dot{\theta} \end{pmatrix} \tag{2}$$

v_x is the motion velocity of the robot on the x-axis (mm/s), v_y is the displacement velocity of the robot on the y-axis (mm/s).

The equations of omnidirectionality (3), (4), (5) allow calculating the individual velocity of each motor from the desired velocity of motion of the robot.

$$w_1 = \left[-v_x \cdot \frac{\sqrt{3}}{2} + \frac{1}{2} \cdot v_y + \left(L \cdot \dot{\theta} \cdot \frac{\pi}{180} \right) \right] \cdot K. \tag{3}$$

$$w_2 = \left[-v_x + \left(L \cdot \dot{\theta} \cdot \frac{\pi}{180} \right) \right] \cdot K \tag{4}$$

$$w_3 = \left[-v_x \cdot \frac{\sqrt{3}}{2} + \frac{1}{2} \cdot v_y + \left(L \cdot \dot{\theta} \cdot \frac{\pi}{180} \right) \right] \cdot K. \tag{5}$$

K is the constant to convert the velocity of mm/s rpm.

$$k = 60 \cdot n / 2\pi \cdot r \tag{6}$$

r is the rim radio (mm).

n is the reduction factor of the gearbox.

w_1, w_2, w_3 is the engine velocity 1, 2 y 3 (rpm).

So that velocity w_1, w_2 y w_3 can be sent to the motor controllers, they must first be converted to pulses per second (PPS) by means of Eqs. (7), (8) y (9).

$$w_{1pps} = w_1 \cdot npev/60 \tag{7}$$

$$w_{2pps} = w_2 \cdot npev/60 \tag{8}$$

$$w_{3pps} = w_3 \cdot npev/60 \tag{9}$$

$npev$ is the number of encoder pulses per revolution.

2.2   Description of Robot Motion

The forward motion of the robot towards the front on the positive x axis and negative x axis is shown in Fig. 1.

v_x Positive the robot advances towards the front.

v_x Negative the robot advances towards the back of the robot (Table 1).

Motion to the left and right of the robot is shown in Table 2 [velocity on the y-axis (v_y)]. The robot advances to the left with a positive velocity and with a negative velocity advances to the right of the robot.

Table 1 Motion of Robotino in x

v_x	Address	Sense of progress	Sense of progress
+	➡	Towards the front (of the robot)	On the positive x axis
−	⬅	Backwards (from the robot)	On the negative x axis

Table 2 Motion of the robot on the axis y

v_y	Address	Sense of progress	Sense of progress
+	⬅	To the left (of the robot)	On the positive axis
−	➡	To the right (of the robot)	On the axis and negative

Table 3 Angular velocity of the robot

$\dot{\theta}$	Address	Direction of the turn	Turn
+	↻	Turn to the left	Counterclockwise direction
−	↻	Turn to the right	Sense to the hands of the clock

Fig. 2 Controller simulation platform

The angular motion of the robot is shown in Table 3, velocity in the clockwise direction is positive and counterclockwise velocity is negative.

$\dot{\theta}$ is the angular velocity of the robot.

In order to implement the fuzzy and proportional controller, the ROBOTINO SIM platform is used as seen in Fig. 2.

Using the platform indicated in Fig. 2 use two optical sensors that will give us the next reading of 0 and 1 as shown in Table 4, these sensors will let you know when the robot is inside or outside the line.

The optical sensor gives a reading of 0 is on the line and when it gives a reading of 1 it is out of line, shown in Table 5.

Table 6 describes the action to be taken when obtaining the result of the readings of the two optical sensors.

Case 1 if the two left and right optical sensors are inside the black line the optical sensor readings for both will be 0, so the robot will follow the straight line as shown in Fig. 3, the path that the robot must follow is described, which is composed of black metal tape.

Figure 3 shows that the two optical sensors are on the line, the robot will follow the straight line.

Case 2 the left sensor is inside and the right sensor is shown in Fig. 4, the action that the robot will take is to turn left as described in Table 6.

Table 4 Optical sensor reading

Reading in robot SIM	
0	Inside the black line
1	Outside the black line

Table 5 Reading of the left and right optical sensors

Optic left	Optic right	Action	Turn	
0	0	➡	Not	The two sensors are inside the black line
0	1	↺	On the left (−)	Right sensor is out of line
1	0	↻	On the right (+)	Left sensor is out of line
1	1	➡	Not	The two sensors are out of line

Case 3 the left sensor is out and the right one is inside the line as shown in Fig. 5, the action that will be taken in turning to the right as described in Table 6.

With the two left and right optical sensors, the decision will be made when it is time to turn to follow the line, in the following part of this article you will see the classic control systems the fuzzy control.

3 Control

A conventional automatic control system that is an interconnection of elements that form a configuration called system, in such a way that the resulting array is able to control itself. A signal is applied to this system r(t) as a way to get an answer or output y(t), and is represented by Fig. 6.

3.1 Classification of Control Systems

In the application of the control systems, the type of control open loop or closed loop must be taken into account.

Open loop control system

The control action is, in a way independent of the output in a system.

Closed loop control system

The control action depends on the output in that system. This system uses a sensor that detects the real response to compare it, with a reference as an input, shown in Fig. 6. For this reason, closed loop systems are called feedback systems (Fig. 7).

r is the voltage reference of the inductive sensor so that the robot is always on the line.

e is the error, u is the output to control the angular velocity of the robot.

Proportional control scheme to control the angular velocity of Robotino (Fig. 8).

Table 6 Adjustable gain values to control the angular velocity

e	kp	u	kp	u	kp	u	kp	u	kp	u	kp	u
0	0.9	0	1.5	0	2	0	2.5	0	3	0	3.5	0
1	0.9	0.9	1.5	1.5	2	2	2.5	2.5	3	3	3.5	3.5
2	0.9	1.8	1.5	3	2	4	2.5	5	3	6	3.5	7
3	0.9	2.7	1.5	4.5	2	6	2.5	7.5	3	9	3.5	10.5
4	0.9	3.6	1.5	6	2	8	2.5	10	3	12	3.5	14
5	0.9	4.5	1.5	7.5	2	10	2.5	12.5	3	15	3.5	17.5
6	0.9	5.4	1.5	9	2	12	2.5	15	3	18	3.5	21
7	0.9	6.3	1.5	10.5	2	14	2.5	17.5	3	21	3.5	24.5
8	0.9	7.2	1.5	12	2	16	2.5	20.5	3	24	3.5	28
9	0.9	8.1	1.5	13.5	2	18	2.5	22.5	3	27	3.5	31.5
9.9	0.9	8.9	1.5	14.8	2	19.9	2.5	24.9	3	29.7	3.5	34.6

Fig. 3 Robot trajectory

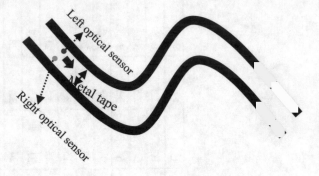

Fig. 4 Left sensor off the line

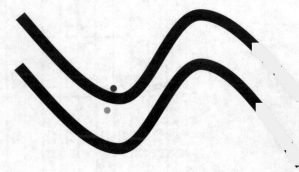

Fig. 5 Right sensor outside the line

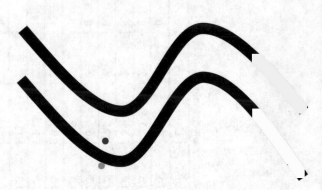

Fig. 6 Control system

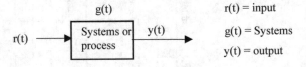

r(t) = input

g(t) = Systems

y(t) = output

$v_D = 0$ is the desired voltage of the inductive sensor, e is the error, $k_p * e$ is the gain adjustable by the error, v_S Inductive sensor voltage.

Where e is the reading of the inductive sensor in voltages on the metal line, kp is the adjustable gain, u is the output of the controller.

Fig. 7 Proportional control
scheme

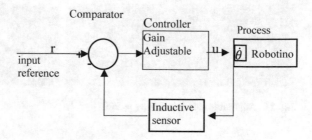

Fig. 8 Proportional control
scheme to control the
angular velocity of Robotino

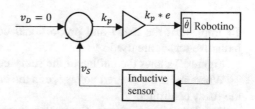

The implementation of the proportional controller to control the angular velocity
of the robot with a constant velocity at 100 mm/s in x and with adjustable gain k =
0.9 did not obtain a good performance when following the line as seen in Fig. 9.

You get a good performance, the robot follows the line if you leave it, shown in
Fig. 10, it should be mentioned that increasing the velocity in x to 200 mm/s does
not get a good performance as shown in Fig. 11.

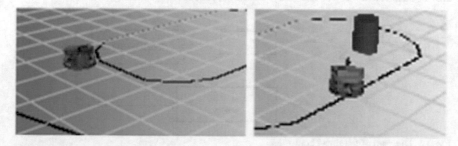

Fig. 9 Adjustable gain k = 0.9, v_x = 100

Fig. 10 Adjustable gain k = 1.5, v_x = 150

Fig. 11 Adjustable gain k = 1.5, $v_x = 200$

3.2 Fuzzy Control

To implement the fuzzy and proportional controller, two optical sensors and an inductive sensor are used.

Figure 12 shows the outline of the fuzzy controller.

Where v_D is the desired voltage, e is the error, v_S inductive sensor voltage, flc is the fuzzy controller.

The Proportional fuzzy Controller is shown in Fig. 13 with the following parameters:

v_D is the desired voltage, e is the error, v_S inductive sensor voltage, kp is the adjustable gain (Fig. 14).

We use 4 membership functions for the error, the first and the last of trapezoidal types and 2 of triangular types, shown in Fig. 15 and for the output that is the angular velocity we also use 4 membership functions two of trapezoidal and two triangular types, shown in Fig. 16. The linguistic values that characterize the linguistic variable are: For the error: very small (VS), small (S), medium (M), long (L).

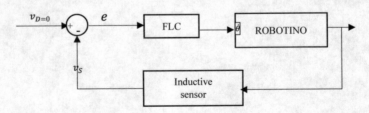

Fig. 12 Diagram of the fuzzy controller

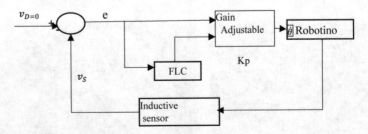

Fig. 13 Diagram of the fuzzy proportional controller

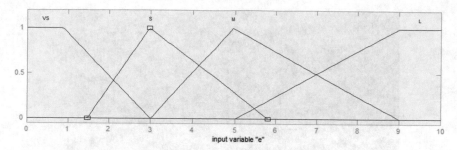

Fig. 14 Fuzzy membership function of e

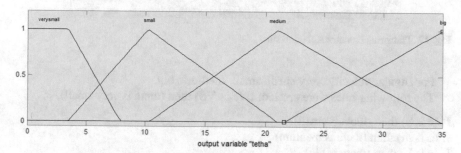

Fig. 15 Fuzzy membership function of theta

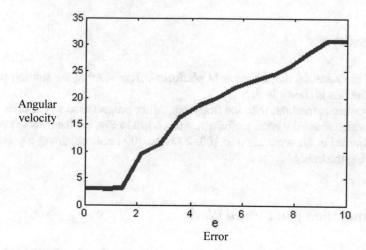

Fig. 16 Fuzzy proportional controller curve

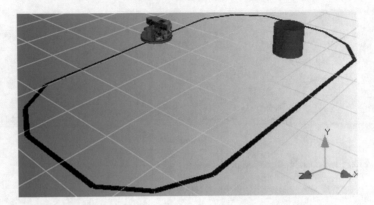

Fig. 17 Platform simulator Robotino sim

For angular velocity: very small, small, medium, big.
The following rules were created: If (e is VS) then (theta is very small).

If (e is S) then (theta is small)
If (e is M) then (theta is medium)
If (e is L) then (theta is big).

4 Results

Start of the robot on the robotic SIM platform before starting the starting point to follow the line is shown in Fig. 17.

Tests were carried out with the fuzzy controller proportional to the rules shown above, which showed a good performance as shown in Fig. 18, the constant velocity were adjusted in the x direction to 100, 200 and 300 mm/s obtaining a good result in tracking the line.

5 Conclusion and Future Work

The difference between the proportional controller and the proportional fuzzy controller is shown in Figs. 8, 9 and 10, where the proportional controller did not give a good result when increasing the velocity in x, on the contrary, the proportional fuzzy controller was able to follow the line increasing the velocity at xa 100, 200 and 300 mm/s, shown in Fig. 18.

As future work we will perform tests with proportional control (kp), control-integral (ki), control-derivative (kd) and finally with the fuzzy-proportional-integral controller that control both the angular velocity and the velocity in x. In addition, we

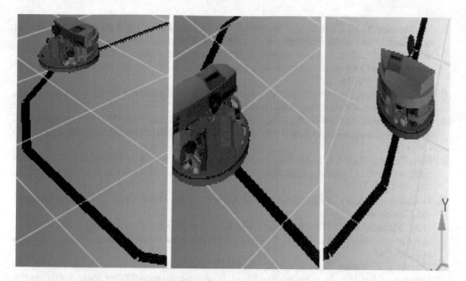

Fig. 18 Result of line tracking by the robot

can perform more experiments with variations in values of some parameters and we could use type-2 fuzzy logic as in [11–16]. In addition, we can deal with other types of applications, like in [17–21].

Acknowledgements The authors would like to express thank to the Consejo Nacional de Ciencia y Tecnologia and Tecnológico Nacional de México/Tijuana Institute of Technology for the facilities and resources granted for the development of this research.

References

1. K.J. Åström, W.K. Ho, Control theory and applications, in *IEE proceedings-D*, vol. 138, no. 2 (Institution of Electrical Engineers, marzo 1980)
2. F.G. Rossomando, C.M. Soria, Identification and control of nonlinear dynamics of a mobile robot in discrete time using an adaptive technique based on neural PID. Neural Comput. Appl. **26**(5), 1179–1191 (2015)
3. M. Pena et al., Fuzzy logic for omni directional mobile platform control displacement using FPGA and bluetooth. IEEE Lat. Am. Trans. **13**(6), 1907–1914 (2015)
4. C.-C. Tsai, X.-C. Wang, F.-C. Tai, C.-C. Chan, Fuzzy decentralized EIF-based pose tracking for autonomous omnidirectional mobile robot, in *2014 International Conference on Machine Learning and Cybernetics* (2014), pp. 748–754
5. R. Choomuang, Hybrid Kalman filter/fuzzy logic based position control of autonomous mobile robot. Int. J. Adv. Robot. Syst. **2**, 197–208 (2005)
6. E.H. Mamdani, N. Baaklini, Prescriptive method for deriving control policy in a fuzzy-logic controller. Electron. Lett. **11**(25–26), 625 (1975)
7. M. Wada, S. Mori, Holonomic and omnidirectional vehicle with conventional tires, in *Proceedings of IEEE International, and undefined 1996*, ieeexplore.ieee.org (1996)

8. M. Masmoudi, N. Krichen, M. Masmoudi, N. Derbel, Fuzzy logic controllers design for omnidirectional mobile robot navigation. Appl. Soft Comput. J. **49**, 901–919 (2016)

9. S. Oltean, M. Dulau, Position control of Robotino mobile robot using fuzzy logic, in *IEEE International Conference on Automation, Quality and Testing, Robotics (AQTR)*, Cluj-Napoca, Romania, May 2010

10. T. Kalmár-Nagy. Real-time trajectory generation for omnidirectional vehicles, in *Proceedings of the American Control Conference* (2002), pp. 286–291

11. C. Leal Ramírez, O. Castillo, P. Melin, A. Rodríguez Díaz, Simulation of the bird age-structured population growth based on an interval type-2 fuzzy cellular structure. Inf. Sci. **181**(3), 519–535 (2011)

12. N.R. Cázarez-Castro, L.T. Aguilar, O. Castillo, Designing type-1 and type-2 fuzzy logic controllers via Fuzzy Lyapunov synthesis for nonsmooth mechanical systems. Eng. Appl. AI **25**(5), 971–979 (2012)

13. O. Castillo, P. Melin, Intelligent systems with interval type-2 fuzzy logic. Int. J. Innov. Comput. Inf. Control **4**(4), 771–783 (2008)

14. G.M. Mendez, O. Castillo. Interval type-2 TSK fuzzy logic systems using hybrid learning algorithm, in *The 14th IEEE International Conference on Fuzzy Systems. FUZZ'05* (2005), pp. 230–235

15. Claudia I. González, Patricia Melin, Juan R. Castro, Olivia Mendoza, Oscar Castillo, An improved sobel edge detection method based on generalized type-2 fuzzy logic. Soft. Comput. **20**(2), 773–784 (2016)

16. Emanuel Ontiveros, Patricia Melin, Oscar Castillo, High order α-planes integration: a new approach to computational cost reduction of General Type-2 Fuzzy Systems. Eng. Appl. AI **74**, 186–197 (2018)

17. P. Melin, O. Castillo, Intelligent control of complex electrochemical systems with a neuro-fuzzy-genetic approach. IEEE Trans. Ind. Electron. **48**(5), 951–955 (2001)

18. E. Rubio, O. Castillo, F. Valdez, P. Melin, C. I. González, G. Martinez. An extension of the fuzzy possibilistic clustering algorithm using type-2 fuzzy logic techniques. Adv. Fuzzy Syst. **2017**, 7094046:1–7094046:23 (2017)

19. Patricia Melin, Alejandra Mancilla, Miguel Lopez, Olivia Mendoza, A hybrid modular neural network architecture with fuzzy Sugeno Integration for time series forecasting. Appl. Soft Comput. **7**(4), 1217–1226 (2007)

20. P. Melin, O. Castillo, *Modelling, Simulation and Control of Non-Linear Dynamical Systems: An Intelligent Approach Using Soft Computing and Fractal Theory* (CRC Press, Boca Raton, 2001)

21. Patricia Melin, Daniela Sánchez, Oscar Castillo, Genetic optimization of modular neural networks with fuzzy response integration for human recognition. Inf. Sci. **197**, 1–19 (2012)

Neural Inverse Optimal Pinning Control of Output Trajectory Tracking for Uncertain Complex Networks with Nonidentical Nodes

Carlos J. Vega and Edgar N. Sanchez

Abstract This chapter presents the development of a control scheme, we named as neural inverse optimal pinning control to achieve output trajectory tracking on uncertain complex networks with nonidentical nodes. A recurrent high order neural network is used to identify the unknown system dynamics of a small fraction of nodes (pinned ones) and by means of this neural model, an inverse optimal controller is designed to synchronize the whole network at an output desired reference. The proposed controller effectiveness is illustrated via simulations. The illustrative example is composed of a network of ten different chaotic nodes.

Keywords Complex network · Neural network · Pinning control · Optimal control · Synchronization

1 Introduction

Complex networks can model many phenomena, e.g.: neural networks; interactions in biological molecules as genes, RNAs, and proteins; social interactions; Internet; power systems; among others [3, 6, 22, 39]. Recently, one of the behaviors which has attracted increasing attention in different fields is synchronization, a typical collective behavior in nature [4, 38, 43]. Synchronization processes are ubiquitous in many systems, such as synchronous communication, encryption systems synchronization, synchronization in biology, flash synchronization, database synchronization and so

This work is supported by CONACYT, Mexico, Project 257200.

C. J. Vega · E. N. Sanchez (✉)
Electrical Engineering Department, Centro de Investigación y de Estudios Avanzados del Instituto Politécnico Nacional, Av. del Bosque 1145, Col. El Bajío, 45019 Zapopan, Jalisco, Mexico
e-mail: sanchez@gdl.cinvestav.mx

C. J. Vega
e-mail: cjvega@gdl.cinvestav.mx

© Springer Nature Switzerland AG 2020
O. Castillo and P. Melin (eds.), *Hybrid Intelligent Systems in Control, Pattern Recognition and Medicine*, Studies in Computational Intelligence 827, https://doi.org/10.1007/978-3-030-34135-0_24

on [2, 5, 23]. A control technique should be designed to ensure synchronization, when it can not be achieved by itself, or if the synchronized state is not the desired one. Among network controllers, the pinning control is very adequate because it only needs to apply local feedback injections to a small fraction of the network nodes (pinned ones) [8, 17, 37].

Mainly, three approaches to analyze complex networks synchronization have been developed; linear systems theory, the master stability function (MSF), and the Lyapunov approach [17, 33, 36, 40–42, 44]. Li et al. [17] derived stabilization conditions for synchronization on networks with identical nodes, linearly and diffusively coupled. In [33], controllability of networks under pinning control schemes was defined via an augmented system approach and the MSF method. In [40], pinning control based on the Lyapunov V-stability approach was investigated; which allows nonidentical nodes. In [42], synchronization for a heterogeneous time-delay complex dynamical network for both continuous- and discrete-time domains was analyzed. In [44], using Lyapunov stability theory, different adaptive synchronization criteria were attained for general complex networks. In [41], pinning-controlled networks were analyzed via a renormalization approach, which included two operations; edge weighting and node reduction. In [36], using Lyapunov stability theory, different pinning synchronization criteria were proposed for a controlled dynamical network with uncertainties. In [18], analytical tools to study general controllability of an arbitrary complex network was developed, identifying the set of driver nodes which can guide the system entire dynamics.

On the other hand, optimal control is a desired goal, which constitutes a hard task to be solved in practise, since, it is required the solution of complex expressions as the partial differential Hamilton-Jacobi-Bellman (HJB) or the Hamilton-Jacobi-Isaacs (HJI) equation. To address this issue, the inverse optimal control appears as an alternative to handle the solution of nonlinear optimal control. In this technique, an optimal control law is stated a priori based on a control Lyapunov function, and then a posteriori, it is determined the minimized performance index, avoiding the solution of HJB or HJI equation [10, 13, 15, 28, 32].

The optimal control formulation requires a model of the system. In practice, such model is not always available. To bypass this drawback, neural networks (NNs) could be used; it has been demonstrated that NNs have adaptation and function approximation properties [12, 16, 21, 24, 34]. This fact has motivated the use of NNs for identification and control of nonlinear systems. Recurrent Higher-Order Neural Networks (RHONN) have been received increasing attention due to its characteristics such that as: its easy implementation, its simple structure and the on-line adjust capacity [1, 25, 27, 30]; NNs have been also employed to synchronization of complex networks [7, 31, 35].

The objective of the present chapter is to extend the recent results of the published paper [35]. These new results are validity for complex networks with nonidentical nodes, ensuring that output trajectory tracking objective is achieved based on the proposed control scheme. Its applicability is evaluated via simulations using a network with ten different nodes. Simulation results are given to illustrate the feasibility of the proposed scheme.

The chapter is organized as follows: In Sect. 2, a mathematical preliminary is provided. Section 3 presents the proposed control scheme. Simulation results are reported in Sect. 4. Finally, conclusions are drawn in Sect. 5.

2 Mathematical Preliminaries

Notations: $diag(\ldots)$ denotes a block-diagonal matrix. A^T and A^{-1} denote the transpose and the inverse of the matrix A, respectively. $\|A\|$ is the Euclidean norm of A. Write $A > 0$ $(A < 0)$ if A is positive (negative) definite. If $f : \mathbb{R}^n \to \mathbb{R}^n$ is a vector field and $h : \mathbb{R}^n \to \mathbb{R}$ is a scalar function, $L_f h$ denotes the directional derivative of $h(x)$ along $f(x)$ given by

$$\frac{\partial h}{\partial x} f(x).$$

$A \otimes B$ represents their Kronecker product.

A brief review on recurrent higher-order neural networks is discussed in [27].

2.1 Recurrent Higher-Order Neural Networks (RHONN)

The RHONN consist of dynamical elements distributed in the form of neurons, which present high-order interactions among them. According to [11], a RHONN is given by

$$\dot{\chi}_i = -\lambda_i \chi_i + \sum_{k=1}^{L} w_{ik} \prod_{j \in I_k} y_j^{d_j(k)}, \qquad i = 1, 2, \ldots, n \tag{1}$$

where χ_i is the ith neuron state, L is the number of higher-order connections, $I_1, I_2, \ldots, I_L$ is a collection of non-ordered subsets of $1, 2, \ldots, m + n$, $\lambda_i > 0$, w_{ik} are the adjustable weights of the network, $d_j(k)$ are nonnegative integers, and y is a vector defined by

$$y = [y_1, \ldots, y_n, y_{n+1}, \ldots, y_{n+m}]^T$$
$$= [S(\chi_1), \ldots, S(\chi_n), S(u_1), \ldots, S(u_m)]^T \tag{2}$$

with $u = [u_1, u_2, \ldots, u_m]$ being the input to the network, and $S(\cdot)$ being a smooth sigmoid function. Equation (2) includes higher-order terms.

Now, define a vector

$$\Upsilon(\chi, u) = \left[\prod_{j \in I_1} y_j^{d_j(1)}, \prod_{j \in I_2} y_j^{d_j(2)}, \ldots, \prod_{j \in I_L} y_j^{d_j(L)} \right]^T \tag{3}$$

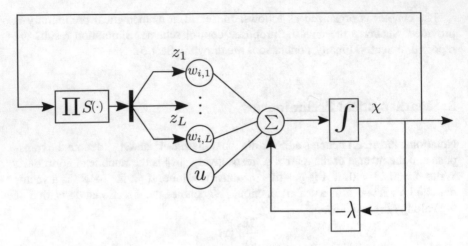

Fig. 1 RHONN scheme

Then, (1) can be written as

$$\dot{\chi}_i = -\lambda_i \chi_i + w_i \Upsilon_i(\chi, u), \qquad i = 1, 2, \dots, n \tag{4}$$

where $w_i = [w_{i,1}, w_{i,2}, \dots, w_{i,L}]^T$.

In this paper, the RHONN scheme used is affine in the control input, as displayed in Fig. 1. This RHONN in a matrix form, is given by

$$\dot{\chi} = -\lambda I_n \chi + W \Upsilon(\chi) + u \tag{5}$$

where $\chi \in \mathbb{R}^n$, $W \in \mathbb{R}^{n \times L}$, $\Upsilon(\chi) \in \mathbb{R}^L$, $u \in \mathbb{R}^n$, and $\lambda > 0$; also it has a Series-Parallel structure i.e., $\Upsilon(\cdot) = \Upsilon(u_e)$, where u_e is an external input, [27].

2.2 Complex Networks

Consider a controlled network consisting of N nodes with linear diffusive couplings, where each node is an n-dimensional dynamical unknown system. The state equations of this dynamical network are given by

$$\dot{\mathbf{x}}_i = f_i(\mathbf{x}_i) + \sum_{j=1, j \neq i}^{N} c_{ij} a_{ij} \mathbf{\Gamma}(\mathbf{x}_j - \mathbf{x}_i), \tag{6}$$

where $\mathbf{x}_i = \left(x_{i_1}, x_{i_2}, \dots, x_{i_n}\right)^T \in \mathbb{R}^n$ for $i = 1, 2, \dots, N$ are the state vector of node i, $f_i : \mathbb{R}^n \to \mathbb{R}^n$ represents the different self-dynamics of node i, constants c_{ij} are the coupling strength between node i and node j, $\Gamma \in \mathbb{R}^{n \times n}$ is the inner coupling matrix that describes the way of linking the components in each pair of connected node vectors $(\mathbf{x}_j - \mathbf{x}_i)$, and the coupling matrix $\mathbf{A} = [a_{ij}] \in \mathbb{R}^{N \times N}$ represents the topological structure of the network. If node i and j for $i \neq j$ are connected, then $a_{ij} = a_{ji} = 1$; otherwise, $a_{ij} = a_{ji} = 0$ for $i \neq j$. If the node degree k_i is defined to be the number of edges connected to node i

$$k_i = \sum_{j=1, j \neq i}^{N} a_{ij} = \sum_{j=1, j \neq i}^{N} a_{ji}, \quad i = 1, 2, \ldots, N.$$

Let the diagonal element of matrix $\mathbf{A} : a_{ii} = -k_i, i = 1, 2, \ldots, N$, which means the linear diffusive coupling, and

$$c_{ii} = \frac{1}{k_i} \sum_{j=1, j \neq i}^{N} a_{ij} c_{ji}$$

for normalization. Then, network (6) in a compact form is

$$\dot{\mathbf{x}}_i = f_i(\mathbf{x}_i) + \sum_{j=1}^{N} c_{ij} a_{ij} \Gamma \mathbf{x}_j \quad i = 1, 2, \ldots, N. \tag{7}$$

2.3 Pining Control Strategy

For network (6), the control objective is to stabilize it onto a homogeneous stationary state $\bar{x}$, which satisfies $f(\bar{x}) = 0$:

$$x_1 = x_2 = \cdots = x_N = \bar{x}.$$

To achieve network synchronization, the pinning strategy is applied on a small fraction of the nodes, which are called pinned ones. For notational simplicity, let these nodes be labeled as $1, 2, \ldots, l$, where $1 \leq l \leq N$. The pinning controllers are given by

$$u_i = g_i(\mathbf{x}_i, \bar{x}), \quad i = 1, 2, \ldots, l. \tag{8}$$

Thus, the pinning controlled network can be described by

$$\begin{aligned}
\dot{\mathbf{x}}_i &= f_i(\mathbf{x}_i) + \sum_{j=1}^{N} c_{ij} a_{ij} \Gamma \mathbf{x}_j + g_i, \quad i = 1, 2, \ldots, l, \\
\dot{\mathbf{x}}_i &= f_i(\mathbf{x}_i) + \sum_{j=1}^{N} c_{ij} a_{ij} \Gamma \mathbf{x}_j, \qquad i = l+1, \ldots, N,
\end{aligned} \tag{9}$$

The following assumption is needed to ensure network synchronization, where

$$D_i = \{\mathbf{x}_i : \|\mathbf{x}_i - \bar{x}\| < \delta\}, \quad \delta > 0, \quad D = \bigcup_{i=1}^{N} D_i.$$

Assumption There is a continuously differentiable Lyapunov function $V(x) : D \subseteq \mathbb{R}^n \mapsto \mathbb{R}_+$ satisfying $V(\bar{x}) = 0$ with $\bar{x} \in D$, such that for each node function $f_i(\mathbf{x}_i)$, there is a scalar θ_i guaranteeing

$$\frac{\partial V(\mathbf{x}_i)}{\partial \mathbf{x}_i} \left(f_i(\mathbf{x}_i) + g_i(\mathbf{x}_i, \bar{x}) + (\theta_i + \psi_i) \, \mathbf{\Gamma} \mathbf{x}_i \right) < 0,$$

$$\forall \mathbf{x}_i \in D_i, \quad \mathbf{x}_i \neq \bar{x}, \tag{10}$$

with constants $\psi_i \geq 0$. $\square$

In (10), θ_i is the passivity degree [40]. This passivity degree is modified by a factor of ψ_i. Define the Lyapunov function for the controlled network:

$$V_N(\mathbf{X}) = \sum_{i=1}^{N} \frac{1}{2} \mathbf{x}_i^T \mathbf{P} \mathbf{x}_i, \tag{11}$$

$$\mathbf{X} = [\mathbf{x}_1^T, \mathbf{x}_2^T, \dots, \mathbf{x}_N^T]^T.$$

By taking its time derivative, one obtains

$$\dot{V}_N(\mathbf{X}) = \sum_{i=1}^{N} \mathbf{x}_i^T \mathbf{P} \left(f_i(\mathbf{x}_i)) + \sum_{j=1}^{N} c_{ij} a_{ij} \mathbf{\Gamma} \mathbf{x}_j + g_i(\mathbf{x}_i, \bar{x}) \right)$$

$$< \sum_{i=1}^{N} \mathbf{x}_i^T \mathbf{P} \left(\sum_{j=1}^{N} c_{ij} a_{ij} \mathbf{\Gamma} \mathbf{x}_j - (\theta_i + \psi_i) \, \mathbf{\Gamma} \mathbf{x}_i \right)$$

$$< \mathbf{X}^T \left(-\mathbf{\Theta} + \mathbf{G} - \mathbf{\Psi} \right) \otimes \mathbf{P} \mathbf{\Gamma} \mathbf{X}, \tag{12}$$

The next result is established.

Theorem 1 [40] *If the coupling strength is sufficiently strong and ψ_i, $i = 1, 2, \dots, l$, is large enough, then*

$$\mathbf{C} = -\mathbf{\Theta} + \mathbf{G} - \mathbf{\Psi},$$

is negative definite with nonzero l, where $\mathbf{\Theta} = diag(\theta_1, \theta_2, \dots, \theta_N) \in \mathbb{R}^{N \times N}$, $\mathbf{G} = (g_{ij}) = (c_{ij} a_{ij}) \in \mathbb{R}^{N \times N}$, $\mathbf{\Psi} \in \mathbb{R}^{N \times N}$ is a diagonal matrix with the first l elements ψ_i, $i = 1, 2, \dots, l$, and other $(N - l)$ elements are all zero.

The above method converts the original stability problem to the study of the negativity property of the matrix $\mathbf{C}$. This analysis method is named as V-stability [40].

3 Neural Inverse Optimal Pinning Control

Neural inverse optimal pinning control is proposed to achieve output trajectory tracking of complex networks with chaotic nonidentical nodes. The control scheme is composed of a RHONN identifier and a controller for each pinned node, where the former is used to build an on-line model for the unknown plant, and the latter to force the unknown node dynamics to achieve output trajectory tracking for the whole network. This scheme is displayed in Fig. 2.

3.1 Control Problem Formulation

In this chapter, the control goal is to achieve that the whole network tracks an output desired trajectory

$$y = h(\mathbf{x}_i), \quad y \in \mathbb{R}^m$$

with the assumption of considering the same number of inputs as outputs for each pinned node, and the pinned nodes given by

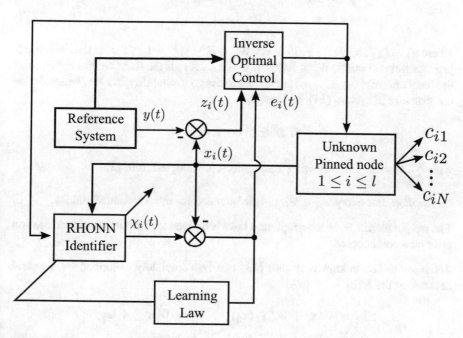

Fig. 2 The proposed control scheme

$$\dot{\mathbf{x}}_i = \hat{f}_i(\mathbf{x}_i) - \sum_{j=1}^{N} g_{ij}\mathbf{\Gamma}\mathbf{x}_j + b_i\mathbf{u}_i, \quad \Rightarrow \quad \begin{aligned} \dot{\mathbf{x}}_{i_1} &= \hat{f}_{i_1}(\mathbf{x}_i) - \sum_{j=1}^{N} g_{ij}\mathbf{\Gamma}\mathbf{x}_{j_1}, \\ \dot{\mathbf{x}}_{i_2} &= \hat{f}_{i_2}(\mathbf{x}_i) - \sum_{j=1}^{N} g_{ij}\mathbf{\Gamma}\mathbf{x}_{j_2} + \mathbf{u}_i, \end{aligned}$$

$$i = 1, \dots, l, \qquad (13)$$

where $\mathbf{x}_{i_1} \in \mathbb{R}^{n-m}, \mathbf{x}_{i_2} \in \mathbb{R}^m, b_i \in \mathbb{R}^{n \times m}, \mathbf{u}_i \in \mathbb{R}^m, h : \mathbb{R}^n \to \mathbb{R}^m$, the vector functions $\hat{f}_{i_1}(\cdot)$ and $\hat{f}_{i_2}(\cdot)$ are unknown, with $\mathbf{x}_i$ being available for measurement.

3.2 Neural Identifier

It is possible to model (13) by an RHONN. Assume that it has a Block Controllable (BC) form. Then, the RHONN is used in a Series-Parallel structure consisting of r blocks, as follows:

$$\begin{aligned} \dot{\boldsymbol{\chi}}_{i_j} &= -\lambda\boldsymbol{\chi}_{i_j} + W_{i_j}\Upsilon_{i_j}(\mathbf{x}_{i_1}, \dots, \mathbf{x}_{i_j}) + W'_{i_j}\mathbf{x}_{i_{j+1}}, \\ \dot{\boldsymbol{\chi}}_{i_r} &= -\lambda\boldsymbol{\chi}_{i_r} + W_{i_r}\Upsilon_{i_r}(\mathbf{x}_i) + W'_{i_r}u_i, \\ j &= 1, \dots, r-1, \quad i = 1, \dots, l, \end{aligned} \qquad (14)$$

where $\boldsymbol{\chi}_i = [\boldsymbol{\chi}_{i_1}, \boldsymbol{\chi}_{i_2}, \dots, \boldsymbol{\chi}_{i_r}]^T \in \mathbb{R}^n$, $W_{i_k} \in \mathbb{R}^{n_k \times L}, k = 1, \dots, r$, is the on-line adjustable weight matrix, $W'_{i_k} \in \mathbb{R}^{n_{k+1} \times L}, k = 1, \dots, r$, is the fixed weight matrix, and the set of numbers $(n_1, \dots, n_r)$ are known as the controllability indexes, which define the structure of system (14), and satisfy

$$n_1 \le n_2 \le \dots \le n_r \le m,$$

with $\sum_{k=1}^{r} n_k = n$. The following assumptions are adopted from [25]:

Assumption For every $w_{ik} \in W_i$, (14) is bounded for every bounded state x_i. □

The weight matrix W_i in the adaptation laws is derived to minimize the identification error on a compact set.

Assumption The unknown system (13) can be completely described by a neural network of the form

$$\begin{aligned} \dot{\mathbf{x}}_{i_j} &= -\lambda\mathbf{x}_{i_j} + W^*_{i_j}\Upsilon_{i_j}(\mathbf{x}_{i_1}, \dots, \mathbf{x}_{i_j}) + W'_{i_j}\mathbf{x}_{i_{j+1}} + \boldsymbol{\omega}_{i_j}, \\ \dot{\mathbf{x}}_{i_r} &= -\lambda\mathbf{x}_{i_r} + W^*_{i_r}\Upsilon_{i_r}(\mathbf{x}_i) + W'_{i_r}u_i + \boldsymbol{\omega}_{i_r}, \\ j &= 1, \dots, r-1, \quad i = 1, \dots, l, \end{aligned} \qquad (15)$$

where W_i^* are the ideal constant weight matrix, the modeling error the term $\omega_i \in \mathbb{R}^n$ representing the modeling error is bounded, which represents the mismatch between the unknown system and the RHONN model with its optimal weight values, and all the other elements are as described above. □

The modeling error is caused by an insufficient number of higher-order terms in the RHONN model and it can be arbitrarily small by selecting appropriately the number L_i of high-order connections.

Now, the identification error is defined as $\mathbf{e}_i = \chi_i - \mathbf{x}_i$, satisfying

$$\dot{\mathbf{e}}_i = -\lambda I_n \mathbf{e}_i + \tilde{W}_i \Upsilon(\mathbf{x_i}) - \omega_i, \tag{16}$$

where $\tilde{W}_i = W_i - W_i^*$.

The learning law is taken from [27]

$$tr\left\{\dot{\tilde{W}}_i^T \tilde{W}_i\right\} = -\gamma \mathbf{e}_i^T \tilde{W}_i \Upsilon(\mathbf{x}_i), \tag{17}$$

which has elements

$$\dot{\tilde{w}}_{ij} = -\gamma e_i \Upsilon_{(\mathbf{x}_i)}, \quad i = 1, 2, \ldots, n, \quad j = 1, 2, \ldots, L.$$

With this adaptation law, the identification error satisfies

$$\sup_{0 \leq t \leq T} \|\mathbf{e}_i(t)\| \leq \mathcal{E},$$

for any $\mathcal{E} > 0$ and any finite $T > 0$. For a detailed stability analysis of (16), see [29].

3.3 Trajectory Tracking Analysis

For trajectory tracking, suppose that the plant has been identified by the RHONN (14) after a finite time T. The identifier cannot reduce the modeling error ω_i to zero; so, ω_i is considered as a disturbances, as follows:

$$\dot{\mathbf{x}}_i \triangleq \dot{\chi}_i + \omega_i, \tag{18}$$
$$\dot{\mathbf{x}}_{i_j} = -\lambda \chi_{i_j} + W_{i_j} \Upsilon_{i_j}(\mathbf{x}_{i_1}, \ldots, \mathbf{x}_{i_j}) + W_{i_j}' \mathbf{x}_{i_{j+1}} + \omega_{i_j},$$
$$\dot{\mathbf{x}}_{i_r} = -\lambda \chi_{i_r} + W_{i_r} \Upsilon_{i_r}(\mathbf{x}_i) + W_{i_r}' u_i + \omega_{i_r},$$
$$j = 1, \ldots, r - 1, \quad i = 1, \ldots, l.$$

In this section, the tracking problem is thus transformed to the stabilization of the error system based on a coordinate transformation to the block-control form. To achieve this transformation, the backstepping technique [14] is applied for each pinned node ($i = 1, 2, \ldots, l$), which is described step-by-step as follows.

Step 1:
the tracking error $\mathbf{z}_1$ is given by

$$\mathbf{z}_1 = \mathbf{x}_1 - y_r.$$

Consider the Lyapunov function candidate as

$$V(\mathbf{z}_1, \mathbf{e}_1, \tilde{W}_1) = \tfrac{1}{2}\mathbf{z}_1^T\mathbf{z}_1 + \tfrac{1}{2}\mathbf{e}_1^T\mathbf{e}_1 + \tfrac{1}{2\gamma}tr\left\{\tilde{W}_1^T\tilde{W}_1\right\} > 0,$$

which its derivative is

$$\begin{aligned}
\dot{V}(\mathbf{z}_1, \mathbf{e}_1, \tilde{W}_1) &= \mathbf{z}_1^T\dot{\mathbf{z}}_1 + \mathbf{e}_1^T\dot{\mathbf{e}}_1 + \frac{1}{\gamma}tr\left\{\dot{\tilde{W}}_1^T\tilde{W}_1\right\} \\
&= \mathbf{z}_1^T(-\lambda\boldsymbol{\chi}_1 + W_1\Upsilon_1(\mathbf{x}_1) + W_1'\mathbf{x}_2 + \boldsymbol{\omega}_1 - \dot{y}_r) - \mathbf{e}_1^T\tilde{W}_1\Upsilon(\mathbf{x}_1) \\
&\quad + \mathbf{e}_1^T(-\lambda I_n\mathbf{e}_1 + \tilde{W}_1\Upsilon(\mathbf{x}_1)).
\end{aligned} \tag{19}$$

It is designed a virtual control $\mathbf{x}_2 = \alpha_1$ to force $\mathbf{z}_1 \to 0$, proposed as

$$\alpha_1 = (W_1')^{-1}(\lambda\boldsymbol{\chi}_1 - W_1\Upsilon_1(\mathbf{x}_1) + \dot{y}_r - k_1\mathbf{z}_1),$$

with $k_1 > 0$. So, (19) becomes

$$\dot{V}(\mathbf{z}_1, \mathbf{e}_1, \tilde{W}_1) = -k_1\|\mathbf{z}_1\|^2 - \lambda\|\mathbf{e}_1\|^2 + \mathbf{z}_1^T\boldsymbol{\omega}_1 + \mathbf{z}_1^T W_1'\mathbf{z}_2.$$

Step j:
The error dynamics for $\mathbf{z}_j$ is derived from

$$\mathbf{z}_j = \mathbf{x}_j - \alpha_{j-1},$$

which represents the error between the actual control and the virtual control. Consider the augmented Lyapunov function candidate as

$$\begin{aligned}
V(\mathbf{z}_1, \ldots, \mathbf{z}_j, \mathbf{e}_1, \ldots, \mathbf{e}_j, \tilde{W}_1, \ldots, \tilde{W}_j) &= V(\mathbf{z}_1, \ldots, \mathbf{z}_{j-1}, \mathbf{e}_1, \ldots, \mathbf{e}_{j-1}, \tilde{W}_1, \ldots, \tilde{W}_{j-1}) \\
&\quad + \frac{1}{2}\|\mathbf{z}_j\|^2 + \frac{1}{2}\|\mathbf{e}_j\|^2 + \frac{1}{2\gamma}tr\left\{\tilde{W}_j^T\tilde{W}_j\right\},
\end{aligned}$$

and

$$
\begin{aligned}
\dot{V}(\mathbf{z}_1, \ldots, \mathbf{z}_j, \mathbf{e}_1, \ldots, \mathbf{e}_j, \tilde{W}_1, \ldots, \tilde{W}_j) = {} & \dot{V}(\mathbf{z}_1, \ldots, \mathbf{z}_{j-1}, \mathbf{e}_1, \ldots, \mathbf{e}_{j-1}, \tilde{W}_1, \ldots, \tilde{W}_{j-1}) \\
& + \mathbf{z}_j^T \left(-\lambda \boldsymbol{\chi}_j + W_j \Upsilon_j(\mathbf{x}_1, \ldots, \mathbf{x}_j) \right. \\
& \left. + W_j' \mathbf{x}_{j+1} + \boldsymbol{\omega}_j - \dot{\alpha}_{j-1} \right) \\
& + \mathbf{e}_j^T \left(-\lambda I_n \mathbf{e_j} + \tilde{W}_j \Upsilon(\mathbf{x}_1, \ldots, \mathbf{x}_j) \right) \\
& - \mathbf{e}_j^T \tilde{W}_j \Upsilon(\mathbf{x}_1, \ldots, \mathbf{x}_j).
\end{aligned}
\tag{20}
$$

The goal at the jth step is to propose a virtual control $\mathbf{x}_{j+1} = \alpha_j$ to stabilize the error $\mathbf{z}_j = 0$, as

$$
\alpha_j = (W_j')^{-1}(\lambda \boldsymbol{\chi}_j - W_j \Upsilon_j(\mathbf{x}_1, \ldots, \mathbf{x}_j) + \dot{\alpha}_{j-1} - k_j \mathbf{z}_j),
\tag{21}
$$

with $k_j > 0$. Replacing (21) in (20) gives

$$
\begin{aligned}
\dot{V}(\mathbf{z}_1, \ldots, \mathbf{z}_j, \mathbf{e}_1, \ldots, \mathbf{e}_j, \tilde{W}_1, \ldots, \tilde{W}_j) = {} & -\sum_{k=1}^{j} (k_k \|\mathbf{z}_k\|^2 + \lambda \|\mathbf{e}_k\|^2) \\
& + \sum_{k=1}^{j} \mathbf{z}_k^T \boldsymbol{\omega}_k + W_j' \mathbf{z}_j^T \mathbf{z}_{j+1}.
\end{aligned}
$$

Step r:

Consider $\mathbf{z}_r = x_r - \alpha_{r-1}$, the new dynamics for (18) in $\mathbf{z}$-variables is

$$
\begin{aligned}
\dot{\mathbf{z}}_j &= -W_{j-1}' \mathbf{z}_{j-1} - k_j \mathbf{z}_j + W_j' \mathbf{z}_{j+1} + \boldsymbol{\omega}_j, \\
\dot{\mathbf{z}}_r &= f_r(\mathbf{x}, \boldsymbol{\chi}, W) + W_r' \mathbf{u} + \boldsymbol{\omega}_r, \\
j &= 1, \ldots, r-1.
\end{aligned}
\tag{22}
$$

Let us define the system

$$
\dot{\boldsymbol{\varepsilon}} = f_\varepsilon(\mathbf{z}, \mathbf{e}, \tilde{W}) + g_\varepsilon(\mathbf{z}, \mathbf{e}, \tilde{W})\mathbf{u},
\tag{23}
$$

where $\boldsymbol{\varepsilon} = [\mathbf{z}^T, \mathbf{e}^T, \tilde{W}^T]^T$. For (23), it is proposed to minimize a cost functional given by

$$
J = \lim_{t \to \infty} \left[\mathfrak{J}(\mathbf{z}(t), \mathbf{e}(t), \tilde{W}(t)) + \int_0^t \left(\mathfrak{l}(\mathbf{z}, \mathbf{e}, \tilde{W}) + \mathbf{u}^T R \left(\mathbf{z}, \mathbf{e}, \tilde{W} \right) \mathbf{u} \right) d\tau \right],
\tag{24}
$$

where $R = R^T > 0$ is a matrix-valued function for all $\mathbf{z}, \mathbf{e}, \tilde{W}$, $\mathfrak{l}(\mathbf{z}, \mathbf{e}, \tilde{W})$ and $\mathfrak{J}(\mathbf{z}, \mathbf{e}, \tilde{W})$ are radially unbounded function [10]. In order to satisfy this requirement, the inverse optimal control technique is applied. Consider the following Control Lyapunov Function (CLF) [10]:

$$V(\mathbf{z}, \mathbf{e}, \tilde{W}) = \frac{1}{2}\mathbf{z}^T\mathbf{z} + \frac{1}{2}\mathbf{e}^T\mathbf{e} + \frac{1}{2\gamma}tr\{\tilde{W}^T\tilde{W}\}. \tag{25}$$

and its derivative is

$$\dot{V} = \mathbf{z}^T\dot{\mathbf{z}} + \mathbf{e}^T\dot{\mathbf{e}} + \frac{1}{\gamma}tr\{\dot{\tilde{W}}^T\tilde{W}\}$$

$$= -W_1(\mathbf{z}, \mathbf{e}) + \sum_{k=1}^{r}\mathbf{z}_k^T\omega_k + W_j'\mathbf{z}_j^T\mathbf{z}_{j+1} + \mathbf{z}_r^T(f_r(\mathbf{x}, \chi, W) + W_r'\mathbf{u}) \tag{26}$$

$$\triangleq -W_1(\mathbf{z}, \mathbf{e}) + L_f V + L_g V\mathbf{u} + L_h V\omega \tag{27}$$

where

$$W_1(\mathbf{z}, \mathbf{e}) = \sum_{k=1}^{j}(k_k\|\mathbf{z}_k\|^2 + \lambda\|\mathbf{e}_k\|^2) > 0$$

$$L_f V = \mathbf{z}_r^T f_r(\mathbf{x}, \chi, W) + W_j'\mathbf{z}_j^T\mathbf{z}_{j+1},$$

$$L_g V = \mathbf{z}_r^T W_r',$$

$$L_h V = \mathbf{z}^T.$$

The control law $\mathbf{u}$, taken from [15], is used:

$$\mathbf{u} = -\frac{1}{2}R^{-1}(\mathbf{z}, \tilde{W})L_g V^T = g^*(\mathbf{z}, \tilde{W}) \tag{28}$$

where

$$g^* = \begin{cases} \left[\dfrac{\phi + \sqrt{\phi^2 + \left(L_g V (L_g V)^T\right)^2}}{L_g V (L_g V)^T}\right]L_g V^T, & \|L_g V\| \neq 0, \\ \\ \qquad\qquad 0, & \|L_g V\| = 0, \end{cases}$$

with

$$\boldsymbol{\phi} = L_f V + |L_h V|\rho^{-1}(\|\mathbf{z}\|), \tag{29}$$

Replacing (28) in (26) gives

$$
\begin{aligned}
\dot{V}\big|_{u=\frac{g^*}{2}} &= -W_1(\mathbf{z}, \mathbf{e}) + L_f V + L_h V \omega \\
&\quad - \frac{1}{2}\left(\phi + \sqrt{\phi^2 + \left(L_g V \left(L_g V\right)^T\right)^2}\right) \\
&= -W_1(\mathbf{z}, \mathbf{e}) + L_h V \omega - |L_h V|\,\rho^{-1}\,(\|\mathbf{z}\|) \\
&\quad + \frac{1}{2}\overbrace{\left(L_f V + |L_h V|\,\rho^{-1}\,(\|\mathbf{z}\|)\right)}^{\phi} \\
&\quad - \frac{1}{2}\left(\sqrt{\phi^2 + \left(L_g V \left(L_g V\right)^T\right)^2}\right) \\
&\leq -W_1(\mathbf{z}, \mathbf{e}) - |L_h V|\left(\rho^{-1}\,(\|\mathbf{z}\|) - \|\omega\|\right) \\
&\quad + \frac{1}{2}\phi - \frac{1}{2}\sqrt{\phi^2 + \frac{1}{2}\left(L_g V \left(L_g V\right)^T\right)^2}.
\end{aligned}
$$

For $\|\mathbf{z}\| \geq \rho(\|\omega\|)$, it is hold

$$
\dot{V}(\mathbf{z}, \mathbf{e}, \tilde{W}) \leq -W_1(\mathbf{z}, \mathbf{e}) - W_2(\mathbf{z}, \tilde{W}) < 0,
$$

where

$$
W_2(\mathbf{z}, \tilde{W}) = -\frac{1}{2}\phi + \frac{1}{2}\sqrt{\phi^2 + \frac{1}{2}\left(L_g V \left(L_g V\right)^T\right)^2} > 0
$$

which is positive definite. Hence, system (23) is input-to-state stable (ISS) [15].

3.4 Inverse Optimality

In this part, inverse optimality is analyzed. In order to demonstrate this fact, Lyapunov function (25) must satisfy the associated HJI equation:

$$
\mathfrak{l}(\mathbf{z}, \mathbf{e}, \tilde{W}) - W_1(\mathbf{z}, \mathbf{e}) + L_f V - \frac{1}{4}L_g V R^{-1}(L_g V)^T + \|L_h V\|\omega = 0. \tag{30}
$$

When $\mathfrak{l}(\mathbf{z}, \mathbf{e}, \tilde{W})$ is selected as $\mathfrak{l}(\mathbf{z}, \mathbf{e}, \tilde{W}) = -\dot{V}$, function (25) is a solution to (30). Hence, $\mathfrak{l}(\mathbf{z}, \mathbf{e}, \tilde{W})$ is written as

$$
\mathfrak{l}(\mathbf{z}, \mathbf{e}, \tilde{W}) = W_1(\mathbf{z}, \mathbf{e}) - L_f V + \frac{1}{4}L_g V R^{-1}(L_g V)^T - \|L_h V\|\omega, \tag{31}
$$

where $\mathfrak{l}(\mathbf{z}, \mathbf{e}, \tilde{W})$ must be positive definite and radially unbounded. Substituting $R^{-1}(\cdot)$ into (31) yields

$$\mathfrak{l}(\mathbf{z}, \mathbf{e}, \tilde{W}) \geq W_1(\mathbf{z}, \mathbf{e}) + W_2(\mathbf{z}, \tilde{W}) > 0,$$

for $\|\mathbf{z}\| \geq \rho(\|\boldsymbol{\omega}\|)$.

In order to establish inverse optimality, let us substitute (31) and $\mathbf{u} = \mathbf{v} - \frac{1}{2}R^{-1}(\cdot)L_g V^T$ into (24) gives

$$
\begin{aligned}
J &= \lim_{t\to\infty} \left[\mathcal{J} + \int_0^t \left(\mathfrak{l}(\mathbf{z}, \mathbf{e}, \tilde{W}) + \mathbf{v}^T R(\mathbf{z}, \mathbf{e}, \tilde{W})\mathbf{v} \right. \right. \\
&\quad \left. \left. - \mathbf{v}^T(L_g V)^T + \frac{1}{4}L_g V R^{-1}(\mathbf{z}, \mathbf{e}, \tilde{W})(L_g V)^T \right) d\tau \right] \\
&= \lim_{t\to\infty} \left[\mathcal{J} + \int_0^t \left(-L_f V + \frac{1}{2}L_g V R^{-1}(L_g V)^T \right. \right. \\
&\quad \left. \left. -(L_g V)\mathbf{v} - \|L_h V\|\boldsymbol{\omega} \right) d\tau + \int_0^t \left(\mathbf{v}^T R\mathbf{v} \right) d\tau \right] \\
&= \lim_{t\to\infty} \left[\mathcal{J} - \int_0^t \left(\sup_{\boldsymbol{\omega}\in\Omega} \left\{ \dot{V}(\mathbf{z}, \mathbf{e}, \tilde{W}) d\tau \right\} \right) + \int_0^t \left(\mathbf{v}^T R\mathbf{v} \right) d\tau \right] \\
&\leq \lim_{t\to\infty} \left[\mathcal{J} - \int_0^t \dot{V}(\mathbf{z}, \mathbf{e}, \tilde{W}) d\tau + \int_0^t \left(\mathbf{v}^T R\mathbf{v} \right) d\tau \right] \\
&\leq V(\mathbf{z}(0), \mathbf{e}(0), \tilde{W}(0)) + \lim_{t\to\infty} \left[\int_0^t \left(\mathbf{v}^T R\mathbf{v} \right) d\tau \right].
\end{aligned}
$$

The minimum value $J^* = \mathfrak{J}^*(\mathbf{z}(0), \mathbf{e}(0), \tilde{W}(0)) = V(\mathbf{z}(0), \mathbf{e}(0), \tilde{W}(0))$ is achieved when $\mathbf{v} \equiv 0$.

3.5 Output Trajectory Tracking for the Whole Network

Next, tracking error stability for the whole network is analyzed. Define $\mathbf{z}_i = \mathbf{x}_i - \mathbf{x}_d$, $\mathbf{x}_d = [y_r, \alpha_1, \ldots, \alpha_{r-1}]^T$, and $\dot{\mathbf{x}}_d = f_d(\mathbf{x}_d)$. The complex network is rewritten as

$$
\begin{aligned}
\dot{\mathbf{z}}_i &= f_i(\mathbf{x}_i) - f_i(\mathbf{x}_d) + \sum_{j=1}^N c_{ij}a_{ij}\boldsymbol{\Gamma}\mathbf{z}_j + b_i\mathbf{u}, \quad i = 1, 2, \ldots, l, \\
\dot{\mathbf{z}}_i &= f_i(\mathbf{x}_i) - f_i(\mathbf{x}_d) + \sum_{j=1}^N c_{ij}a_{ij}\boldsymbol{\Gamma}\mathbf{z}_j, \qquad i = l+1, \ldots, N,
\end{aligned}
\tag{32}
$$

Consider the following Lyapunov function for the controlled network:

$$V_N(\mathbf{Z}) = \sum_{i=1}^{N} \frac{1}{2} \mathbf{z}_i^T \mathbf{P} \mathbf{z}_i, \tag{33}$$

$$\mathbf{Z} = [\mathbf{z}_1^T, \mathbf{z}_2^T, \dots, \mathbf{z}_N^T]^T.$$

and its derivative is given as

$$\dot{V}_N(\mathbf{Z}) = \sum_{i=1}^{N} \mathbf{z}_i^T \mathbf{P} \left(f_i(\mathbf{x}_i) - f_i(\mathbf{x}_d) + \sum_{j=1}^{N} c_{ij} a_{ij} \mathbf{\Gamma} \mathbf{z}_j + b_i \mathbf{u} \right)$$

$$< \sum_{i=1}^{N} \mathbf{z}_i^T \mathbf{P} \left(\sum_{j=1}^{N} c_{ij} a_{ij} \mathbf{\Gamma} \mathbf{z}_j - (\theta_i + \psi_i) \mathbf{\Gamma} \mathbf{z}_i \right)$$

$$< \mathbf{Z}^T \left(-\mathbf{\Theta} + \mathbf{G} - \mathbf{\Psi} \right) \otimes \mathbf{P} \mathbf{\Gamma} \mathbf{Z}, \tag{34}$$

where $\mathbf{\Theta} = diag(\theta_1, \theta_2, \dots, \theta_N) \in \mathbb{R}^{N \times N}$, $\mathbf{G} = (g_{ij}) = (c_{ij} a_{ij}) \in \mathbb{R}^{N \times N}$, $\mathbf{\Psi} \in \mathbb{R}^{N \times N}$ is a diagonal matrix with the first l elements ψ_i, $i = 1, 2, \dots, l$, and other $(N - l)$ elements are all zero.

4 Simulation Results

To verify the proposed control applicability, consider that the complex network is scale-free [3] with 10-nodes and four types of nodes, which are described by Chua's system (Node 1, 2, 3 and 4) [20]

$$\begin{aligned}
\dot{x} &= a_C \left(y - x - f_C(x) \right) \\
\dot{y} &= x - y + z \\
\dot{z} &= -b_C y - C_C z,
\end{aligned} \tag{35}$$

with

$$f_C(x) = \begin{cases} x > 1, & -m_{b_C} x - m_{ac} + m_{bc} \\ x < -1, & -m_{b_C} x + m_{ac} - m_{bc} \\ x = 1, & -m_{ac} x \end{cases},$$

where $a_C = 10$, $b_C = 15$, $c_C = 0.0385$, $m_{ac} = -1.27$, and $m_{h_C} = -0.68$

Lorenz's system (Node 5, 6 and 7) [19] is formulated as

$$
\begin{aligned}
\dot{x} &= a_L(y - x) \\
\dot{y} &= x(b_L - z) - y \\
\dot{z} &= xy - c_L z,
\end{aligned}
\tag{36}
$$

where $a_L = 10$, $b_L = 28$ and $c_L = 8/3$.

Chen's system (Node 8 and 9) [9] is given by

$$
\begin{aligned}
\dot{x} &= a_{Ch}(y - x) \\
\dot{y} &= (c_{Ch} - a_{Ch})x - xz + c_{Ch}y \\
\dot{z} &= xy - b_{Ch}z,
\end{aligned}
\tag{37}
$$

where $a_{Ch} = 35$, $b_{Ch} = 3$ and $c_{Ch} = 28$.

And Rössler's system (Node 10) [26] is written as

$$
\begin{aligned}
\dot{x} &= -(y + z), \\
\dot{y} &= x + a_R y, \\
\dot{z} &= b_R + zx - c_R),
\end{aligned}
\tag{38}
$$

where $a_R = 0.2$, $b_R = 0.2$ and $c_R = 5.7$.

The coupling strengths are constants $c_{ij} = c = 70$, $\mathbf{\Gamma} = diag(1, 1, 1)$, and the output $y = y_r$ is selected as

$$
y_r(t) = sin(4\pi t) - 6.
$$

The network matrix $\mathbf{A}$ is

$$
A =
\begin{bmatrix}
-8 & 1 & 1 & 1 & 1 & 1 & 1 & 1 & 0 & 1 \\
1 & -5 & 1 & 1 & 0 & 1 & 0 & 0 & 1 & 0 \\
1 & 1 & -4 & 0 & 1 & 1 & 0 & 0 & 0 & 0 \\
1 & 1 & 0 & -3 & 0 & 0 & 1 & 0 & 0 & 0 \\
1 & 0 & 1 & 0 & -2 & 0 & 0 & 0 & 0 & 0 \\
1 & 1 & 1 & 0 & 0 & -3 & 0 & 0 & 0 & 0 \\
1 & 0 & 0 & 1 & 0 & 0 & -2 & 0 & 0 & 0 \\
1 & 0 & 0 & 0 & 0 & 0 & 0 & -1 & 0 & 0 \\
0 & 1 & 0 & 0 & 0 & 0 & 0 & 0 & -1 & 0 \\
1 & 0 & 0 & 0 & 0 & 0 & 0 & 0 & 0 & -1
\end{bmatrix}.
$$

Figure 3 visualizes this network. The controller is defined as (28) and applied to Node 1. For simulations, the following RHONN is used to model the pinned node

Fig. 3 Scale free network
with 6 nodes

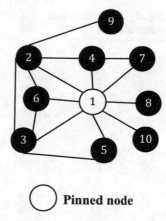

◯ **Pinned node**

$$\dot{\chi}_{1_1} = -5\chi_{1_1} + W_1 \Upsilon_1(\mathbf{x}_1)$$
$$\dot{\chi}_{1_2} = -5\chi_{1_2} + W_2 \Upsilon_2(\mathbf{x}_1)$$
$$\dot{\chi}_{1_3} = -5\chi_{1_3} + W_3 \Upsilon_3(\mathbf{x}_1) + u$$

with

$$\Upsilon_1(\mathbf{x}_1) = [tanh(x_{1_1}); tanh^2(x_{1_1})],$$
$$\Upsilon_2(\mathbf{x}_1) = [tanh(x_{1_1}); tanh(x_{1_2})],$$
$$\Upsilon_3(\mathbf{x}_1) = [tanh(x_{1_2}); tanh(x_{1_3}); tanh(x_{1_2})tanh(x_{1_3})].$$

Simulations are done using MATLAB. They are performed as follows. From $t = 0s$ to $t = 2s$, the network runs without interconnections and control actions ($c_{ij} = 0$, $\mathbf{u}_1 = 0$). From $t = 2s$, the coupling strengths are selected as $c_{ij} = c = 70$, and the proposed control law is turned on, and the complex network is forced to track the desired output trajectory.

Figure 4 presents the control input signal $u_i(t)$ applied to the single pinned node. Figure 5 displays the states evolution of the entire network. Figure 6 exhibits the time

Fig. 4 Control input signal
for the pinned node

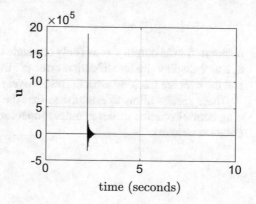

time (seconds)

Fig. 5 Time evolution of
network states

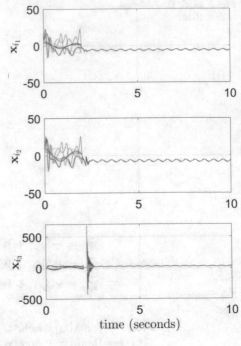

Fig. 6 Time evolution of
Node 1 $\mathbf{x}_1$

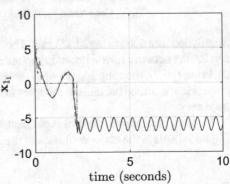

response for the output $y = x_1$ (only one state of the whole network). Finally, Figs. 7,
8, and 9 display the identification error $\mathbf{e}_{i_i}$, the neural network weights for node 1,
and the average tracking error $\bar{z}_1$, respectively.

These results allow to established that the proposed neural inverse optimal pinning control scheme achieves output trajectory tracking successfully for the whole
complex network.

Fig. 7 Time evolution of
identification error

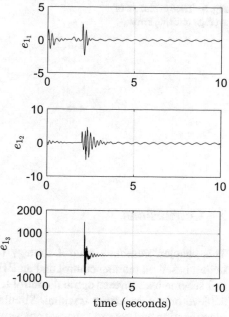

Fig. 8 Time evolution of
neural network weights for
Node 1

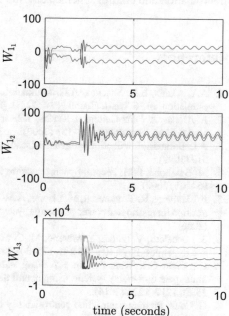

Fig. 9 Time evolution of average tracking error

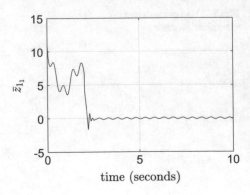

time (seconds)

5 Conclusions

This chapter presents a control strategy for complex networks with nonidentical nodes, based on pinning control and an RHONN for identification. This novel control scheme uses inverse optimal control technique combined with backstepping to achieve output trajectory tracking. Simulation results illustrate that the proposed identification and control scheme achieves the desired goals.

References

1. A.Y. Alanis, E.N. Sanchez, A.G. Loukianov, M.A. Perez, Real-time recurrent neural state estimation. IEEE Trans. Neural Netw. **22**(3), 497–505 (2011)
2. A. Arenas, A. Díaz-Guilera, J. Kurths, Y. Moreno, C. Zhou, Synchronization in complex networks. Phys. Rep. **469**(3), 93–153 (2008)
3. A.L. Barabási, R. Albert, Emergence of scaling in random networks. Science **286**(5439), 509–512 (1999)
4. M. Barahona, L.M. Pecora, Synchronization in small-world systems. Phys. Rev. Lett. **89**(5), 054101 (2002)
5. F. Blaabjerg, R. Teodorescu, M. Liserre, A.V. Timbus, Overview of control and grid synchronization for distributed power generation systems. IEEE Trans. Ind. Electron. **53**(5), 1398–1409 (2006)
6. S. Boccaletti, V. Latora, Y. Moreno, M. Chavez, D.U. Hwang, Complex networks: structure and dynamics. Phys. Rep. **424**(4), 175–308 (2006)
7. C.P. Chen, G.X. Wen, Y.J. Liu, F.Y. Wang, Adaptive consensus control for a class of nonlinear multiagent time-delay systems using neural networks. IEEE Trans. Neural Netw. Learn. Syst. **25**(6), 1217–1226 (2014)
8. G. Chen, Pinning control and controllability of complex dynamical networks. Int. J. Autom. Comput. **14**(1), 1–9 (2017)
9. G. Chen, T. Ueta, Yet another chaotic attractor. Int. J. Bifurcat. Chaos **9**(7), 1465–1466 (1999)
10. P. Kokotović, M. Arcak, Constructive nonlinear control: a historical perspective. Automatica **37**(5), 637–662 (2001)
11. E.B. Kosmatopoulos, M.A. Christodoulou, P.A. Ioannou, Dynamical neural networks that ensure exponential identification error convergence. Neural Netw. **10**(2), 299–314 (1997)

12. E.B. Kosmatopoulos, M.M. Polycarpou, M.A. Christodoulou, P.A. Ioannou, High-order neural network structures for identification of dynamical systems. IEEE Trans. Neural Netw. **6**(2), 422–431 (1995)
13. M. Krstic, H. Deng, *Stabilization of nonlinear uncertain systems* (Springer, New York, 1998)
14. M. Krstic, I. Kanellakopoulos, P.V. Kokotovic et al., *Nonlinear and Adaptive Control Design*, vol. 222 (Wiley, New York, 1995)
15. M. Krstic, Z.H. Li, Inverse optimal design of input-to-state stabilizing nonlinear controllers. IEEE Trans. Autom. Control **43**(3), 336–350 (1998)
16. A.U. Levin, K.S. Narendra, Control of nonlinear dynamical systems using neural networks. II. Observability, identification, and control. IEEE Trans. Neural Netw. **7**(1), 30–42 (1996)
17. X. Li, X. Wang, G. Chen, Pinning a complex dynamical network to its equilibrium. IEEE Trans. Circ. Syst. I Regul. Pap. **51**(10), 2074–2087 (2004)
18. Y.Y. Liu, J.J. Slotine, A.L. Barabási, Controllability of complex networks. Nature **473**(7346), 167 (2011)
19. E.N. Lorenz, Deterministic nonperiodic flow. J. Atmos. Sci. **20**(2), 130–141 (1963)
20. T. Matsumoto, A chaotic attractor from Chua's circuit. IEEE Trans. Circ. Syst. **31**(12), 1055–1058 (1984)
21. K.S. Narendra, S. Mukhopadhyay, Adaptive control using neural networks and approximate models. IEEE Trans. Neural Netw. **8**(3), 475–485 (1997)
22. M.E. Newman, The structure and function of complex networks. SIAM Rev. **45**(2), 167–256 (2003)
23. L.M. Pecora, T.L. Carroll, Synchronization in chaotic systems. Phys. Rev. Lett. **64**(8), 821 (1990)
24. A.S. Poznyak, W. Yu, E.N. Sanchez, J.P. Perez, Nonlinear adaptive trajectory tracking using dynamic neural networks. IEEE Trans. Neural Netw. **10**(6), 1402–1411 (1999)
25. L.J. Ricalde, E.N. Sanchez, Output tracking with constrained inputs via inverse optimal adaptive recurrent neural control. Eng. Appl. Artif. Intell. **21**(4), 591–603 (2008)
26. O.E. Rössler, An equation for continuous chaos. Phys. Lett. A **57**(5), 397–398 (1976)
27. G.A. Rovithakis, M.A. Christodoulou, *Adaptive Control with Recurrent High-Order Neural Networks: Theory and Industrial Applications* (Springer Science & Business Media, 2012)
28. E.N. Sanchez, F. Ornelas-Tellez, *Discrete-Time Inverse Optimal Control for Nonlinear Systems* (CRC Press, Boca Raton, 2017)
29. E.N. Sanchez, L.J. Ricalde, Chaos control and synchronization, with input saturation, via recurrent neural networks. Neural Netw. **16**(5), 711–717 (2003)
30. E.N. Sanchez, L.J. Ricalde, R. Langari, D. Shahmirzadi, Rollover prediction and control in heavy vehicles via recurrent high order neural networks. Intell. Autom. Soft Comput. **17**(1), 95–107 (2011)
31. E.N. Sanchez, D.I. Rodriguez-Castellanos, G. Chen, R. Ruiz-Cruz, Pinning control of complex network synchronization: a recurrent neural network approach. Int. J. Control Autom. Syst. 1–10 (2017)
32. R. Sepulchre, M. Jankovic, P.V. Kokotovic, *Constructive Nonlinear Control* (Springer Science & Business Media, 2012)
33. F. Sorrentino, M. di Bernardo, F. Garofalo, G. Chen, Controllability of complex networks via pinning. Phys. Rev. E **75**(4), 046103 (2007)
34. J.A. Suykens, J.P. Vandewalle, B.L. de Moor, *Artificial Neural Networks for Modelling and Control of Non-linear Systems* (Springer Science & Business Media, 2012)
35. C.J. Vega, E.N. Sanchez, Neural inverse optimal pinning control for synchronization of complex networks with nonidentical chaotic nodes, in *2018 International Joint Conference on Neural Networks (IJCNN)* (IEEE, 2018), pp. 1–5
36. L. Wang, Y.X. Sun, Robustness of pinning a general complex dynamical network. Phys. Lett. A **374**(13–16), 1699–1703 (2010)
37. X.F. Wang, G. Chen, Pinning control of scale-free dynamical networks. Phys. A Stat. Mech. Appl. **310**(3), 521–531 (2002)

38. X.F. Wang, G. Chen, Complex networks: small-world, scale-free and beyond. IEEE Circ. Syst. Mag. **3**(1), 6–20 (2003)
39. D.J. Watts, S.H. Strogatz, Collective dynamics of small-world networks. Nature **393**(6684), 440–442 (1998)
40. J. Xiang, G. Chen, On the v-stability of complex dynamical networks. Automatica **43**(6), 1049–1057 (2007)
41. J. Xiang, G. Chen, Analysis of pinning-controlled networks: a renormalization approach. IEEE Trans. Autom. Control **54**(8), 1869–1875 (2009)
42. L. Xiang, Z. Chen, Z. Liu, F. Chen, Z. Yuan, Pinning control of complex dynamical networks with heterogeneous delays. Comput. Math. Appl. **56**(5), 1423–1433 (2008)
43. J. Zhou, T. Chen, Synchronization in general complex delayed dynamical networks. IEEE Trans. Circ. Syst. I Regul. Pap. **53**(3), 733–744 (2006)
44. J. Zhou, J.S. Lu, J. Lü, Pinning adaptive synchronization of a general complex dynamical network. Automatica **44**(4), 996–1003 (2008)

Printed in the United States
By Bookmasters